A Field Guide to the
Neogene Sedimentary Basins of the
Almería Province, SE Spain

Edited by

A.E. MATHER, J.M. MARTÍN,

A.M. HARVEY & J.C. BRAGA

b

Blackwell
Science

© 2001 by
Blackwell Science Ltd
Editorial Offices:
Osney Mead, Oxford OX2 0EL
25 John Street, London WC1N 2BS
23 Ainslie Place, Edinburgh EH3 6AJ
350 Main Street, Malden
MA 02148-5018, USA
54 University Street, Carlton
Victoria 3053, Australia
10, rue Casimir Delavigne
75006 Paris, France

Other Editorial Offices:
Blackwell Wissenschafts-Verlag GmbH
Kurfürstendamm 57
10707 Berlin, Germany

Blackwell Science KK
MG Kodenmacho Building
7–10 Kodenmacho Nihombashi
Chuo-ku, Tokyo 104, Japan

Iowa State University Press
A Blackwell Science Company
2121 S. State Avenue
Ames, Iowa 50014-8300, USA

First published 2001
Set by Graphicraft Limited, Hong Kong
Printed and bound in Great Britain by
MPG Books Ltd., Bodmin

DISTRIBUTORS

Marston Book Services Ltd
PO Box 269
Abingdon, Oxon OX14 4YN
(*Orders*: Tel: 01235 465500
Fax: 01235 465555)

USA
Blackwell Science, Inc.
Commerce Place
350 Main Street
Malden, MA 02148-5018
(*Orders*: Tel: 800 759 6102
781 388 8250
Fax: 781 388 8255)

Canada
Login Brothers Book Company
324 Saulteaux Crescent
Winnipeg, Manitoba R3J 3T2
(*Orders*: Tel: 204 837 2987)

Australia
Blackwell Science Pty Ltd
54 University Street
Carlton, Victoria 3053
(*Orders*: Tel: 3 9347 0300
Fax: 3 9347 5001)

A catalogue record for this title
is available from the British Library

ISBN 0-632-05919-2

Library of Congress
Cataloging-in-Publication Data has been
applied for

For further information on
Blackwell Science, visit our website:
www.blackwell-science.com

Contents

Contributors

Editors

Anne. E. Mather *Department of Geographical Sciences, University of Plymouth, Drake Circus, Plymouth, Devon PL4 8AA, UK*

José M. Martín *Departamento de Estratigrafía y Paleontología, Universidad de Granada, Campus Fuentenueva, 18002 Granada, Spain*

Adrian M. Harvey *Department of Geography, Roxby Building, University of Liverpool, Liverpool PO Box 147, L69 3BX, UK*

Juan. C. Braga *Departamento de Estratigrafía y Paleontología, Universidad de Granada, Campus Fuentenueva, 18002 Granada, Spain*

Contributors

Roy Alexander *Department of Geography, Chester College, Parkgate Road, Chester CH1 4BJ, UK*

Juan M. Fernández-Soler *Departamento de Mineralogía y Petrología, Universidad de Granada, Granada E-18071, Spain*

María Teresa Gómez-Pugnaire *Departamento de Mineralogía y Petrología, Facultad de Ciencias, Universidad de Granada, 18002 Granada, Spain*

Peter Haughton *Department of Geology, University College Dublin, Belfield, Dublin 4, Ireland*

Diane Spivey *RPS Environmental Consultants, 3 Linenhall Place, Chester CH1 2LP, UK*

Martin Stokes *Department of Geological Sciences, University of Plymouth, Drake Circus, Plymouth, Devon PL4 8AA, UK*

Chapter 1
Introduction to the Field Guide

ANNE E. MATHER, JOSÉ M. MARTÍN,
ADRIAN M. HARVEY & JUAN C. BRAGA

1.1 Aim of this field guide

This field guide focuses on the Neogene and Quaternary evolution of the sedimentary basins of the Almería Province, south-east Spain, providing an integrated approach to the geology and geomorphology. The guide assumes a basic knowledge of geology and geomorphology and should be of interest to earth scientists at all levels, from the keen amateur through to undergraduate or postgraduate students and academics visiting the region. It touches on some areas of geology and geomorphology which are associated with specialist terminology (e.g. sequence stratigraphy and igneous and metamorphic geology). It is beyond the scope of this volume to expand on these areas in detail, but a list of texts which do so is given in the Appendix.

The emphasis of this guide is initially to provide a general background to the area based on the Sorbas Basin, where the Neogene depositional sequence is most continuous (Chapter 2). More specialist trips then examine the structure and development of the basins (Chapter 3), the shallow marine sedimentation (Chapter 4) the continental sedimentation (Chapter 5) and, finally, the uplift, dissection and geomorphological evolution of the basins (Chapter 6).

The modern geometry of the Neogene sedimentary basins is shown in Fig. 1.1.1, and the main excursion routes in Fig. 1.1.2.

1.2 Background to the study area

Almería is the easternmost province of Andalucía in southern Spain. It is at a latitude of 37°N, bordered by the Mediterranean Sea, some 200 km from the north coast of Africa.

The province lies within the Betic Cordillera, an Alpine mountain chain resulting from the collision of the European/Iberian and African plates from Late Mesozoic to Middle Cenozoic times. During the Miocene, compression was replaced by extensional and strike-slip tectonics, and the area

I

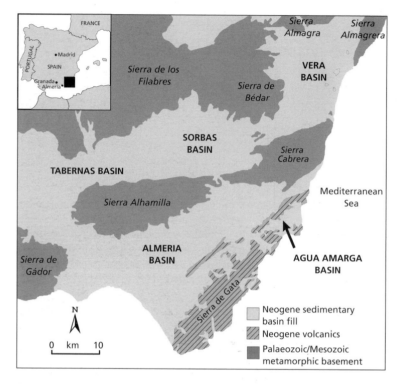

Fig. 1.1.1. The sedimentary basins and sierras of the Almería region.

has undergone substantial uplift since the Pliocene. The region is still tectonically active. The city of Almería was virtually destroyed by an earthquake in 1522, and the small town of Vera suffered a similar fate during the
last century. The Andalucían earthquake of 25 December 1884 affected the
area near Alhama de Granada and reached intensity IX on the Mercalli
scale, killing 800 people (López-Arroyo *et al.* 1980). It was similar in scale
to the earthquake which hit Kobe, Japan in January 1995 (7.2 on the
Richter scale).

The sedimentary basins were defined during the Tortonian (late
Miocene), by the development of left-lateral strike-slip faults cross-cutting
the basement schists. The sedimentary infill of the basins was dominantly
marine until the Pliocene, then terrestrial until the Quaternary, when
regional uplift caused a switch from deposition to erosion, creating the
modern landforms.

Almería Province is a superb area in which to study the relationships
between tectonics, sedimentary geology and geomorphology. The basin-fills

(a)

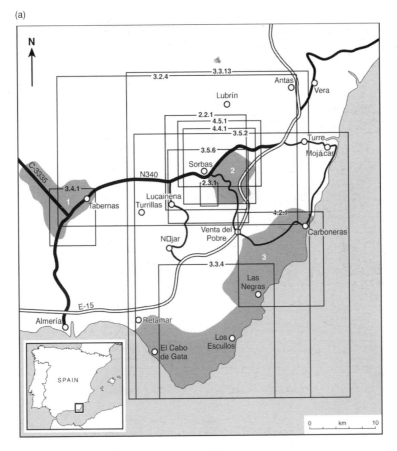

Fig. 1.1.2. Main excursion routes. (a) Chapters 2–5.

include an enormous variety of sedimentary rocks, reflecting the rapid environmental changes that occurred during the Neogene and Quaternary. The rocks range from deep water turbidites, through shallow marine rocks including carbonate reefs, evaporitic gypsum (related to the Messinian salinity crisis) and a whole suite of nearshore and shoreline facies, to low- and high-energy fluvial deposits. The rocks show a wide range of deformation from syn-sedimentary soft sediment deformation to brittle strike-slip faulting. The rocks are well exposed as a result of the rapid incision of the river systems during the Quaternary.

A spectacular suite of landforms has resulted from the incision of the developing drainage, under a dominantly dry Quaternary climatic regime. The erosional landforms include pediments, canyons and badlands; the

(b)

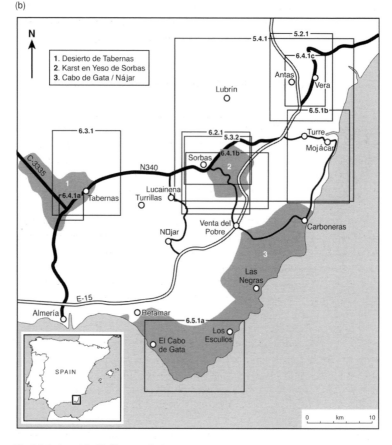

Fig. 1.1.2. (*cont'd*). (b) Chapters 5–6.

depositional forms include alluvial fans and braided rivers, and magnificent pedogenic calcretes. The region offers the best developed tectonically active and semiarid landscape in western Europe.

1.3 Climate

The region enjoys the driest climate in western Europe, with mean annual precipitation of less than 300 mm, most of which falls in autumn or winter, in relatively short duration but high-intensity storms. Daytime temperatures range from *c.* 15°C in January to *c.* 40°C in July and August. Frosts are rare, except in the sierras. The climate is most favourable for field-work in spring and autumn. June, July and August are hot (*c*. 40°C), often with poor visibility as a result of hazy conditions. Summer also has the disadvantage of being the peak tourist season with generally higher prices.

1.4 Accommodation, travel and general facilities

Accommodation is available in most of the key settlements (Fig. 1.1.2), with small hotels or 'hostals' in Tabernas, Sorbas and Vera, but these cannot cater for large groups. Sorbas benefits from a field centre which can handle groups of up to 40. The coastal towns, especially Almería, Mojácar and Carboneras, have hotels and other accommodation which can cater for larger parties, together with other facilities such as car hire. Some useful points of contact are included in the Appendix.

The area can be reached by train or bus to Almería. The easiest method of transport within Europe is to take advantage of cheap charter flights to Almería or Alicante. Almería is the nearest airport, but Alicante (to the north of the region, located along the Costa Blanca) has a wider range of flights and cheaper car hire. Driving time from these airports to the centre of the field area is approximately 30 min from Almería and 2.5 h from Alicante.

Coaches can be hired from tour operators, but generally greater flexibility will be enjoyed with minibuses or hire cars. Many of the tracks and minor roads are unsuitable for large vehicles and parking of such vehicles can be a problem. Suitability of access to field locations and special permission requirements for access to protected areas are given in Table 1.1.1. Figure 1.1.2 details the distribution of protected areas (both Paraje and Parque Natural) and to aid the planning of your trip. Permission to visit the protected areas can be obtained by writing to the Medio Ambiente (see Appendix for contact details), detailing the purpose of your visit, people involved and proposed itinerary.

1.5 Map coverage

The best topographic maps (1 : 50 000) are those of the Mapa Militar de España series. A new series of excellent 1 : 25 000 maps is available for a limited part of the area. Much map coverage is now available in CD ROM format (see Appendix for contact details). The region is also covered by at least two sets of air photos and LANDSAT Thematic Mapping imagery. The geological maps of the region are of variable quality, but undergoing revision. Maps relevant to this guide are outlined in Table 1.1.1. Addresses from where they may be obtained are given in the Appendix. Grid

Table 1.1.1. Details of map coverage and stop accessibility (all stops are accessible by car or minibus)

Excursion	Duration (days)	1 : 50 000 Topographic/ Geology Sheets
2.2 Transect of the Sorbas Basin	1	Sorbas
2.3 A hike to Cantona view point	0.5	Sorbas
3.2 The basement	1	Tabernas
		Macael
		Vera
3.3a The calc-alkaline volcanics of the Cabo de Gata	0.5	Cabo de Gata
		El Pozo de los Frailes
3.3b The volcanics of the Almería, Níjar/ Carboneras and Vera Basins	0.5	Cabo de Gata
		Almería
		Carboneras
		Vera
3.4 The turbidites of Tabernas†	1	Tabernas
		Almería
3.5a Plio-Pleistocene deformation: the Carboneras and Palomares Fault Zone	1	Mojácar
		Carboneras
		Cabo de Gata
3.5b Plio-Pleistocene deformation: the Sierra Cabrera Northern Boundary Fault and adjacent areas of the Sorbas Basin	0.5	Sorbas
4.2 Temperate water carbonates of the Agua Amarga Basin	1	Carboneras
4.3 Tropical carbonates of Níjar	1	Carboneras
		Almería
4.4 Tropical carbonates of Sorbas	1	Sorbas
4.5 Evaporites and stromatolites of Sorbas	1	Sorbas
5.2 Late Pliocene Gilbert-type fan deltas of the Vera Basin	1	Vera
5.3 Mio-Pliocene marine to continental transition of the Sorbas Basin	1	Sorbas
5.4 Plio-Pleistocene alluvial environments of the Sorbas and Vera Basins	1	Sorbas
		Vera
6.2 Drainage evolution and river terraces of the Sorbas Basin	1	Sorbas
6.3 Geomorphology of the Quaternary alluvial fans and related features of the Tabernas Basin†	1	Tabernas
6.4 Badlands†	1	Tabernas
		Sorbas
		Vera
6.5 The landforms of the coastal zone	1	Cabo de Gata
		El Pozo de los Frailes
		Carboneras
		Sorbas
		Mojácar
		Garrucha

* Permission to visit the protected areas can be obtained by writing to the Medio Ambiente (see Appendix for contact details), detailing the purpose of your visit, people involved and proposed itinerary (for A: Parque de Cabo de Gata–Níjar; B: Paraje Karst en Yeso de Sorbas and C: Paraje Desierto de Tabernas); or for D: Cortijo Urra Field Centre, Urra, Sorbas, by contacting the owner. See Appendix for correspondence addresses.

1 : 25 000 Topographic Sheet	*Permission required	Stops to be omitted for coaches
Sorbas & Polopos	none	none
Polopos	none	none
	none	5
Fernán Pérez	all (A)	none
Morrón de Genoveses		
El Pozo de los Frailes		
Campohermoso	1 (A)	1
	2,3,4,5,6 (C)	none
Castillo de Macenas	4,6 (A)	none
El Agua del Medio		
Fernán Pérez		
Carboneras		
Polopos	none	2
El Agua del Medio		
Sorbas		
Carboneras	all (A)	3
Níjar	none	none
Campohermoso		
Sorbas	none	none
Polopos		
Sorbas	1,2 (B)	none
	none	4
Sorbas	none	2,3,4,6
Polopos		
Sorbas	none	3,5,6
Polopos		
Sorbas	1,6 (B)	2
Polopos	5 (D)	
	6,7 (C)	3,4
Sorbas	2b (C)	3
Polopos	4 (B)	
Morrón de Genoveses	1,2,3 (A)	none
El Pozo de los Frailes		
El Agua del Medio		
Garrucha		
Castillo de Macenas		

† Please note that excursions which visit the west of the Tabernas Basin (3.4, 6.3 & 6.4) may have restricted access to some stop localities due to the construction of the new Granada–Almería Autovia along the route of the N324. The affected stops are highlighted at the beginning of each excursion.

references throughout this guide are taken from the topographic map series. (Note that the Instituto Geológico y minero de España (IGME) geology maps use a different grid system.) Note that in the description of stop transfers **KM** (e.g. **KM** 485) is used to indicate kilometre markers along the roadside. Distance is given as km (e.g. 12 km).

Chapter 2
Introduction to the Neogene Geology of the Sorbas Basin

ANNE E. MATHER, JUAN C. BRAGA,
JOSÉ M. MARTÍN & ADRIAN M. HARVEY

2.1 Introduction

Thick, continental to marine sedimentary sequences occur throughout the Neogene basins of Almería Province (southern Spain). The Sorbas Basin retains both the most complete and well-exposed sedimentary succession and a diverse range of Quaternary landscapes. It is thus used here to illustrate the regional stratigraphy and Quaternary landscape evolution, which is less continuously or well-exposed in the adjacent Neogene basins. This section is based on a transect across the Sorbas Basin from south to north, essentially following the stratigraphy from the oldest rocks up the sequence through the basin-fill, and at the same time following the Quaternary dissection of the basin and associated landforms.

The Sorbas Basin is a narrow, east–west elongate, intramontane basin of Neogene age within the Betic Cordillera of Almería Province. It is bounded by the Sierra de los Filabres to the north and by the Sierras Alhamilla and Cabrera to the south (see Fig. 1.1.1). These reliefs comprise metamorphic rocks from the Internal Betic Zone (see Section 3.2).

The Neogene infill of the basin comprises a series of units separated by unconformities (Fig. 2.1.1). The lowermost unit consists of conglomerates and sands, probably Middle Miocene (?Serravallian) in age. The overlying upper Tortonian (Upper Miocene) unit is made up of platform carbonates—including some coral patch reefs—and conglomerates that pass basinwards into conglomerates, sandstones and silty marls deposited in submarine fans (Haughton, 1994). The Messinian sequence, in turn, consists of five units (Martín & Braga, 1994).

1 A temperate carbonate unit (Azagador Member of Völk, 1967; Fig. 2.1.2), composed of bioclastic calcarenites and sandstones, locally mixed with conglomerates, with abundant bryozoans, bivalves, coralline algae and benthic Foraminifera, and minor brachiopods, solitary corals, barnacles and gastropods (Wood, 1996). These platform deposits grade upwards and laterally into marls (lower Abad Member of Völk, 1967) mainly dated as Messinian (Iaccarino et al. 1975; Serrano, 1979). The Tortonian–Messinian

9

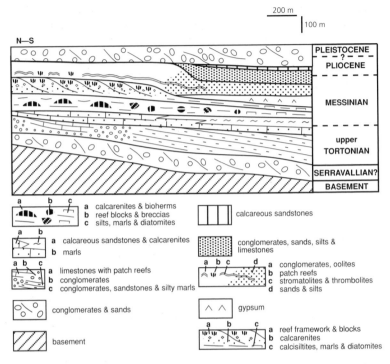

Fig. 2.1.1. Neogene and Quaternary lithostratigraphy of the Sorbas Basin. (Modified after Martín & Braga, 1994.)

boundary (7.1 Ma; Berggren *et al.* 1995) is recorded in the centre of the basin near the base of the marls (Sierro *et al.* 1993; Gautier *et al.* 1994). This temperate carbonate unit is considered in Excursion 4.2 but in the Agua Amarga Basin.

2 A bioherm unit consisting of tropical platform carbonates, locally mixed with siliciclastics, with intercalated lensoid, coral patch reefs and *Halimeda* mounds (Braga *et al.* 1996a; Martín *et al.* 1997), changing basinwards to marls that intercalate with diatomites that contain abundant planktonic Foraminifera. The coiling change in *Neogloboquadrina acostaensis* indicates an age of 6.2 Ma for this unit (using the Berggren *et al.* 1995 time-scale; Braga & Martín, 1996a). This unit is considered in Excursion 4.4.

3 A unit of prograding fringing reefs composed of *Porites* corals encrusted by stromatolites (Riding *et al.* 1991a; Braga & Martín, 1996a). The top of these reefs is an erosional surface with signs of subaerial exposure. This reef unit passes basinwards into marls with intercalated diatomites, containing abundant calcareous nanoplankton and planktonic Foraminifera of Messinian age (Toelstra *et al.* 1980; Sierro *et al.* 1993). Magnetostratigraphy (Gautier *et al.* 1994) indicates the start of Chron 3R (5.9 Ma; Berggren *et al.* 1995) at the top of the marls. The two Messinian reef units correspond to the Cantera Member and its basinal lateral equivalent, the upper Abad Member marls (in the sense of Völk, 1967) (Figs 2.1.1 and 2.1.2). These bioherm and fringing reef units are considered in Excursions 4.3 and 4.4.

4 Selenite gypsum deposits (evaporite unit) (Yesares Member of Ruegg, 1964; Fig. 2.1.2) overlie a basin-scale, major unconformity on top of the fringing-reef unit and laterally equivalent marls (Riding *et al.* 1991a; Martín & Braga, 1994; Riding *et al.* 1998). They onlap the eroded reef slopes (Braga & Martín, 1992), occurring as banks (up to 20 m thick) of vertically arranged, twinned, selenite gypsum crystals, separated by silt–marl interbeds (up to 3 m thick) (Dronkert, 1977). Sulphur and strontium isotope data indicate a marine origin for this gypsum (Playà *et al.* 1997). The uppermost gypsum banks intercalate with beds (up to 5–6 m thick) of marine silts and marls containing benthic and planktonic Foraminifera, including *Globorotalia gr. miotumida sensu* Sierro *et al.* (1993), *Neogloboquadrina acostaensis* and *Globigerina multiloba* that indicate a Messinian age (Riding *et al.* 1998). This unit is considered in Excursion 4.5.

5 The last Messinian unit (Sorbas Member of Ruegg, 1964; Fig. 2.1.2) corresponds mainly with siliciclastic sands and conglomerates (Roep *et al.* 1979, 1998), with local oolitic carbonates and coral (*Porites*) patch reefs, changing laterally basinwards to silts and marls and intercalating giant (up to 15 m in diameter) microbial (stromatolite–thrombolite) domes (Martín *et al.* 1993; Braga *et al.* 1995; Braga & Martín, 2000). The marls in this unit once again contain *Globorotalia gr. miotumida sensu* Sierro *et al.* (1993), *Neogloboquadrina acostaensis* and *Globigerina multiloba* (Riding *et al.* 1998). These deposits directly onlap, at the margins of the basin, the eroded Messinian reefs.

The Pliocene to Quaternary sediments comprise two main units.

1 The Cariatiz Formation of Mather (1991) (Fig. 2.1.2). This is composed of alluvial fan conglomerates along the northern and western margins of the basin (Moras Member of Mather, 1991) which pass basinwards to coastal plain silts and sands (Zorreras Member of Ruegg, 1964). The Cariatiz Formation sedimentation was punctuated by two basin-wide ostracod-rich carbonate units representing the deposits of a shallow, brackish-water lake (Mather, 1991) and is capped by a basin-wide marine unit. The Cariatiz Formation is considered to be Pliocene in its upper part (Martín & Braga, 1994), and Messinian in its lower part (Roep *et al.* 1979; Ott d'Estevou, 1980). These sediments are considered in Excursion 5.3.

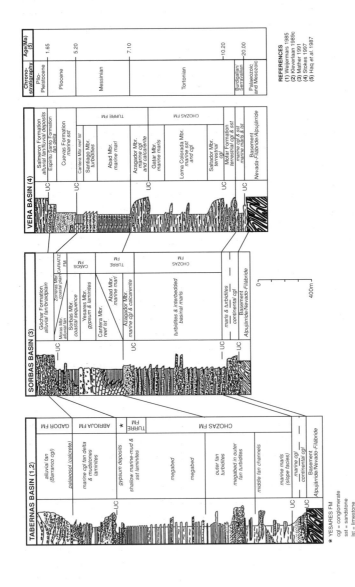

Fig. 2.1.2. Simplified geological sections of the Tabernas, Sorbas and Vera basins. (Modified after Weijermars, 1985; Kleverlaan, 1989c; Mather & Harvey, 1995; Stokes, 1997.)

2 A sequence of fluviatile conglomerates, the Plio-Pleistocene Góchar Formation (Ruegg, 1964; Fig. 2.1.2). These conglomerates mark the final infill of the basin and the dominance of continental sedimentation in the form of braided fluvial systems and alluvial fan deposits (Mather, 1991; Mather & Harvey, 1995). These sediments are considered in Section 5.4.

This Neogene stratigraphy can also be recognized in the other basins in the Almería area (Fig. 2.1.2). Nevertheless, when referring specifically to single basins, some variations from this 'Sorbas stratigraphy' can be found (Fig. 2.1.2). In the case of the Tabernas Basin (the western continuation of the Sorbas Basin; Fig. 1.1.1) the upper Tortonian submarine-fan deposits are especially well represented, with scarce occurrences of Messinian deposits (temperate carbonates, reefs, evaporites; Kleverlaan, 1987, 1989a,b). In the Vera Basin, immediately to the east of the Sorbas Basin (Fig. 1.1.1), there are thick marly sequences of upper Tortonian and Messinian age (Barragan, 1997) and a large Gilbert-type delta developed in the Pliocene (Postma & Roep, 1985; Stokes, 1997). In the Almería Basin (Fig. 1.1.1) there are abundant, shallow-marine, Pliocene deposits (Aguirre, 1995). In the Agua Amarga Basin, in the volcanic Cabo de Gata area (Fig. 1.1.1), a temperate carbonate unit of lower Tortonian age is intercalated between volcanic rocks dated radiometrically as 9.6 and *c.* 8 Ma, respectively. In addition, a Messinian sequence, very similar to that of the Sorbas Basin (including the Azagador Member temperate carbonates, the reef units and a unit similar to the Sorbas Member), occurs on top of the youngest volcanic rocks (Braga *et al.* 1996b). The Tortonian temperate carbonates of the Agua Amarga Basin are considered in Excursion 4.2.

Continued uplift stimulated a change from net aggradation to net erosion in the Pleistocene, creating the landscape seen today (considered in Chapter 6). The incision of the river systems is recorded in the deep, erosional canyons developed in the more resistant lithologies, such as the Sorbas Member, and the development of badlands in the weaker lithologies, such as the Abad Member marls. Steep slopes accompanying the incision, combined with the seismic activity and the action of high-magnitude, low-frequency rainstorm events, have led to a wide range of mass movement features including rotational slips, landslides and topples.

Within the Almería region as a whole there are three groups of Quaternary deposits (Chapter 6): (i) river terrace and other alluvium; (ii) alluvial fans and other hillfoot deposits; (iii) coastal sediments. Within the Sorbas Basin, excellently exposed river terraces record the sequence of Quaternary dissection through the Neogene basin-fill succession. There are small mountain-front alluvial fans on the northern margins of the basin, but Quaternary alluvial fans are better developed in the Tabernas and Almería basins. Quaternary coastal sediments do not occur in the Sorbas Basin, as

14Chapter 2

they are restricted to the modern coastal zone. For a fuller description of the Quaternary sequences and geomorphology see Section 6.1.

2.2 Excursion: Transect of the Sorbas Basin (one day)

ANNE E. MATHER, JUAN C. BRAGA,
JOSÉ M. MARTÍN & ADRIAN M. HARVEY

Introduction

This excursion introduces the basic Neogene basin stratigraphy and geomorphology for the Almería region using the Sorbas Basin as a type example. The excursion starts at the south of the basin, in the Sierra Alhamilla, and follows a south–north transect, finishing in the Sierra de los Filabres which delimits the northern margin of the basin (Fig. 2.2.1).

Transfer to Stop 1

This stop is at KM 12.3 on the Al-104 from Sorbas to Venta del Pobre and Carboneras (Fig. 2.2.1).

Stop 1: Peñas Negras (848 013; Polopos 1 : 25 000)

The road-cut on the east side of the road exposes Tortonian rocks, and shows well-developed, channelled turbidites.

To the west there is a general view of the southern margin of the basin. To the south and SW are the mountains of the Sierra de Alhamilla comprising low-grade, blackschists of the upper part of the Nevado–Filábride nappe complex, overlain by brown/purple Triassic metacarbonates and phyllites of the Alpujárride nappe complex. The mountain front faults can be identified by slivers of red, purple and other brightly coloured Triassic lithologies. The mountains are faulted against the Tortonian marls and turbidites, which dip steeply towards the north, and are exposed in the foreground and at the base of the large escarpment to the NW. The steep, low ridges to the west and NW are formed by the more resistant sandstone and conglomerate units within the Tortonian.

Looking northwards (Fig. 2.2.2) the Messinian deposits are visible above the upper Tortonian marls, comprising bioclastic, temperate carbonates (Azagador Member), followed by marls, the upper part of which are the lateral, basinal equivalents of Messinian reefs (Cantera Member). In the highest hill (Cerrón de Hueli) Messinian gypsum (Yesares Member) occurs on top of these marls. The gypsum is capped by the Sorbas Member (Fig. 2.2.2), which contains stromatolites.

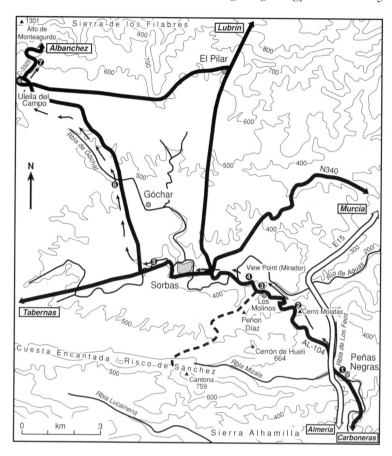

Fig. 2.2.1. Location map of the field-trip stops for Excursion 2.2.

The ancestral Aguas–Feos valley flowed south from here across the mountains; this course is today exploited by the route of the motorway. Terrace fragments record the Pleistocene evolution (Harvey & Wells, 1987; Harvey *et al.* 1995) south of Peñas Negras; the modern Río Aguas now flows to the east into the Vera Basin. The Rambla de Mizala, a tributary from the west, now forms the main headwater of the Feos. It developed as a subsequent stream, exploiting the strike of weak lithologies in the Tortonian rocks, and capturing earlier north-flowing basinal drainage. The continuation of this process can be observed from Cantona, at the head of the valley (788 016), where steep, east-draining gullies are about to capture a less steep, north-flowing drainage. This is explored further in Excursion 6.2.

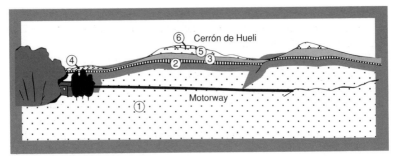

Fig. 2.2.2. Stop 1, Peñas Negras view to the north of: 1, Upper Tortonian silty marls, turbidite sandstones and conglomerates; 2, Azagador Member temperate carbonates; 3, Abad Member marls (undifferentiated); 4, Cantera Member reefs; 5, Yesares Member gypsum; and 6, stromatolitic carbonates and sands from the Sorbas Member.

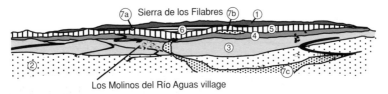

Fig. 2.2.3. Stop 2, Cerro Molatas view to the north: 1, Nevado–Filabre basement of Sierra de los Filabres; 2, Azagador Member temperate carbonates; 3, grey marls from the lower Abad Member; 4, yellow marls with intercalated diatomites from the upper Abad Member; 5, Yesares Member gypsum; 6, post-evaporitic Messinian (Sorbas and Zorreras Members undifferentiated); and 7, fluviatile Quaternary deposits: (a,b) ancient river terraces; and (c) present-day, Río de Aguas alluvial deposits (see also Fig. 4.1.3).

Transfer to Stop 2

Take the road north towards Sorbas. At Cerro Molatas (832 045), on the apex of a right-hand bend, is a dramatic view down the lower Aguas valley. This is Stop 2.

Stop 2: Cerro Molatas (832 045; Polopos 1 : 25 000)

The view south from Stop 2 takes in the southern part of the basin and includes the older part of the basin-fill observed at Stop 1. Towards the NW the outcrop of the gypsum can be traced, forming an escarpment (Fig. 2.2.3). To the north and east the same escarpment can be seen across the canyon of the Río Aguas. The gypsum has a maximum thickness of 130 m and crops out over an area of 25 km^2 in this part of the Sorbas Basin.

Below the gypsum, in a belt stretching from the east, through Los Molinos below you, and to the west is the lower ground of the Abad

Member marl. This terrain is deeply dissected by gullies and badlands. Note also the topple failures where the underlying Abad Member marls have been eroded from beneath the gypsum.

To the east of the motorway at 836 054, the Azagador Member can be seen dipping into the lower Aguas valley. At this point the Azagador Member consists of locally conglomeratic, bioclastic, calcareous sands and calcarenites with abundant bivalves, coralline algae, bryozoans, echinoids and brachiopods. The upper yellowish marly sequence (Abad Member) is the basinal lateral equivalent of the Messinian reefs (Cantera Member). The Azagador Member rests unconformably on Tortonian marls, exposed in the hillside below Cerro Molatas. Note the planar slides and associated tensional failures where the Azagador Member calcarenite has moved down-dip over the Tortonian marls. For a fuller description of the Quaternary geology and landform development see Excursion 6.2, Stop 6.

Transfer to Stop 3

Continue in the same direction along the road, after passing through the village of Los Molinos del Río Aguas. Climbing the hairpin bends on to the top of the gypsum escarpment you will come to a cross-roads where the road flattens out. Take the dirt track to the right (the turning to the left leads to the gypsum quarry). If in a coach, park here. If in a car or minibus, proceed down the dirt track to the first grove of sparse olive trees. To the south is a hill. Stop 3 is at the top of this hill.

Stop 3: Los Molinos Mirador (819 056; Sorbas 1 : 25 000)

This site provides panoramic views north into the centre of the basin, and south across the Los Molinos area and the site of the Aguas–Feos river capture (Figs 6.2.5 and 6.2.6). In the Pliocene and earlier Pleistocene this river flowed through the gap to the north of Cerro Molatas, but was captured by the lower Aguas at *c.* 100 ka (Harvey *et al.* 1995), eroding headwards from the east along the outcrop of the weak Abad Member marl. For a fuller description of the Quaternary geology and landform development see Excursion 6.2, Stop 1.

The ridge here is formed by the Yesares Member gypsum. A little down the ridge to the SW are exposures of *in situ* selenite crystals in growth position. At the top of the hill, resting unconformably on the gypsum, are two conglomerate units. The older unit, resting on brecciated marl, is not in depositional position and contains dewatering structures (reorientated clasts). These features probably reflect deformation, associated with underlying gypsum cave collapse. This resembles similar disturbed deposits, which occur on a larger scale at Peñón Díaz, 2 km to the SW (810 045). We interpret these rocks to represent the preserved base of the Góchar Formation. The second conglomerate unit is *in situ* within a channel, cut into the first set and the underlying gypsum. We interpret these to be an eroded remnant of a Quaternary gravel terrace (Terrace A, Harvey *et al.* 1995).

The view north from Stop 3 takes in the central part of the Sorbas Basin and includes the upper part of the basin-fill sedimentary succession. The low ground in the foreground, the valley of the Río Aguas, follows the strike of the Sorbas Member. Beyond, the reddish rocks with the two thin, white carbonate units are the Zorreras Member. This is capped by the Góchar Formation conglomerates. The mountains in the distance are the Sierra de los Filabres.

Transfer to Stop 4

Rejoin the road and turn right. As you drive north you will drive down off the gypsum escarpment. At KM 3 park at the stone-built view point (Mirador). At 813 060 (on the south side of the road) you can observe deformed sections of Pleistocene river terraces (Terrace B, Harvey & Wells, 1987). This theme is followed in Excursion 6.2.

Stop 4: Urra Mirador (813 061; Sorbas 1 : 25 000)

Here the Sorbas Member consists of siltstones, claystones and marls, with lesser amounts of sandstones. The Zorreras Member, visible in the distance (Fig. 2.2.4), is made up of red clays and silts, with minor sands, that are locally quarried. Interbedded with the clays and silts are two beds of white, ostracod-rich limestones of lacustrine origin.

Note the extensive flat topography with red soils of Terrace C to the east. This is an extensive terrace level related to the ancestral Aguas–Feos, which used to flow through the Aguas–Feos gap at this stage. To the north, in the middle, near distance, note the extensive brown sediments which make up Terrace D surfaces inset into the yellow marls of the Sorbas Basin. The large, white building to the NW (Cortijo Urra Field Study Centre) sits

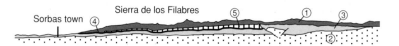

Fig. 2.2.4. Stop 4, Urra Mirador view to the north: 1, Nevado–Filabre basement of Sierra de los Filabres; 2, sandstones, siltstones and claystones from the Sorbas Member; 3, clays, silts, sands and limestones from the Zorreras Member; 4, marine bioclastic sandstones from the Lower Pliocene; and 5, Plio-Pleistocene Góchar Formation conglomerates. Vertical scale is exaggerated by 1.5.

on top of one such surface. These sediments (examined in Excursion 6.2, Stop 5) represent low-energy, ponded sediments of Terrace D of the Río Aguas (Mather *et al.* 1991) which may relate to deformation along an active strike-slip structural lineament, the Infierno–Marchalico lineament (described in Excursion 3.5b, Stop 2).

Transfer to Stop 5

From the Mirador proceed north along the road for *c.* 2.5 km to the junction with the N340. Turn left along the main road. Drive past the town of Sorbas, which sits on coastal deposits of the Sorbas Member. The road follows an abandoned (post river capture, Terrace D) meander of the Río Aguas. Follow the road for *c.* 1 km past the main entrance to Sorbas Town (on the right). Park on the right-hand shoulder adjacent to the 'Taller Mecánico'.

Stop 5: 'Taller Mecánico', west of Sorbas village (768 064; Sorbas 1 : 25 000)

Along the bed of the Rambla de los Chopos, which is an upstream tributary of the Río de Aguas, at *c.* 1 km to the west of Sorbas there is an excellent section of prograding beach/barrier deposits (Roep *et al.* 1979; 1998; Dabrio *et al.* 1985; Fig. 2.2.5). The lower part consists of trough cross-bedded calcareous sandstones. They correspond to shoreface deposits. The upper part of the section is made up of low-angle, parallel-laminated calcareous sandstones burrowed at their top. They constitute the foreshore and backshore sediments which prograde to the east on top of the

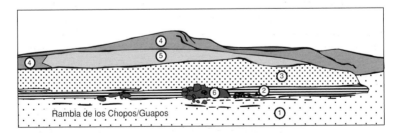

Fig. 2.2.5. Stop 5, Rambla de los Chopos view to the north: 1, beach deposits (low-angle, seaward-dipping, parallel-laminated and trough cross-bedded calcareous sands); 2, lagoonal deposits (laminated clays with mudcracks and bird-foot imprints); 3, aeolian dune deposits (low-angle, landwards-dipping, parallel-laminated fine sands); 4, Plio-Pleistocene Góchar Formation conglomerates; 5, Quaternary river terrace; and 6, fallen blocks from the cliff (collapse instigated by the 1973 flash flood in the region; see Thornes, 1974 for details on the regional impact of the flood event).

shoreface deposits. Locally they include calcareous breccias interpreted as fragments of eroded beachrock (Roep *et al.* 1979). These deposits are overlain by fine-grained sands and laminated clays with mud cracks and bird-foot tracks (Roep *et al.* 1979; Dabrio *et al.* 1985). They represent aeolian dunes (Fig. 2.2.5).

To the west the sequence is unconformably overlain by conglomerates which appear to thicken and crop out down to the river bed, displaying intercalated red sands and showing evidence of deformation. These sediments are part of the Góchar Formation.

The top of the section is capped unconformably by a well-cemented conglomerate of Sierra de los Filabres provenance (amphibole–mica schists— see fallen blocks in the river bed). The conglomerate is well imbricated and fluvial in origin and includes blocks derived from the Góchar Formation. This conglomerate marks one of the Quaternary terraces (level C, precapture terrace of Harvey & Wells 1987; Harvey *et al.* 1995) set into both the Góchar Formation conglomerates and the Sorbas Member.

What is interesting to note is the lack of the Zorreras Member at this locality. This in part can be accounted for by erosional loss. However, careful tracing of the Góchar–Zorreras contact which occurs in the section on the north side of the river, although partially obscured by large block topples of Góchar Formation conglomerates, indicates that the Zorreras Member thins to zero in this vicinity as a function of palaeorelief (see Excursion 5.3).

Transfer to Stop 6

Continue along the N340 to the west. Take the first turn on the right towards Uleila del Campo. Follow the road for 5 km. After crossing a bridge over the Rambla de Góchar, park at the side of the road. Note the conglomerates exposed in the apex of the meander bend to the west. This is Stop 6.

Stop 6: Rambla de Góchar-type locality of the Góchar Formation (753 110; Sorbas 1 : 50 000)

Stop 6 exposes a succession of Góchar Formation conglomerates, overlying Messinian reef limestones (Cantera Member), exposed east of the bridge, and shoreline facies of the Sorbas Member, exposed west of the bridge. The top of the section forms approximately the end-Góchar Formation depositional surface, the final stage of basin-filling. Since the end-Góchar the rambla has trenched only *c.* 25 m into this surface, which represents the total Quaternary dissection in this part of the basin (contrast this with the greater incision at Stops 2 and 3). The dissection stages are marked by a series of rather ill-defined river terraces with red soils, on the east side of the rambla.

On the south side of the rambla, about 100 m upstream from the bridge, carbonate rocks are exposed. These comprise rippled sands and algal carbonates at the top of the Sorbas Member. They are cut out abruptly to the west by a NNW–SSE orientated fault which downthrows the top of the Sorbas Member a few metres to the floor of the rambla.

The overlying deposits form a more or less complete sequence through the Góchar Formation (Fig. 2.2.6a). They comprise conglomerates, whose clast content is dominated by amphibole–mica schists from the Nevado–Filábride nappe in the Sierra de los Filabres to the north. Palaeocurrents are dominantly to the SSE. The sequence reflects a prograding fan–fluvial sequence, part of the Góchar System of the Góchar Formation (Mather 1991; Mather & Harvey 1995; explored in Excursion 5.4). The lowest parts comprise poorly sorted sandstones displaying oblique alignments of clast-rich zones. These could represent: (i) weakly defined, collapsed erosional structures from channels cut in a poorly consolidated, wet medium; or (ii) shear surfaces within a mudflow. Rare shell fragments and the internal mould of the gastropod *Iberus* have been found in this unit. This deposit may be the lateral equivalent of the top Cariatiz Formation marine unit (Fig. 2.1.2). This basal sandstone is cut into by small, laterally migrating channels which are typically single thread (Fig. 2.2.6b). Further up-sequence the channels become much wider but shallower, suggesting a higher energy braidplain environment (Fig. 2.2.6a).

Three prominent, laterally continuous soils are present in the sequence. The topmost is accessible at the beginning of the river cliff (Fig. 2.2.6a). A scramble up to the section reveals a well-developed soil profile. The soil profile appears to lack a pedogenic carbonate horizon, the partial cementation is by secondary carbonate penetrating root casts and cracks from the overlying gravels.

Transfer to Stop 7

Return to the road. Drive north towards Uleila del Campo. At the junction turn right onto the C-3325 (signed for Cantoria). Continue for about 800 m to the second left-hand bend (712 167). Just beyond the bend there is a large gravel area on the right, suitable for parking several cars or a coach. This is Stop 7.

Stop 7: *Sierra de los Filabres, east of Uleila (712 167; Macael 1 : 50 000)*

This location completes the south–north transect of the basin and affords a view back south across the Sorbas Basin, and east along the Filabres

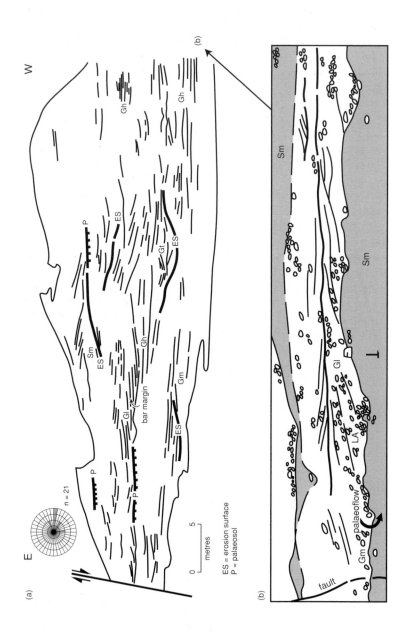

mountain front. The local bedrock, exposed in the road-cut, comprises Nevado–Filábride complex, grey amphibole–mica schists (the lithology that dominates the clasts of the Plio-pleistocene Góchar Formation conglomerates locally, and the modern rambla sediments, as seen in Stop 6).

On the north rim of the basin, the basal Messinian Azagador Member calcarenites and reefs rest directly on the basement rocks. These form the low ridge visible *c.* 3 km south of this view point. Locally they are capped by Góchar Formation conglomerates. These rocks are better exposed along the C-3325 road *c.* 4 km SW of Uleila. To the east, Quaternary alluvial fans with associated red soils rest on the basement at the mountain front and occupy the low ground to the north of the low Azagador Member ridge.

In the distance to the south the southern rim of the basin can be seen. To the SSE you are looking up the dip slope of the gypsum escarpment towards El Cerrón de Hueli (813 023). Beyond and to the left lies the Sierra Cabrera. To the SW you are looking across the undissected centre of the basin towards the dip slope of the Messinian reef (Cantera Member) escarpment. Beyond lies the Sierra Alhamilla.

If time permits, a broader panoramic view can be obtained from higher up the road towards the col over the Filabres, or, better still, from the shrine on top of the Alto de Monteagudo (705 187). To reach the shrine, proceed another *c.* 7 km up the C-3325 around the back of the ridge crest, near to KM 31 (alternative viewpoint), take the dirt road (usually passable with care by car) on the left up to the shrine.

2.3 Excursion: A hike to Cantona view point (half day)

ADRIAN M. HARVEY

As an alternative or additional half-day introductory excursion, a hike to the top of Cantona (Fig. 2.3.1), the highest point in the area (759 m), is well worth while as an introduction to the broader-scale regional geology and geomorphology.

Fig. 2.2.6. Detail of the Góchar Formation type section at Stop 6. (Drawings modified after Mather, 1991.) (a) General view of the sedimentary architecture of the Góchar Formation type section nearest the road showing bars, multiple fill channels, erosion surfaces and palaeosol. Palaeocurrents are dominantly to the east. (b) Example of a small, meandering channel with lateral accretion surfaces (LA) indicating migration towards the left. The channel is cut into massive sands. These contain very rare shell fragments and may be the margins of the marine unit which marks the top of the Cariatiz Formation. Sm, massive sand; Gm, massive conglomerate; Gh, horizontally bedded conglomerate; Gl, low-angle bedded conglomerate; Gt, trough cross-bedded conglomerate.

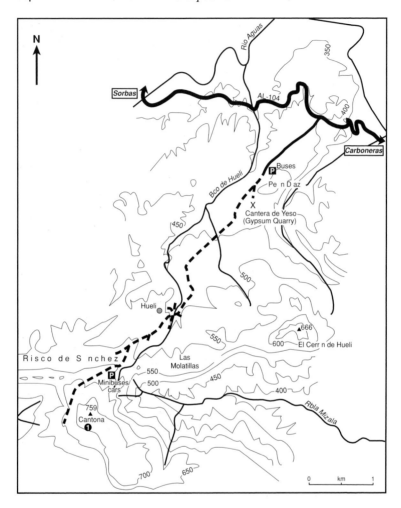

Fig. 2.3.1. Location map for Cantona hike, Excursion 2.3.

Directions to Stop 1

The access roads to the Cantona hike beyond the quarry turn-off (816 055; Sorbas 1 : 25 000) are not suitable for buses. Take the Al-104, Sorbas to Venta del Pobre road to KM 3.6. This point is on the gypsum plateau about 1.5 km west of the village of Los Molinos del Río Aguas. There, turn SW onto the access road to the gypsum quarries (signed Cantera de Yeso). At the quarry entrance take the dirt road on the right of the quarries, heading SW towards the abandoned hamlet of Hueli. After *c.* 2 km bear left, following the road through some ruined houses, then after *c.* 2.5 km from the gypsum quarries the road crosses the Barranco de Hueli on a tight right-hand bend. There is an irrigation tank here beside the Barranco. The abandoned hamlet of Hueli is above you to the west. Unless you have a high-clearance vehicle it is best to park here (796 026).

En route from the gypsum quarry to Hueli the track has been traversing the dip slope of the gypsum escarpment, which culminates in the hill of El Cerrón de Hueli (813 023), and is capped by rocks of the Terminal Carbonate complex (Sorbas Member).

Continue on foot (or in a high-clearance vehicle), taking the track which curves to the NW from the Barranco, then curves SW below Hueli. Follow the track SSW, on the west side of a low hill (note that this section of the track is not shown on the 1 : 25 000 map). Beyond the hill the track emerges onto a low plateau before beginning to climb up the Azagador Member dip slope.

You are now stratigraphically below the Yesares Member gypsum, towards the southern margins of the Sorbas Basin, in the zone where patch reefs of the Cantera Member interdigitate with marls of the Abad Member. The low hills are patch reefs (see Excursion 4.4, Stop 1).

From the plateau take the right-hand track up the dip slope for another 500 m to a clump of small trees. Park here (at 783 017). Beyond here the track is totally impassable for all vehicles. Continue on foot. The track continues to climb up the eastern flank of a ridge towards a col.

In this area, note the unconformity between the gently northward-dipping Azagador Member carbonates (stratigraphically below the patch reefs), and the underlying, steeply dipping Tortonian marls. Note also extensive, structurally controlled, pipe-induced gullying along the track.

At the col (782 014) clamber up the ridge to the left (east) of the track to reach the summit of Cantona (marked by a triangulation pillar).

Stop 1: Panoramic view from the summit of Cantona (altitude 759 m; 784 012; Polopos 1 : 25 000)

Cantona itself is on Messinian reefs of the Cantera Member, where they cross from the southern margins of the Sorbas Basin over the Alhamilla uplift, to the northern margins of the Almería Basin (to the south) (see Excursion 4.3).

The Sierra de Alhamilla basement continues to the SW, and comprises mostly dark schists of the uppermost part of the Filábride complex, flanked by Triassic dolomites of the Alpujárride nappe (see Section 3.2). The mountain front coincides with a major fault system (see Excursion 3.5b), identifiable by the slivers of multicoloured, predominantly purple rocks. The village of Lucainena, below the very steep mountain front *c.* 8 km to your west, was developed as a mining centre, exploiting iron ores from mineralization associated with the basement. The same rock and topographic patterns can be used to trace the Sierra Alhamilla eastwards. Beyond them, with a similar structure, the high mountains are the Sierra Cabrera. On a clear day in the far NNW the (usually) snowcapped Sierra Nevada (Nevado–Filábride complex) is visible, and to their south are the Sierra Gádor (dominantly Alpujárride complex), forming the western boundary of the Tabernas Basin. The high mountains along most of the northern skyline are the Sierra de los Filabres (Nevado–Filábride complex; see Excursion 3.2), the major sediment source area for the Tabernas and Sorbas basins.

The low mountains and hills to the SE include the Cabo de Gata volcanic rocks. The low ridge beyond the 'plastic' agriculture of the Almería Basin to the south marks where the volcanics have been brought in along the Carboneras Fault System (see Excursion 3.5a). Beyond them are the volcanics of the Cabo de Gata itself (see Excursion 3.3a).

The sedimentary basins occupy the lower ground between the mountain ranges. To the north the Sorbas Basin is in front of you, the Tabernas Basin to the left (west) and the Vera Basin far to the right (east). The Almería Basin is to the south. The pale, marly, Tortonian rocks outcrop along the southern margins of the Sorbas and Tabernas basins, exposed along the west–east orientated valley of the Rambla de Lucainena to your west and the similar strike valley of the Rambla de Mizala to your east. They dip steeply to the north and strike obliquely NE–SW, an alignment picked out by the low sandstone hills in the centre of the Lucainena valley. To the south they are truncated by the mountain-front faults. They are overlain unconformably by the Azagador Member carbonates, which dip gently north and form the pronounced escarpments bounding the north sides of both the Lucainena and Mizala valleys.

Capping the escarpments, to the west at Risco de Sánchez (765 016) and to the east at Las Molatillas (800 020), are patch reefs of the Cantera Member (see Excursion 4.4). The reefs continue through Cantona, and can be traced SE resting directly on basement schists. To the south they are downfaulted, but continue into the northern margin of the Almería Basin (see Excursion 4.3), forming a major scarp around the southern margin of the Sierra de Alhamilla. The lateral equivalent of the reefs in the centre of the Sorbas Basin is the Abad Member marl, visible as the gullied white terrain *c.* 8–10 km away to the NE, where it is exposed along the lower Aguas valley. Above the Abad Member marl is the Yesares Member gypsum (see Excursion 4.5) forming the pronounced scarp north of the Aguas valley (8–10 km to the NE), and El Cerrón de Hueli, the hill to the ENE (at 814 023).

Terrain formed on the upper part of the Messinian basin-fill can be identified. Due north, beyond Hueli (792 028) and towards the village of Sorbas, the creamy coloured country marks the outcrop of the shallow, marine Sorbas Member (see Excursion 4.5), and away to its right the reddish-coloured terrain marks the outcrop of the Zorreras Member (see Excursion 5.3). The relatively level terrain away to the north of Sorbas and in the western part of the basin to the NW, together with the reddish terrain nearer to you in the NW, is formed on the Góchar Formation, representing the last phase of basin-filling (see Excursion 5.4).

The topographic contrasts between the basins reflect the different interactions between uplift and dissection during the evolution of the drainage network. The original drainage of the Sorbas Basin developed as centripetal drainage towards the basin centre (Mather, 1991; Mather & Harvey, 1995). The master drainage exited south, along the line of the modern Feos valley, as a transverse course across the low basement saddle between the Sierras Alhamilla and Cabrera (see Section 6.1), into the Almería Basin. This course was originally superimposed from the retreating end-Cariatiz Formation sea, and became antecedent with uplift of the Alhamilla–Cabrera axis. This zone is visible *c.* 10 km to the east, as a low in the Sierra Alhamilla ridge. Another original transverse drainage was the Lucainena–Alias system, which crossed the Sierra Alhamilla just to the south of Cantona. The modern canyon is visible *c.* 3 km to your south, beyond the fault zone.

The original north–south master drainage was captured near Los Molinos by the aggressive Lower Aguas, eroding headwards along the outcrop of the weak Abad Member marl (see Excursion 6.2). This area is visible as the eroded badland area *c.* 10 km to the NE, where the view extends down the course of the Aguas into the Vera Basin. Other captures have modified the original drainage network. Uplift along the southern margins of the Sorbas and Tabernas basins stimulated the development of subsequent drainage into the weak Tortonian rocks. The Rambla Mizala and the upper part of the Lucainena drainages developed in this way. Headwards erosion of the upper Lucainena was taking place during the Plio-Pleistocene, isolating the Sorbas drainage from older source areas within the Alhamillas (Mather 1993a, 2000a; Excursion 5.4). At the far end of the Lucainena valley, a remnant of the earlier basinal drainage is visible in the form of the shallow, concave, pediment slope profile that extends from the Alhamillas into the upper part of the basin. The capture process is still ongoing. The small valley that drains NNE from just below the summit of Cantona is part of the Sorbas drainage system. However, capture is imminent by the steep gullies which form the headwaters of the Rambla Mizala, which are cutting back into that drainage (788 017) to the NE.

From this view point the contrast between the deeply incised erosional nature of the southern part of the Sorbas Basin and the much less dissected northern part is evident. The Sorbas Basin contrasts with less elevated adjacent basins, where local relief is not so great. Visible to the south, the extensive plastic greenhouses are situated on coalescent Quaternary alluvial fans in the centre of the Almería Basin. The upper part of the Tabernas Basin, although undergoing dissection in the Pliocene or early Pleistocene, has not yet been affected by Late Pleistocene tectonically induced dissection. There are extensive coalescent Quaternary alluvial fans (see Excursion 6.3), in that area, just discernible from this view point as smooth, low-gradient surfaces *c.* 12 km to your WNW, adjacent to the tower of the Solar Research Station. The badlands of the deeply dissected western end of the Tabernas Basin (see Excursion 6.4) are not visible from here.

Chapter 3
The Development of the Neogene Basins

ANNE E. MATHER

3.1 Introduction

The sedimentary basins within the Almería Province are typically fault-bounded against sierras composed dominantly of metamorphosed Palaeozoic (or older) rocks (Priem *et al.* 1966; Nieto *et al.* 1997). Many of these faults are still active and are seismically monitored (University of Granada Geophysics web page, http://www.ugr.es/iag/iagpds.html). These structures have affected sedimentation in the basins throughout their evolution (Weijermars, 1991; Mather & Westhead, 1993). To the south of one of the major fault zones in the region, the Carboneras Fault Zone, volcanic rocks are prolific (see Fig. 1.1.1). Although some Pliocene volcanic activity has been recorded in the Cartagena area (Bellon *et al.* 1983), within the Almería Province the activity is predominantly of Tortonian age and was synchronous with marine sedimentation. It is the aim of this section of the guide to place the sedimentary basins in context by providing: (i) an introduction to the basement (the provenance for the later siliciclastic Neogene fill) and early tectonic evolution of the region (see Section 3.2); (ii) an introduction to the volcanic rocks which are locally intercalated with the early marine basin-fill (see Section 3.3); (iii) an example of the interaction of the earliest basin-fill and tectonics in the Tabernas Basin (see Section 3.4); and (iv) examples of the range of sedimentary deformation features and landform modification associated with regional Plio-Pleistocene fault activity (see Section 3.5).

The regional tectonic framework

The Betic Cordillera formed at the southern margin of the Iberian craton during the collision of the European/Iberian and African plates from the Late Mesozoic to Middle Cenozoic (De Jong, 1991; Monié *et al.* 1991) and together with the Moroccan Rif form the western limit of the Alpine Mediterranean belt (Fig. 3.1.1). The Hercynian basement of the Iberian craton is overlain by Triassic to Early Tertiary sedimentary rocks (Hermes, 1978; García-Hernández *et al.* 1980), which were deformed during the Miocene to form the External Zone of the Betic Cordillera. These comprise

29

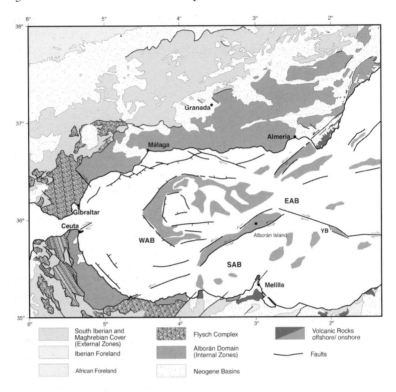

Fig. 3.1.1. Simplified geological sketch map of the Alborán Sea. EAB, Eastern Alborán Basin; SAB, Southern Alborán Basin; WAB, Western Alborán Basin; YB: Yusuf Basin (Comas *et al.* 1996).

a Pre-Betic Zone of predominantly platform carbonate sediments and a Sub-Betic Zone of deeper marine sediments. The Internal Zones comprise a stack of thrust sheets of more highly deformed Palaeozoic, Mesozoic and possibly older rocks, including low- to high-grade metamorphics. The Internal and External Zones are separated by Cretaceous to Early Miocene flysch sediments that mainly crop out in the western part of the Cordillera (Martín-Algarra & Vera, 1982; Fontboté, 1984).

In the Internal Zone the lowest pile of nappes, the Nevado–Filábride 'nappe complex', is composed predominantly of schists, the lowest part of which lacks strong evidence of horizontal displacement and could be attached to the underlying basement (Weijermars, 1991). The movement of the overlying units is complex (Weijermars, 1991). It has been suggested that they may have moved as much as 50–80 km from the south, if a dominant transport direction towards the north–NW is used, as suggested by Platt *et al.* (1983). These Nevado–Filábride rocks underwent a complex sequence of metamorphism related to subduction between 85 and 80 Ma (Weijermars, 1991). Overlying the Nevado–Filábride complex are two groups of thrust sheets, the Alpujárride and Maláguide complexes. The Alpujárride Group comprise mainly Triassic sedimentary and low-grade metamorphic rocks. The Maláguide Group, which has been largely removed by erosion from the Almería Betics, comprises dolomites and clastics of a wide age-range.

Although clearly related to the convergence of the Iberian and African plates, the detailed origin of the Betic Cordillera has been the subject of much controversy. Central to the controversy are the relations between the Betic Cordillera, the Rif of Morocco and intervening Alborán Sea (Fig. 3.1.1). The Betics and the Rif are thrust outward, away from the Alborán Sea. The Alborán Basin is underlain by relatively thin continental crust and an anomalously low-velocity upper mantle. Models include extensional collapse of thickened continental lithosphere (Platt & Vissers, 1989) and radial nappe spreading from a mantle diapir in the Alborán Sea (Weijermars, 1985). The lithology, metamorphism and deformation recorded within the nappe units and its regional significance are examined more fully in Section 3.2.

Following emplacement of the nappes, interaction between the African and Iberian plates then became dominated by strike-slip and extensional fault movement. The extensional tectonics (Middle Miocene–Tortonian) are thought to be related to the formation of the Alborán Sea (Comas *et al.* 1992), of which the Almería and Sorbas Basins were initially part, before becoming independent in the Tortonian. These fault systems dominate the neotectonic patterns of south-east Spain (Bousquet, 1979; Sanz de Galdeano, 1990) and have created a system of blocks and basins, with Betic basement locally uplifted in mountain blocks or locally downfaulted to form sedimentary basins. The local patterns of extensional or compressional tectonics are complex, reflecting interaction of fault blocks along the strike-slip and related fault systems (Keller *et al.* 1995).

The neotectonic patterns of the Almería region are dominated by the SW–NE-trending Carboneras Fault System and the north–south-trending Palomares Fault System (Weijermars, 1987). Other major faults are the east–west-trending Alhamilla and Cabrera bounding faults (Sanz de Galdeano, 1987). The uplifted mountain blocks comprise to the north the Sierra de los Filabres and in the centre the Sierras Alhamilla and Cabrera. To the SE of the Carboneras Fault Zone is the Neogene volcanic zone of the Cabo de Gata. Between the uplifted zones lie the Neogene sedimentary basins; south of the Filabres are the Tabernas, Sorbas and Vera Basins, and south of the Sierras Alhamilla and Cabrera is the Almería Basin

truncated in the south by the Carboneras Fault Zone from the Cabo de
Gata (Fig. 1.1.1).

Early stages of basin-filling

The Sorbas and Tabernas Basins were created as an east–west-orientated
half-graben, resulting from the switch in the Middle Miocene from compressional to local extensional tectonics. The oldest basin-fill deposits are
?Serravallian terrestrial conglomerates which crop out on the northern
flank of the Sierra Cabrera, where the clasts are Alpujárride, and in the
Tabernas Basin, where the clasts are Nevado–Filábride. In the Tabernas
Basin these terrestrial conglomerates, mostly proximal alluvial fan debris-
flows, rapidly give way to marine debris-flows, apparently in a fan-delta
environment (Kleverlaan 1989b; Doyle *et al.* 1997). Rapid subsidence ensued
into the Tortonian and thick sequences of muds, turbidites and subaqueous
debris-flows accumulated (Kleverlaan, 1987, 1989a,b; Haughton, 1994).

The main sediment source during the Tortonian was the uplifted
Filábres. At this stage of basin evolution the Sierra Alhamilla was largely
submerged, as is indicated by the presence of turbidites in the Huebro
graben, in the heart of the Sierra de Alhamilla. The Tortonian sediments
are best represented in the Tabernas Basin, where they include thick muds,
turbidites and debris-flows, with deposition dominated by several Nevado–
Filábride-fed submarine fan systems. These are examined in Section 3.4.
Continued deformation of the basin margins led to the development of
spectacular slump structures, and basin-wide megabeds, interpreted as
seismites (Kleverlaan, 1987, 1989a,b). Tortonian sediments in the Sorbas
Basin include turbiditic sandstones and debris-flow conglomerates within
a generally marly sequence, but in the Vera and Almería Basins tend to be
muddy with debris-flows.

South of the Carboneras Fault Zone the Cabo de Gata volcanics are
largely of Tortonian age (11–8 Ma, see Section 3.3). The volcanic rocks are
a suite of calc-alkaline rocks, ranging in composition from pyroxene
andesite to rhyolite. Within the volcanic field there are several calderas,
which are associated with gold, alunite and base metal mineral deposits.
The volcanic centres, which were probably emergent, are onlapped by thin
sequences of shallow marine carbonates with intercalated tuffs.

Towards the end of the Tortonian, regional compression replaced
the extension which had created the basins. The Tortonian sediments
became strongly folded by north–south to NNW–SSE compression (Sanz
de Galdeano & Vera, 1992) and the Sierras Alhamilla and Cabrera were
uplifted and their northern bounding faults became reverse faults. NE–
SW-trending basement faults brought about some separation between the
Sorbas Basin and the Tabernas and Vera Basins. This is apparent in the

marked unconformity which exists between the Messinian and Tortonian deposits of the Sorbas Basin. In the Tabernas Basin the unconformity is much less marked. Interactions between the tectonics and sedimentation during this period are demonstrated in Section 3.4.

Late stages of basin-filling

Continued tectonism throughout the Neogene and Quaternary, coupled with fluctuating global sea-levels, led to the continued isolation of the sedimentary basins and a transition from marine to continental depositional environments associated with the inversion of the basins (see Chapters 4 and 5). During the Quaternary, fault activity offset landforms and influenced regional erosional and depositional patterns (see Chapter 6). Section 3.5 explores the impact of the regional deformation on the more recent sedimentary basin development (Pliocene, Pleistocene and Holocene) by examining the impact of the main fault systems on the contemporaneous sediments and landforms. These data are used to provide information on rates of deformation over both Neogene and Quaternary time-scales.

3.2 The basement geology of the Almería Province

MARIA T. GÓMEZ-PUGNAIRE

Introduction

The Internal Zones (Fig. 3.2.1) are considered to be an allochthonous tectonic element comprising several complexes including rocks that have undergone one or more metamorphic processes. These complexes are (from top to bottom): (i) the Dorsalian and Pre-Dorsalian complexes; (ii) the Maláguide complex (Blumenthal, 1927; Durand-Delga, 1968); (iii) the Alpujárride complex (Van Bemmelen, 1927); and (iv) the Nevado–Filábride complex (Egeler & Simon, 1969). The Dorsalian and Pre-Dorsalian complexes consist of a number of small, dispersed, tectonic units that crop out along the contact of the Internal Zones and the Campo de Gibraltar flysch units. From a palaeogeographical point of view these units may be related to either the Alpujárride or the Maláguide complexes.

The Maláguide complex

The most extensive outcrops of the Maláguide complex (Blumenthal, 1935) appear in the western part of the Betic Cordillera, in the province of Málaga, in Sierra de las Estancias (Almería) and Sierra Espuña (Murcia) (Fig. 3.2.1). From a stratigraphic point of view the units belonging to this complex

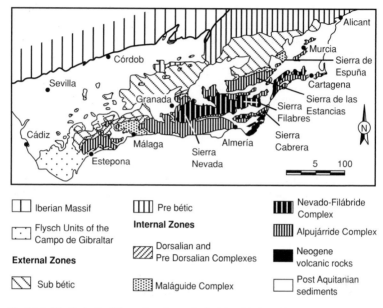

Iberian Massif

Flysch Units of the Campo de Gibraltar

External Zones

Sub bétic

Pre bétic

Internal Zones

Dorsalian and Pre Dorsalian Complexes

Maláguide Complex

Nevado-Filábride Complex

Alpujárride Complex

Neogene volcanic rocks

Post Aquitanian sediments

Fig. 3.2.1. The Betic Cordillera of south-east Spain (after Jabaloy 1993).

consist of a pre-Permian basement and a Triassic–Tertiary cover. The contacts separating the Maláguide complex from the underlying Alpujárride are low-angle normal faults (Aldaya *et al.* 1991; Galíndo-Zaldívar *et al.* 1989).

The Maláguide basement consists of lower unfossiliferous, fine-grained, graphite-bearing mica schists (Mon, 1969) and overlying fossiliferous sediments ranging from the Silurian to the Carboniferous (Soediono, 1971; Geel, 1973; Herbig, 1983, 1984). The Maláguide cover comprises Permo-Triassic red sandstones, Jurassic–Cretaceous shallow-water carbonates and pelagic limestones, and Tertiary nummulite limestones and marine siliciclastic deposits (Azema, 1961; Soediono, 1971; Geel, 1973).

The Alpujárride complex

The Alpujárride complex extends from Estepona (Málaga) to Cartagena (Murcia; Fig. 3.2.1). It consists, like the Maláguide complex, of two different series attributed to a pre-Permian basement and an essentially Mesozoic cover. However, the differences from the Maláguide complex are considerable, as the Palaeozoic and Mesozoic series have undergone Alpine (and probably older) metamorphism. Palaeontological data are scarce in a substantial part of the series and the age in some formations is only partially known.

The Alpujárride complex consists of several tectonic units that have been grouped into lower, middle and upper nappes (Junta de Andalucía, 1985) according to their metamorphic grade and lithology. The metamorphic grade increases from the lower to upper nappes, although in each individual nappe the metamorphic grade increases towards the bottom. From a lithological point of view, the upper nappes include thick sequences of Palaeozoic basement, in the middle nappes the basement sequence is thinner and in the lower nappes the basement materials traditionally attributed to the Alpujárride complex are absent.

Lithology

A complete lithological sequence of the Alpujárride complex consists of three main units (in ascending order): (i) dark metamorphic rocks (pre-Permian basement); (ii) ?Permian–Triassic pale phyllites and quartzites with minor carbonate; and (iii) a thick sequence of Triassic carbonates (Fig. 3.2.2).

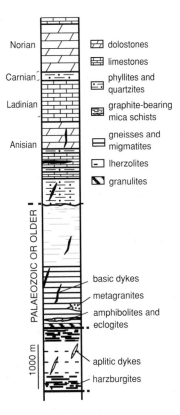

Fig. 3.2.2. Lithostratigraphic column of the Alpujárride complex.

Pre-Permian basement

The pre-Permian basement consists of up to 1000 m of graphite-bearing mica schists, migmatites, locally with eclogite and granulite intercalations and quartzites. These rocks show low- to intermediate-pressure metamorphism of different intensity depending on the profile although, in general, the intensity decreases eastwards.

In the westernmost part of the Alpujárride complex, on top of the graphite-bearing mica schists and quartzites, crop out one of the most interesting and largest Alpine-type peridotite bodies in the world, the 'Ronda peridotites'. They consist of lherzolites and harzburgites with minor amounts of dunite and peridotite (Obata, 1980).

Cover

The Alpujárride cover comprises a basal formation of phyllites, quartzites and metaconglomerates up to 1000 m thick, with thin carbonate, calcschist and gypsum layers intercalated towards the top. Permian–Middle Triassic age is attributed to the entire formation on the basis of its stratigraphic position (Egeler & Simon, 1969; Aldaya *et al.* 1979). This is overlain by carbonate rocks up to 2000 m in thickness. These are shallow-platform to basinal marine dolostones and limestones (Delgado *et al.* 1981). Dasycladacean algae and conodonts indicate an Anisian–Norian (Rhaetian) age for these carbonates (Kozur *et al.* 1985; Braga & Martín, 1987; Martín & Braga, 1987). Younger Alpujárride rocks are Jurassic–Cretaceous pelagic marls, radiolaritic marls and marly limestones.

Poorly recrystallized dykes and sills of basic rocks with very well-preserved igneous structure are locally abundant (e.g. in Sierra de Almagro) as intrusions in the carbonates and phyllites (López Sánchez-Vizcaíno *et al.* 1991).

Metamorphic evolution

The first event (M_1) of the Alpine metamorphism affecting the Alpujárride cover developed under high-pressure, low-temperature conditions (see Table 3.2.1; Bakker *et al.* 1989; Azañón & Goffé, 1997). Higher pressure and higher temperature conditions have been estimated in basement blackschist rocks (Torres-Roldán, 1981; Tubía & Gil-Ibarguchi, 1991; Azañón *et al.* 1996; García-Casco & Torres-Roldán, 1996) but the metamorphic evolution of the basement might not be unequivocally correlatable with the Alpine metamorphism in the Mesozoic cover. $^{40}Ar/^{39}Ar$ data indicate an age of 25 Ma for the end of the high-pressure event (Monié *et al.* 1991). However, recent SHRIMP U–Pb analyses in zircons indicate a younger age (19.9 ± 1.9 Ma) for the peak of this event (Sánchez-Rodríguez & Gebauer, 2000).

The second event (M_2) was produced by near-isothermal decompression, which yielded lower pressure assemblages overprinting the previous high-pressure minerals (Loomis 1972; Torres-Roldán, 1979; Bakker *et al.* 1989; Vissers *et al.* 1995; García-Casco & Torres-Roldán, 1996; Platt *et al.* 1996; Azañón *et al.* 1997). The age of the end of this metamorphic event has been estimated by ^{40}Ar/^{39}Ar dating at 19 Ma (Monié *et al.* 1991). The third event (M_3) developed in a low-temperature, low-pressure regime under greenschist facies conditions.

In the basement blackschists a progressive metamorphic evolution has been described from the biotite zone to the sillimanite–K-feldspar zone. Local migmatite layers indicate that partial melting conditions were reached. The age of this metamorphism is, at least in part, Hercynian (Sánchez-Rodríguez, 1998; Zeck & Whitehouse, 1999).

Deformational history and related structures

The following interpretation of the structure and tectonic evolution of the Alpujárride complex reflects the most commonly accepted hypothesis about the main deformational processes. However, crucial topics, such as the emplacement of the Ronda peridotites, and the number and sequence of the compressive and extensional episodes, are still controversial.

D_1 deformational phase

The oldest structure of the Alpujárride complex, an axial planar foliation (S_1), is recognizable only very locally as a folded and relict minor structure or as inclusions in the porphyroblasts, because it was largely obliterated by the main S_2 deformation. This deformational event was coeval with the high-pressure, low-temperature metamorphism (M_1).

D_2 deformational phase

The main structures are penetrative axial-plane isoclinal folds (F_2), an associated L_2 mineral lineation, and some mylonitic bands showing associated S-C fabrics (Álvarez, 1987; Tubía, 1988; Azañón *et al.* 1997). The condensed metamorphic mineral zones and their subparallelism with the S_2 tectonic fabric and the lithological sequences are interpreted as the result of vertical shortening produced by tectonic extension during a crustal thinning process. This extensional deformation took place during the exhumation of the Alpujárride complex (Bakker *et al.* 1989; Galindo-Zaldívar *et al.* 1989; Platt & Vissers, 1989; De Jong, 1991; González-Lodeiro *et al.* 1996; Azañón *et al.* 1997; among others).

D_3 deformational phase

Large-scale (km) F_3 folds affected the main S_2 foliation and associated L_2 lineation. A crenulation cleavage, S_3, developed in the hinge zones of small-scale F_3 folds. These folds, the most conspicuous structures in the Alpujárride complex, produced inversions and duplications in the metamorphic sequence (Azañón *et al.* 1997). However, other authors consider these duplications to be the result of ductile thrusting (Cuevas, 1990; Tubía *et al.* 1992).

Later deformational phases

Three Early to Middle Miocene extensional episodes have been distinguished with the following related structures: (i) extensional crenulation cleavage at the contact between the Alpujárride and the Nevado–Filábride complexes (Platt & Vissers, 1989); (ii) low-angle normal faults that affected the same contact (Cuevas, 1990; García-Dueñas *et al.* 1992; Vissers *et al.* 1995); and (iii) large, normal brittle faults (Galíndo-Zaldívar *et al.* 1989).

Nevado–Filábride complex

The Nevado–Filábride complex crops out as a large tectonic window under the Alpujárride complex. It occurs essentially in the Sierra de los Filabres and in the Sierra Nevada, where it builds the highest peaks of the Iberian Peninsula, and in smaller sierras in the south-eastern part of the Betic Cordillera (Fig. 3.2.1). The entire Nevado–Filábride succession has undergone one or more metamorphic processes.

Lithology

The Nevado–Filábride complex has been subdivided into two distinct major lithological units separated by an original unconformity (Fig. 3.2.3).
1 The *basement* (Egeler, 1963) is made up of a monotonous succession (> 4000 m thick) of dark, graphite-bearing mica schists alternating with quartzites, that increase upwards in thickness and abundance. Thin beds of dark marbles occur locally. Metagranite bodies crop out at the top in the easternmost part of the Sierra de los Filabres.
2 The *cover* (Egeler, 1963) consists of three formations (from bottom to top):
   (a) Tahal schists (Nijhuis, 1964). These are light-coloured, feldspathic quartzites and mica schists, up to 1500 m thick. Relics of sedimentary structures indicate a shallow-marine origin for this formation (De Jong & Bakker, 1991).

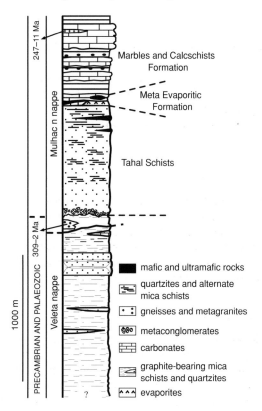

Fig. 3.2.3. Lithostratigraphic column of the Nevado–Filábride complex.

(b) Meta-Evaporitic Formation (Gómez-Pugnaire *et al.* 1994). This comprises up to 80 m of brecciated marbles, calcschists and fine-grained scapolite-bearing metapelites with gypsum pseudomorphs, barite and halite relics.

(c) Marble and Calcschist Formation (Voet, 1967). Laminated and massive calcitic and dolomitic marbles with some intercalated layers of mica schists and calcschists, up to 1000 m in thickness, in the Cóbdar–Macael area (López Sánchez-Vizcaíno, 1994).

Metamorphosed igneous rocks

These rocks constitute only a small proportion of the complete Nevado–Filábride complex but they are very significant for reconstructing the tectonic and metamorphic evolution of the complex.

The *metabasites* are intruded only in the cover materials as dykes of various size, from several centimetres to 10 m in width (Muñoz, 1986; Bodinier *et al.* 1987; Gómez-Pugnaire & Fernández-Soler, 1987; Bakker *et al.* 1989). There is a transition from undeformed metabasites preserving the igneous textures to rocks which are sheared and foliated. Gabbros with no metamorphic recrytallization can be found enclosed by eclogites and/or amphibolites in the less deformed igneous bodies (Voek, 1967; Morten *et al.* 1987; Franz *et al.* 1994). Aluminium-rich xenoliths in the gabbros indicate that the intrusion of the basic magmatism took place in the continental crust (Gómez-Pugnaire & Muñoz, 1991).

The *ultramafic rocks* are usually transformed into serpentinites forming lens-shaped bodies, 10–30 m thick, although exceptionally they may reach up to 300 m in thickness. Relics of the ultramafic rocks can locally be found with the earliest olivine–enstatite high-pressure metamorphic assemblage still preserved (Trommsdorff *et al.* 1998; Puga *et al.* 1999).

Igneous rocks of *acid* composition occur in both the cover and the basement, locally still preserving the unmetamorphosed granite protolith and original intrusive features, such as sharp contacts with the host metasediments and contact metamorphic rocks (Nijhuis, 1964; Voet, 1967; López Sánchez-Vizcaíno, 1994; Nieto, 1995). They are tourmaline-rich quartz–feldspar gneisses with a pronounced foliation and augen structures.

Age of the basement and cover

Fossil dates confirm a Palaeozoic (Middle Devonian; Lafuste & Pavillon, 1976) and older (Precambrian; Gómez-Pugnaire *et al.* 1982) age for dispersed sites within the basement. The age of the emplacement of the metagranites which were intruded in the basement was radiometrically (Rb–Sr whole-rock isochron) established as 269 ± 6 Ma (Priem *et al.* 1966) and recently (Sm/Nd method) as 307 ± 34 Ma (Nieto *et al.* 1997), which implies a pre-Upper Carboniferous age for the host rocks. Permo-Triassic and older ages have been determined from the igneous acidic intrusions found in the cover (Priem *et al.* 1966; Puga, 1976; Gómez-Pugnaire *et al.* 2000). Radiometric ages of the metabasites intruded in the cover range from 213 ± 2.5 Ma ($^{40}Ar/^{39}Ar$ method, Nieto *et al.* 1997) to 146 ± 3 (Rb–Sr whole-rock isochron; Hebeda *et al.* 1980).

The traditional assumption of a Permo-Triassic age for the Nevado–Filábride cover was based upon lithostratigraphic correlation with biostratigraphically dated Alpujárride sections (Fallot *et al.* 1960; Egeler & Simon, 1969). However, the radiometric ages of the interlayered gneisses indicate a Palaeozoic or older age for the entire Nevado–Filábride metasediments (Gómez-Pugnaire *et al.* 2000).

Metamorphic evolution

Alpine metamorphism in the cover series

The Alpine metamorphism in the Nevado–Filábride complex took place in three main stages that evolved from high-pressure/low-temperature to low-pressure/low-temperature regimes (Nijhuis, 1964; Puga, 1976; Gómez-Pugnaire, 1981; Vissers, 1981; Gómez-Pugnaire & Fernández-Soler, 1987; Bakker *et al.* 1989; De Jong, 1991). The maximum pressure–temperature conditions for each stage are summarized in Table 3.2.1.

1 The *first metamorphic stage* developed in a high-pressure and relatively low-temperature environment. During this stage, eclogites and blueschists formed from the metabasites and kyanite–talc–phengite assemblages in the evaporitic sediments (Gómez-Pugnaire *et al.* 1994). This metamorphic stage presumably developed syn- to post the S_1 deformational phase during a continent–continent collision event (see below).

2 The *second metamorphic stage* implies a continuous, nearly isothermal, decrease in pressure with respect to the first metamorphic event, probably related to the decompression produced by uplift (see references above).

3 The *third stage* is a low-temperature/low-pressure retrograde stage, which developed under greenschist facies conditions; this stage may be synchronous with the D_3 deformation event and is, for most authors, the end of the Alpine metamorphism. Nevertheless, a new reheating phase after the third stage has been proposed (Vissers, 1981; Bakker *et al.* 1989; De Jong, 1991). This late stage supposedly took place at temperature conditions similar to those of the second metamorphic event but at lower pressures (about 3–4 kbar).

Table 3.2.1. Pressure and temperature conditions in the Alpine metamorphic evolution of the Alpujárride and Nevado–Filábride complexes

Event	Pressure (kbar)	Temperature (°C)	Deduced from
Alpujárride complex			
First (M_1)	6–12	300–480	Cover
?	15	600–650	Basement
Second (M_2)	3	550	
Nevado–Filábride complex			
First (M_1)	12–20	550–650	Metabasites
	max. 16–18	about 650	Meta-evaporites
Second (M_2)	6–8	600–650	
Third (M_3)	< 4	350–400	

Metamorphism in the basement

Two types of basement series are distinguished on the basis of their different mineral assemblages and metamorphic evolution.

1 The basement series of the so-called Mulhacén nappe (Puga *et al.* 1974) is overlain by the rocks attributed to the Alpine cover. These basement rocks contain a low-pressure mineralogical assemblage which is not thermodynamically compatible with the high- to intermediate-pressure Alpine metamorphic assemblages found in the rocks of the overlying cover and therefore represents the mineral relics of a pre-Alpine metamorphic process (Puga, 1976; Gómez-Pugnaire & Sassi, 1983). In addition, the low-pressure minerals show the effect of a higher-pressure overprint that can be correlated with the Alpine event.

2 The basement in the so-called Veleta nappe (Puga *et al.* 1974) consists of a monotonous mica schist and quartzite sequence which occurs extensively at the bottom of the Nevado–Filábride complex with no clear indication of polyphase metamorphism (Gómez-Pugnaire & Franz, 1988). In spite of its large outcrop and thickness (up to 4000 m), the variations in mineralogical composition are small. The estimated pressure–temperature conditions are less than 500°C and 2–4 kbar (Gómez-Pugnaire & Franz, 1988).

Mineralogical features similar to those of the pre-Alpine metamorphism in the Mulhacén nappe suggest a pre-Alpine age for the metamorphism in the entire Veleta nappe (Gómez-Pugnaire & Franz, 1988). The lower pressure and temperature metamorphic conditions in the deepest rocks of the Nevado–Filábride complex indicate that the Mulhacén nappe was thrust over the Veleta nappe during its exhumation, after the main Alpine metamorphic stages recorded in the Mulhacén nappe.

Deformational history and related structures

The Nevado–Filábride rocks have undergone several ductile deformational events followed by brittle or ductile–brittle ones (Platt & Vissers, 1980; Vissers, 1981; Martínez-Martínez, 1986; De Jong, 1991; Galíndo-Zaldívar, 1993; Jabaloy, 1993; Soto, 1993). The hypotheses concerning the Nevado–Filábride evolution, number, geometric features, and geodynamic meaning of the different deformational phases are still controversial, but there is general agreement about the number and style of the main deformational structures.

Structures older than the main foliation

The intense overprint produced during the main deformation (D_2) almost completely obliterated the previous structures, making it very difficult to elucidate their geometry and kinematics. However, tight to isoclinal folds and associated axial-plane foliation (S_1 schistosity) can be observed locally (Nijhuis, 1964; Puga, 1976; Gómez-Pugnaire, 1981; Vissers, 1981). These folds vary in size from millimetres (as inclusions in the porphyroblasts) to a few metres. There is general agreement that the D_1 deformation phase occurred essentially at the end of the high-pressure metamorphic event.

Main deformational phase

During this deformational phase S_2 foliation was developed throughout the complex. It is an axial-plane crenulation foliation of close, tight and isoclinal folds which may range from large-scale major folds to outcrop-scale microfolds. A thick, ductile mylonitic foliation with associated brittle extensional crenulation cleavage is observed towards the contact with the Alpujárride complex, indicating a large-scale shear zone along the entire contact between the two complexes (Platt *et al.* 1984; Álvarez, 1987; Jabaloy *et al.* 1993; González-Lodeiro *et al.* 1996).

At present it is widely accepted that away from the shear zone the main blastesis of the second, intermediate-pressure, metamorphic event is basically synchronous with the S_2 foliation (Gómez-Pugnaire, 1981; Jabaloy, 1993), whereas the mylonitic foliation was produced afterwards. However, there is no agreement on the number of mylonitic foliations and their relative timing of formation. Some authors propose several shear zones, some of them away from the contact with the Alpujárride complex involving only Nevado–Filábride rocks (García-Dueñas *et al.* 1988). Moreover, whereas some authors include the mylonitic foliation in the D_2 deformational phase as an evolution of the previous non-mylonitic structures (Jabaloy, 1993; González-Lodeiro *et al.* 1996), others consider it to have developed during a subsequent deformation phase here named D_3 (Vissers, 1981; Bakker *et al.* 1989).

D_3 deformational phase

The main foliation and associated folds are deformed by millimetre- to kilometre-scale folds developing an axial-plane crenulation cleavage, which includes a variety of morphologies depending on the previous deformational structures as well as on the composition of the rock. F_3 and S_3 are not penetrative everywhere in the complex but are locally the most evident structures. The F_2 and F_3 fold systems which are approximately homoaxial, form interference patterns. According to some authors, the F_3 folds are the result of a compressive episode that produced a thrusting event responsible for the present arrangement of the tectonic units (Vissers, 1981; Bakker *et al.* 1989; De Jong, 1991; Soto, 1993). This phase took place under greenschist facies conditions during the last stages of uplift of the Nevado–Filábride complex.

Latest structures

These structures include large-scale, low-angle normal faults affecting the current contact between the Alpujárride and Nevado–Filábride complexes (Bakker *et al.* 1989; De Jong, 1991; González-Lodeiro *et al.* 1996). Brittle structures (joints, microfaults) and large, east–west-trending open folds, such as the Sierra Nevada dome in the core of which the Nevado–Filábride complex is exposed, also occur (Galíndo-Zaldívar, 1993). These open folds affected the Nevado–Filábride and Alpujárride contact and small gravitational structures developed as a result of slope instability (García-Dueñas & Comas, 1971; Galíndo-Zaldívar, 1993; Jabaloy, 1993).

Tectonic interpretation

The following tectonic reconstruction of the Internal Zone of the Betic Cordillera reflects the most commonly accepted hypothesis about the main deformational processes. There are, nevertheless, still several controversial topics, such as the age of the metamorphism in the basement series, the peridotite emplacement, and the number and sequence of the compressional and extensional episodes.

It is generally accepted that the D_1 deformational phase and associated high-pressure metamorphism took place during a crustal thickening episode that was the result of a continental collision produced by the convergence of the African and Eurasian plates. The radiometric dates of the high-pressure metamorphism in the Nevado–Filábride complex indicate that convergence began about 51 Ma (Monié *et al.* 1991), although earlier dates have been suggested (De Jong, 1991). During this first thrusting episode, the stack of the tectonic units created a thickened continental crust that was underthrust by more than 40 km. The higher pressure–temperature conditions of the Mulhacén rocks are regarded as being the consequence of their lower position in the tectonic pile, underlying the lower high-pressure metamorphic rocks belonging to the Alpujárride complex. However, this location may also be a result of the lower original stratigraphic position of the Mulhacén nappe with respect to the Alpujárride rocks (Gómez-Pugnaire *et al.* 2000). The original thrusts are not identifiable at present because of the overprint produced by later penetrative structures.

A pre-Alpine emplacement of the Ronda peridotites was proposed in some of the earliest studies (Kornprobst, 1976). Radiometric dates obtained in the host metasediments indicate an age of 22 Ma (Loomis, 1975; Priem *et al.* 1979; Michard *et al.* 1983) for the last low-pressure metamorphic recrystallization and, consequently, an Alpine age was assumed for the emplacement of the peridotites. Recent radiometric dates obtained in the migmatitic gneisses underlying the peridotites suggest, however, a Palaeozoic age (313 Ma, Sánchez-Rodríguez, 1998; 329 Ma, Acosta, 1998) for the partial fusion process during the regional metamorphism. This dating has once again raised the question of the date of the peridotite emplacement and of the regional metamorphism in the Alpujárride basement materials.

The presence of graphite pseudomorphs after diamond (Pearson *et al.* 1989; Davies *et al.* 1993) implies a deep lithospheric or asthenospheric origin for the peridotites. In the earliest models, emplacement was thought to be the result of some form of mantle diapirism (Van Bemmelen, 1969; Loomis, 1975; Torres-Roldán, 1979) that also supposedly produced the gravitational emplacement of the Betic complexes. In the most recent models, however, the Ronda peridotites represent a piece of lithospheric mantle of the subducted lithospheric slab that was uplifted during the continental collision and afterwards was tectonically emplaced in the continental crust.

A rapid decompression in the metamorphic pressure–temperature path (M_2) is coeval with the D_2 and D_3 deformational events in the Nevado–Filábride and Alpujárride complexes (Vissers, 1981; Gómez-Pugnaire & Fernández-Soler, 1987; Bakker *et al.* 1989; García-Casco & Torres-Roldán, 1996; Azañón *et al.* 1997), suggesting rock exhumation during extensional tectonics. This extensional episode evolved from ductile (the mylonitic foliation) to ductile–brittle (the extensional crenulation cleavage) and finally brittle (low-angle normal faults) conditions from the Burdigalian to the Middle Miocene (Monié *et al.* 1991; García-Dueñas *et al.* 1992; Watts *et al.* 1993). During this episode, vertical shortening of the previously thickened crust took place (Platt & Vissers, 1989; Galíndo-Zaldívar, 1993; Jabaloy, 1993). This thinning must be the origin of the extensional structures found at the contact between the Alpujárride and Nevado–Filábride complexes (Cuevas & Tubía, 1990; Balanyá *et al.* 1993, 1997; Azañón *et al.* 1996).

The origin of this extensional process has been related to detachment (Blanco & Spakman, 1993; Zeck, 1996), delamination (De Jong, 1991; Comas *et al.* 1992; García-Dueñas *et al.* 1992; Tubía *et al.* 1997) or convective removal (Platt & Vissers, 1989; Van der Wal & Vissers, 1993; Vissers *et al.* 1995) of part of the cold, gravitationally unstable lower lithosphere. These tectonic mechanisms induced uplift and heating in the previously subducted continental crust and mantle rocks. This extensional period continued producing successive extensional structures with different transport directions, ending about 16–17 Ma ago (De Jong, 1991; Monié *et al.* 1991).

Low-angle extensional fault systems developed during the latest extensional events. A crustal shortening event has, however, been proposed prior to them. Kilometre-scale folds (F_3), supposedly cut by low-angle faults, are thought to be related to a compressive episode that resulted in recumbent folds and nappes that inverted the lithological sequences (Crespo-Blanc *et al.* 1994; Azañón *et al.* 1997; Balanyá *et al.* 1997).

From the Middle Miocene onwards, the tectonic evolution of the belt is much better represented in the sediments deposited in the basins, which are treated in more detail in other chapters of this field guide.

3.2 Excursion: The Nevado–Filábride basement of the Sierra de los Filabres (one day)

MARIA T. GÓMEZ-PUGNAIRE

Introduction

The best-represented complex in the sierras surrounding the Sorbas and Almería Basins is the Nevado–Filábride, in which nearly the entire lithological sequences can be observed (Fig. 3.2.3). For a one-day field-trip in which the pre-Neogene basement can be well seen, we have selected the area from Tabernas to Los Gallardos in the Sierra de los Filabres. The Alpujárride complex has poor outcrops in this sierra and is not included in this field guide. Nevertheless, anyone who is interested in the Triassic phyllites, quartzites and dolostones from the Alpujárride can see them in the northern side of Sierra Cabrera, along the road from the Urbanización Cortijo Grande to the Urbanización Sierra Cabrera, as well as on the road from Níjar to Lucainena. The Maláguide complex is only represented in this area by scarce, dispersed and very small outcrops.

Directions to Stop 1

Stop 1 is located on the road from Tabernas to Tahal, 10.3 km north of Tabernas. At this point the road crosses an older road at a bend where it is possible to park. The new road is not shown on the 1 : 50 000 map, but it is represented approximately in Fig. 3.2.4.

Stop 1: Road section (593 114; Tabernas 1 : 50 000)

At this stop the dark rocks of the Veleta nappe basement (Nevado–Filábride complex) can be observed. They are foliated graphite-bearing mica schists consisting mainly of quartz, white micas and (oxy)chlorite, usually with chloritoid, plagioclase and garnet. Rounded black crystals of chloritoid (< 1 mm) and garnet crystals (red, very small, locally oxidized grains, 2–3 mm in size) may be observed. Oxychlorites appear as yellow–brown flakes very similar to biotite. The only variation observed in the lithological sequence is the occurrence of interstratified plagioclase-bearing quartzite from a few millimetres to 1 m in thickness. In general, schistosity is well developed and parallel to the original bedding, as indicated by alternating mica schists and quartzites.

Both schistosity (S_2) and a weakly developed lineation (L_2) can be observed at this section, both of which formed in a single deformational phase (D_2). In the outcrop, especially where there are quartzite layers, foliation is developed parallel to the axial plane of isoclinal folds (F_2), from millimetre-scale to 1–2 m. The geometry of the folds also causes this foliation to be parallel to the fold limbs and to the original bedding (S_0). In the hinges of the folds the schistosity intersects the bedding and gives rise to lineation parallel to the fold axes.

Tranfer to Stop 2

Travel north for approximately 1 km on the same road.

Stop 2: New road (600 119; Tabernas 1 : 50 000)

A metre-scale fold in quartzites can be observed. The S_2 foliation is here refolded by a crenulation (F_3) that is perfectly visible in the hinge zone of the large fold and the axis of the microfolds, and also observable in the lower limb. Note that S_2 and the axial planes of F_3 are parallel, indicating that the two deformations are homo-axial.

Transfer to Stop 3

Retrace the route southwards for 500 m. Take the turn-off leading to Senes. Drive through Senes and carry on 4 km beyond the settlement, on the road from Senes to Tahal. This road is not represented on the 1 : 50 000 map but it is drawn approximately in Fig. 3.2.4. The stop is easily identifiable by the change in rock colour, from black to light brown. From the turn-off almost to Stop 3, all the outcrops along the road are of the same dark grey mica schists with variable crenulation development.

Stop 3: Roadside (610 176; Macael 1 : 50 000)

A narrow layer (20–30 m) of light-coloured chloritoid–garnet mica schists occurs in this outcrop and can be traced over several metres. These schists have the same mineralogical composition as the underlying black mica

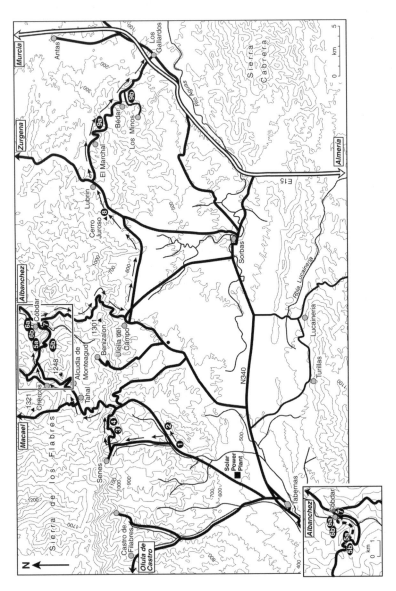

Fig. 3.2.4. Location map of the field-trip stops for Excursion 3.2.

schists. Both lithologies are from the basement of the Mulhacén nappe. The light schists have been interpreted, because of the identical mineralogical composition in dark and light rocks, as a metamorphosed weathering zone of the dark mica schists that produced oxidation of the graphite (Vissers, 1981).

Two phases of deformational structures can be observed, particularly evident in the contact between dark and light mica schists. A second generation of folds (F_3) refolds the S_2 structures that have similar features to those described for Stops 1 and 2. An axial planar crenulation cleavage, with a centimetre-scale variation in intensity, may be the most prominent structure locally (Fig. 3.2.5). The homo-axial character of the two deformation events is clearly observable in the 0.5 m-scale quartz vein located at the contact of the dark and light mica schists. A mineral lineation (L_3), as well as the folded L_2, can be observed in the hinge zones of the F_3 folds. The geometry displayed by this contact suggests that it is deformed by large-scale folds (F_4).

To observe the lithological and metamorphic differences between rocks outcropping at Stop 3 (Mulhacén nappe basement) and Stops 1 and 2 (Veleta nappe basement), retrace your steps to a point 500 m before Stop 3 (605 173; Macael 1 : 50 000). At this spot the contact between the Veleta nappe and the overlying basement of the Mulhacén nappe is visible. There is a ductile–brittle shear zone that juxtaposes the more quartzitic and less recrystallized rocks of the Veleta nappe and the dark mica schists described above.

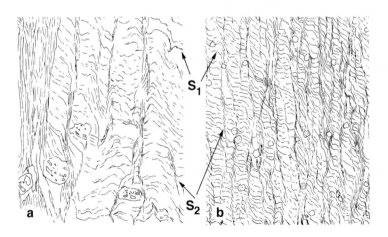

Fig. 3.2.5. Two stages of S_2 crenulation cleavage development sketched from the outcrop at Stop 3. Width of each area is 1 cm. (Modified after Vissers, 1981.)

Transfer to Stop 4

Continue along the road (northwards) for 150–200 m past Stop 3.

Stop 4: Roadside (610 178; Macael 1 : 50 000)

The rocks cropping out above the light mica schists are the Tahal schists, which are the basal formation of the Nevado–Filábride cover. The succession cropping out near Chercos Viejo village (apart from a few carbonate and metabasite intercalations) belongs to the Tahal Schist Formation. Although this stop is a good place to see the relationships between basement and cover, the best section to make structural and mineralogical observations in the Tahal schists is along the road from Cóbdar (Stop 7) to Uleila del Campo.

This is a rather monotonous sequence of light psammitic schist with many intercalations of quartzites from millimetres to several metres in thickness. In adjacent sections the lithologies vary in quartz content from mica schists to quartzites.

The mineralogical composition recognizable with the naked eye consists of quartz, white mica, chlorite, plagioclase and local amphibole. Chlorite is particularly easy to recognize because of its green colour and large crystal size (up to 1 mm). White, irregular plagioclase porphyroblasts up to a few millimetres across are conspicuous. Occasional idioblastic red garnet and dark green chloritoid and amphibole can be observed along the road up to Tahal.

The sequence is folded by centimetres- to hundreds of metres-scale folds that affect the structures related to the D_2 deformation event. These folds also developed an axial planar crenulation cleavage, but in these competent lithologies the S_3 schistosity is less penetrative and transposition of S_2 and the bedding is rarely observed. Where competent alternating quartzite layers occur, the F_3 folds are open and the crenulation cleavage underwent refraction across the adjoining competent and incompetent layers, resulting in fans that converge towards the inner arcs of the folds. The schistosity fan structures are widespread.

Transfer to Stop 5

Continue along the road to the junction at 629 171. Turn left to Tahal. Take the road from Tahal to Alcudia de Monteagud and from there to Chercos. Take a small paved road 1.5 km before Chercos, signed for Cóbdar. It is inaccessible for large buses, but vans and minibuses can get through. If in a bus go directly to Stop 6. This stop is about 6.5 km from Chercos, at the point where the present road crosses the older, abandoned one (Fig. 3.2.4). The section to be visited lies from this point down to the river. Vehicles are best parked by the river.

Stop 5a: Road from Chercos to Cóbdar (689 236; Macael 1 : 50 000)

At this stop the metapelites belonging to the meta-evaporitic formation can be observed. They are dark grey, massive rocks showing white, round or almond-like scapolite crystals of variable size (ranging from < 1 mm to 2–3 mm, Fig. 3.2.6). These rocks are very similar to the metabasite dykes cutting across them and can be difficult to distinguish in the outcrop because of the lack of preferred mineral orientation (foliation or lineation) and the very fine grain size of the matrix (0.02–0.05 mm). At this stop the distribution of the scapolite crystals is very irregular. Tourmaline may be visible, together with scapolite, in the outcrop. It occurs as elongated black crystals < 2 mm in size, showing as irregular a distribution as the scapolite crystals. The local differences in the fluid composition during metamorphism (NaCl- or B, F-richer fluids) may be responsible for the sharp variations of some components. It is worth noting that the rocks contain polycrystalline aggregates, not visible in the outcrops, consisting of quartz with many inclusions of barite and anhydrite, that show the typical shapes of gypsum crystals.

Along the section, strongly brecciated marbles alternate with the meta-evaporites and both are intruded by several dykes of porphyritic

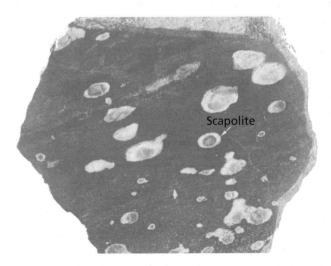

Fig. 3.2.6. Hand sample (Stop 5a) of biotite-rich metapelite showing rounded scapolite crystals. Zonation in the crystals is caused by the cores being altered to carbonate. Crystal indicated by arrow is 2 mm in diameter.

eclogitized basic rocks. The metabasites and meta-evaporites are especially difficult to distinguish, but the green colour, the larger grain size, and the rectangular shape of the plagioclase phenocrysts in the former are diagnostic. The intrusion of the dykes in these rocks is evidence of the continental setting of the Nevado–Filábride magmatism.

Some 800 m NW along the road from Stop 5a is Stop 5b.

Stop 5b: Alternative stop to see the meta-evaporites with well-developed scapolite crystals (684 238; Macael 1 : 50 000)

In addition to the meta-evaporites, irregular white kyanite + talc + white mica pseudomorphs (< 1 mm) after gypsum also occur in this section. This is the highest pressure assemblage registered in the metasediments of the Nevado–Filábride cover.

On the road from Chercos to Cóbdar, marbles and calcschists crop out. The white marbles belonging to this formation are worked in quarries visible from the road, with one of the largest being located over Cóbdar village, about 1 km eastwards. The quarries increase in number and size from Chercos to Macael, where the greatest volume of marble is extracted in Spain. The Macael marble is a coarse-grained, almost pure calcitic marble alternating with yellow–brownish and white–grey banded dolomitic marble. In the outcrops opaque and white mica flakes are visible to the naked eye.

Transfer to Stops 6 and 7

On the same road go eastwards until you arrive at Cóbdar. At the bridge, near the swimming pool, is the best place to park. The streets of the village are very narrow and it is better to go on foot. The best place to make the observations are on the hill immediately north of the sports centre. To get there, follow the gravel road towards the cemetery, go past the sports centre and take a hairpin bend around a small house. The road then doubles back on itself, striking east. Just where it begins to go east, there is a ravine straight ahead (Stop 6a; 701.5245; Macael 1 : 50 000) and a pathway off to the west. Take the pathway until you are situated right above the sports centre (Stop 6b; 700 245).

Stop 6a: North of Cóbdar village (702 245; Macael 1 : 50 000)

This location is the main place in the Nevado–Filábride complex where unmetamorphosed relics of igneous rocks can be observed and where crustal xenoliths embedded in basic rocks have been recognized.

Dark brown gabbros are cross-cut by aphanitic, porphyritic and doleritic dykes (Fig. 3.2.7). These rocks have not undergone metamorphism and

Fig. 3.2.7. Gabbro from Stop 6a (G) cut by very fine-grained, porphyritic dykes (P).
In the outcrop phenocrysts of rectangular plagioclase appear white and idioblastic olivine
appears reddish due to alteration are present. Width of photo is 2 m.

the mineralogical composition is almost completely igneous. Yellow–green
olivine, black pyroxene and white plagioclase grains can be identified in the
outcrop.

From Stop 6a traverse *c.* 100 m due west to Stop 6b.

Stop 6b: North of Cóbdar village (700 245; Macael 1 : 50 000)

At this locality, abundant, large (up to 5 cm), rectangular white crystals are
widespread (Fig. 3.2.8). They consist of pseudomorphs after andalusite
crystals that may be identified in the core (the pink colour) surrounded by a
corona of green spinel. These andalusite crystals represent crustal inclusions (xenoliths) enclosed by the basic rocks during their emplacement.
The andalusite alteration—and partial melting—was produced after their

Fig. 3.2.8. Andalusite xenoliths (up to 5 cm) in an unmetamorphosed gabbro (Stop 6b).

incorporation into the hot igneous material. The xenoliths indicate the continental emplacement of the Nevado–Filábride magmatism.

Transfer to Stop 7

Return to the bridge and walk up the Cóbdar to Uleila del Campo–Albanchez road in a NE direction. Stop 7 runs from the bridge at Cóbdar for about 1 km along this road.

Stop 7: Cóbdar Bridge (from 702 243 to 705 244; Macael 1 : 50 000)

Massive amphibolitized eclogites, showing perfectly preserved igneous textures, crop out. Along this section ophiolites and gabbroic rocks as well as porphyritic, very fine-grained rocks appear. Although the relationships between these two types of rock are not clearly distinguishable as a result of the metamorphic overprint, they are the same gabbros cut by dykes observed at the previous stop. In spite of their igneous texture, these rocks have been completely eclogitized and the entire mineral assemblage is metamorphic. The most evident minerals are the epidote that constitutes the pseudomorphs after igneous plagioclase (the white rectangular or irregular minerals), the small grains of yellow–green omphacite and the reddish garnet crystals. The dark green amphibole is widespread in the matrix and also occurs as larger crystals, together with white plagioclase, filling the later veins.

At approximately 200 m from Cóbdar, a small gabbro outcrop (as in Stop 6) appears completely surrounded by eclogitic rocks. The gabbro probably represents the protolith of the surrounding eclogites that escaped the metamorphic recrystallization when the fluids from the adjacent metasediments were exhausted, as the sharp contact (clearly visible in this outcrop) with the eclogites suggests. This site is easily located by the notable colour differences between the eclogitized rocks (dark green) and the igneous relics (dark brown).

Transfer to Stop 8

Follow the road from Cóbdar to Uleila del Campo, and up to 2 km past the cross-roads of the Cóbdar to Albanchez and Cóbdar to Uleila del Campo roads (Fig. 3.2.4). The metabasites and the upper carbonate formation of the Nevado–Filábride complex crop out along the road. From this point to Uleila del Campo (from 706 250 to 708 161) the road goes through the Tahal schists described at Stop 4. Stop 8 is on the road from Uleila del Campo to Lubrín (Fig. 3.2.4). The stop is at a large, open area, located about 11.5 km from Uleila del Campo.

Stop 8: Cerro del Jaroso (803 180; Vera 1 : 50 000)

The outcropping rocks are also part of the cover of the Nevado–Filábride complex overlying the Tahal schists and the meta-evaporite formation. At this stop, layers up to 2 m thick of white and bluish grey calcitic and light brown dolomitic marbles with intercalated dark green amphibolites appear (Fig. 3.2.9). These rocks crop out continuously over the next 2 km until Venta de la Huertecica (817 185). The white and brownish varieties of marbles are almost pure carbonate rocks whereas in the grey variety, white mica flakes and elongated amphibole crystals are abundant and visible to the naked eye. The amphibolites display foliation parallel to the contact with the carbonate layers and some of them also show well-developed banding defined by strongly orientated light green to white lenses and spindles. The spindles, consisting of epidote + white mica, are pseudomorphs

Fig. 3.2.9. Metabasite dyke showing an accumulation of elongated relics of igneous plagioclase in the centre of the dyke (Stop 8). Plagioclase phenocrysts are now altered to epidote and white mica. Also note chilled margin visible at the bottom of the photograph, without plagioclase phenocrysts.

after igneous phenocrysts of plagioclase that clearly disappear towards the contacts with the alternating marbles. This reflects chilled margins of intrusive basic igneous material. These features indicate that the basic rocks were intruded as sills into the carbonates.

The rocks are folded by open, similar and in some places isoclinal, centimetre- and kilometre-scale folds (F_3), that affected the main foliation (S_2). S_2 is the only foliation visible in the rocks. Some of these large folds can be observed on the sides of the ravine parallel to the road.

A serpentinite body occurs at the highest part of the hill at Stop 8 (803 180; Fig. 3.2.4). The climb is not easy, but it is the only accessible body of serpentinite in the area. Between the road and the serpentinite there is a small outcrop of garnet-bearing (up to 3 cm) amphibolites richer in amphibole than similar rocks at this stop and lacking epidote spindles. Overlying these rocks are metre-scale outcrops of basic rocks metasomatized by Ca-rich fluids and transformed into rodingites. These rocks consist of a massive garnet and clinopyroxene matrix cross-cut by veins of idioblastic garnet. The serpentinites are massive light green rocks with a poorly developed foliation parallel to the main regional foliation (S_2). Their present conformable occurrence is probably the result of the deformational phases that affected the cover of the Nevado–Filábride complex.

Fig. 3.2.10. Tourmaline-rich gneiss (Stop 9) showing augen texture and well-developed platy foliation. The augen consist of K-feldspar. Width of photo is 1.2 m.

Transfer to Stop 9

From Cerro del Jaroso drive to Bédar and Los Gallardos. There are two alternative localities with easily accessible good outcrops of the granite body intruded into the basement of the Mulhacén nappe. The two sections are located near Bédar village (Fig. 3.2.4), the first one (Stop 9a) 1.5 km from El Marchal on the road from Lubrín to Bédar, and the second one (Stop 9b) 1.1 km from Bédar on the road from Bédar to Los Gallardos.

Stop 9: (a: 876 190–879 181; Vera 1 : 50 000)
(b: 917 158; Vera 1 : 50 000)

At Stops 9a and 9b a body of granite intruded into the Palaeozoic basement of the Nevado–Filábride complex can be seen, although the original granite mineralogy and textures were almost completely transformed into gneiss during the Alpine metamorphism.

There are two textural types: (i) 'augen' gneisses; and (ii) banded gneisses. The first type consists of large lenticular crystals or crystalline aggregates of K-feldspar set in a matrix showing extremely penetrative foliation (Fig. 3.2.10). The second type is even-grained gneisses characterized by alternating dark and light layers parallel to the foliation. The differentiation in layers is caused by the preferential concentration (metamorphic segregation) of melanocratic (tourmaline, green and brown biotite, chlorite, magnetite) and leucocratic (plagioclase, quartz, white mica, K-feldspar) minerals. All the transitions between these two types can be observed. The most conspicuous mineralogical features in the outcrops are the presence of large tourmaline crystals, several millimetres in length, and large flakes of green or brown biotite and white micas.

The rocks are highly deformed, with mylonitic foliation and well-developed extensional lineation being the most evident structures.

3.3 Volcanics of the Almería Province

JUAN M. FERNÁNDEZ-SOLER

Introduction

The volcanic region of south-east Spain lies in the eastern sector of the Betic Cordilleras, mainly within the Internal or Betic Zones, with only some outcrops of ultrapotassic rocks located in the External Zones. In this volcanic zone there is a wide representation of facies typical of active continental margins, including rocks of the calc-alkaline series, the high-K calc-alkaline series and the shoshonitic series, together with a well-developed zone of ultrapotassic rocks (lamproites).

Types of volcanism

Most of the volcanism (Figs 3.3.1 and 3.3.2) developed in the Middle–Late Miocene, producing four main groups of lithologies (López-Ruiz & Rodríguez-Badiola, 1980; Hernandez *et al*. 1987).

1 A typical calc-alkaline series, from basaltic andesite to rhyolite, is present in the Cabo de Gata with a few small outcrops near Aguilas (e.g. Bordet, 1985; Cunningham *et al*. 1990; Toscani *et al*. 1990; Fernández-Soler, 1992). The rocks range in age from about 15 to 7 Ma (Bellon *et al*. 1983; Di Battistini *et al*. 1987).

2 A peraluminous volcanic province, consisting of scattered volcanic centres in Almería and Murcia provinces, which formed by anatectic processes of crustal materials, and interaction with magmas of mantle origin (e.g. Zeck, 1970). These rocks belong mostly to the high-K, calc-alkaline and shoshonitic series, and frequently contain igneous and metamorphic enclaves. Their peraluminous character is manifested in the common presence of minerals such as cordierite, garnet, andalusite and spinel. Their ages range from 11 to 6.8 Ma (Hernandez *et al*. 1987).

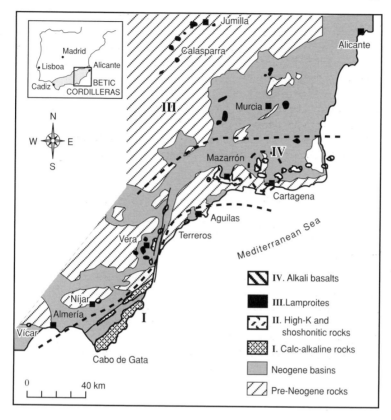

Fig. 3.3.1. Distribution of Neogene volcanic zones in south-east Spain. I, Calc-alkaline rocks of the Cabo de Gata Group; II, high-K calc-alkaline and shoshonitic rocks (peraluminous volcanics) of the Níjar–Vera Mazarrón group; III, ultrapotassic rocks (lamproites) of Murcia and Vera; IV, alkali basalts of Cartagena.

3 An ultrapotassic (lamproite) zone (e.g. Venturelli *et al.* 1984; Mitchell & Bergman, 1991), formed by small volcanic centres dispersed across a wide area in the provinces of Almería and Murcia.

4 Some small centres of Pliocene alkaline basalts and basanites in the area of Cartagena (e.g. Dupuy *et al.* 1986). These basalts contain abundant inclusions of peridotite and lower-crust granulite, and have been dated at 2.6–2.8 Ma (Bellon *et al.* 1983).

In the Alborán Sea (Fig. 3.1.1), well-log and geophysical data indicate that magmatic activity has been common since the earliest Middle Miocene (Comas *et al.* 1992). Major periods of volcanic activity are recorded in the

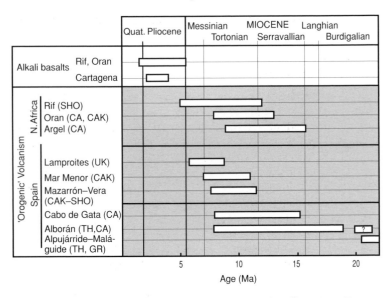

Fig. 3.3.2. Chronology of Neogene magmatic activity in the Betic–Rifean orogen. Data mainly from Hernandez (1983) and Bellon *et al.* (1983), Torres Roldán *et al.* (1986), Di Battistini *et al.* (1987), Hernandez *et al.* (1987), Hoernle *et al.* (1999) and our unpublished data. Magmatic affinities are indicated as follows: CA, calc-alkaline; CAK, high-K calc-alkaline; GR, granitoid dykes; SHO, shoshonite; TH, tholeiitic; UK, ultrapotassic.

Langhian–Serravallian, Serravallian–Tortonian and Messinian (Comas *et al.* 1992). Samples dredged from the Alborán sea-floor include two pyroxene andesites and basaltic andesites, and also some dacites and tuffaceous materials similar, at first appearance, to the calc-alkaline rocks of Cabo de Gata (Molin, 1980). Recent sampling by submersible has shown a widespread development, in the Alborán sea-floor, of a low-K, tholeiitic basaltic series (similar to back-arc basin basalts) associated with calc-alkaline rhyolitic volcanism, and also the presence of high-alumina basalts in some locations. Some volcanic pebbles have also been collected from Ocean Drilling Program Leg 161 Sites 977 and 978. They range from basalt to rhyolite, belonging to the tholeiitic, calc-alkaline and shoshonitic series (Hoernle *et al.* 1999). $^{40}Ar/^{39}Ar$ ages range between 6.1 and 12.1 Ma. In addition, the Alborán islet, the only emergent point of the Alborán sea-floor, comprises a layered sequence of augite–hypersthene basaltic and andesitic rocks (Hernández-Pacheco & Ibarrola, 1970). According to available geochemical data, the rocks belong to a low-K, tholeiitic series, which, together with their trace element signature, indicates an affinity to back-arc basin suites.

Significance of the volcanism

The origin of volcanism in south-east Spain is a matter of debate. Earlier models suggested a relation to an oceanic plate which subducted northwards under the Alborán Basin from an ancient suture in North Africa (Araña & Vegas, 1974). This model was later modified by López-Ruiz & Rodríguez-Badiola (1980), who explained the occurrence of calc-alkaline, high-K calc-alkaline and shoshonitic rocks by the subduction of the Alborán sea-floor itself. This would be consistent, in their opinion, with the observation that potassium contents increase with the depth of the subduction zone. This hypothesis is, however, incompatible with the geological evolution of the Betic–Rif orogen and with the appearance in North Africa of similar volcanic occurrences. In other hypotheses (e.g. Puga, 1980) the relation was made to an older (Cretaceous) subduction of oceanic crust.

More recent hypotheses, based on the observed relationship between volcanism, regional fracture systems (NE–SW and NW–SE), the evolution of the Gibraltar Arc, and the development of the Neogene basins, as well as the geochemical features of the magmatic products, seem to exclude a direct relation of volcanism to any active subduction of oceanic crust. According to Hernandez *et al.* (1987) and De Larouzière *et al.* (1988), the Betic–Rifean magmatism is genetically related to a large-scale 'trans-Alborán' strike-slip fault system.

The origin of most of the Neogene volcanism appears to be related to the development of the Alborán Sea basin, formed as a consequence of extensional processes along low-angle normal faults, acting as a 'back-arc basin' behind the Gibraltar Arc, and producing an extremely thin crust under the Alborán Sea (e.g. Platt & Vissers, 1989; Turner *et al.* 1999). The magmatism would then be related to an asymmetric extensional tectonic regime, after the Alpine collision. These extensional processes could be related to a delamination or detachment of the root of the lithosphere thickened during the continental block collision. Melting is triggered by the huge, fast decompression as well as by the ascent of hot asthenospheric mantle over the delaminated slab (e.g. Platt & Vissers, 1989). The presence of a detached, vertical, SW–NE-striking lithospheric slab is confirmed by seismic tomography studies (Blanco & Spakman, 1993). According to Zeck (1996), the slab would correspond to a portion of the subducted oceanic lithospheric slab, which became detached after the collision of the Iberian plate and the continental Betic–Ligurian lithosphere.

**Calc-alkaline volcanism of the Cabo de Gata (Excursion 3.3a,
Stops 1–6; Excursion 3.3b, Stop 1)**

The Cabo de Gata chain, 10 km wide by 40 km long, is the most extensive
volcanic area in the Betic Cordillera (Fig. 3.1.1). It is separated from the
Almería Basin by the Carboneras Fault, a sinistral transcurrent fault
zone that displaced the volcanic chain from its original position in the
Alborán Sea. Similar calc-alkaline volcanic rocks have also been exposed
by the tectonic activity within a 1.5-km-wide ridge along the trace of the
Carboneras Fault System (La Serrata de Níjar).

Other small outcrops of calc-alkaline andesites are found about 40 km
to the north of the Cabo de Gata field, near the localities of Aguilas
and Terreros. These rocks are associated with Langhian–Serravallian
sediments, and have been dated at 14.1 Ma by Bellon *et al.* (1983). Some
andesite breccias are found about 40 km to the west, near Vícar, and
several layers of volcaniclastic rocks are intercalated in the Upper Miocene
sequences (Ott d'Estevou, 1980; Pascual-Molina, 1997).

The Cabo de Gata series comprises basaltic and pyroxene andesites,
pyroxene–hornblende andesites, pyroxene dacites, hornblende dacites,
cummingtonite-bearing dacites, and biotite rhyolites (rhyodacites) as the
main petrographical groups (SiO_2: 54–75%). Geochemically, these groups
overlap. Their major and trace composition is typical of that of other
orogenic, medium-K calc-alkaline series, with high Al_2O_3 and CaO,
moderate FeO and K_2O, and low Na_2O, TiO_2 and P_2O_5, moderate incompatible element contents, and very low compatible element contents.
Remarkably, $^{87}Sr/^{86}Sr$ ratios in the mafic terms are high (0.7083–0.7139)
and $^{143}Nd/^{144}Nd$ values are low (0.512144–0.51239; data from Toscani
et al. 1990; Fernández-Soler, unpublished data) and $\delta^{18}O$ values are high
(8–11‰; López-Ruiz & Wasserman, 1991).

Geochemical calculations and petrographical features indicate that the
most basic magmas were produced by partial melting of a subduction-
affected mantle section, possibly contaminated by material derived from
the continental crust. Later evolution of these melts was by fractional
crystallization and contamination by crustal rocks, producing the most
acidic members (rhyolites) from andesitic magmas. A large fraction of the
Cabo de Gata dacites originated by mingling and mixing of the andesitic
and rhyolitic magmas in shallow magma chambers (Toscani *et al.* 1990;
Fernández-Soler, 1992).

There are two main volcanic groups.

1 The Lower Volcanic Group (Middle Miocene to Tortonian; *c.* 15–10 Ma).

2 The Upper Volcanic Group (Upper Tortonian; *c.* 9–7.5 Ma).

Both groups contain all the petrological aspects of the calc-alkaline series,
from basaltic andesites to rhyolites, although andesites and dacites are the
most abundant.

The volcanic rocks are covered by carbonate rocks from the Upper Miocene. Much of the volcanism was produced below or near sea-level, and this is manifested in the formation of peculiar volcanic facies. Sedimentary marine intercalations are present in some localities, especially on top of the lower volcanic group (Braga *et al.* 1996b; Betzler *et al.* 1997).

The predominant type of volcanic activity was the extrusion of dacitic domes, which formed large fields of emission points linked by SE–NW and SW–NE fractures. In addition, some lava flows and strombolian emissions of basaltic andesites and pyroxene andesites can also be recognized. The acidic rocks also include frequent pyroclastic layers. The most remarkable are ignimbritic eruptions related to the formation of the Rodalquilar Caldera complex, a large volcanic structure that was affected by hydrothermal, acid-sulphate alteration, resulting in the production of Au mineralization which has been exploited intermittently since the last century (Rodalquilar mines; Lodder, 1966; Sanger von Oepen *et al.* 1989; Rytuba *et al.* 1990; Arribas, 1993; Arribas *et al.* 1995).

The Rodalquilar Caldera is an oval, 4 km × 8 km, collapse structure (Fig. 3.3.3). The caldera-fill mainly comprises two pyroclastic units: the Cinto and the Lázaras Ignimbrites (Rytuba *et al.* 1990). Each of these units was produced as the result of a highly explosive eruption, successively associated with the collapse of the Rodalquilar Caldera and the Lomilla Caldera, a small caldera nesting within the former (Fig. 3.3.3). Outflow ignimbrites extended to the north (Las Negras sector) and to the south, where they make up most of the La Rellana plateau. Associated with these processes was the resurgence and tilting of the basin-fill as a consequence of the intrusion of subsequent andesitic domes. This also caused intense hydro-thermal alteration of the volcanic rocks and the development of gold–alunite mineralization, exploited in the Cerro Cinto mines (Arribas *et al.* 1995).

Below the Rodalquilar Caldera complex, the Lower Volcanic Group (older than *c.* 10 Ma) in the southern Cabo de Gata chain is made up of three main units (from bottom to top).

1 A lower unit of rhyolitic ignimbrites, clearly recognizable by its white colour, associated with some rhyolitic–dacitic domes. This unit mainly crops out in localities along the coast.

2 Amphibole andesites and pyroxene andesites (breccias, domes, flows, ignimbrites, etc.) that make up most of the southern part of the Cabo de Gata chain. Most of them are intensely affected by hydrothermal alteration and contain Pb–Zn–Au deposits (Santa Bárbara mines).

3 Scarcely altered amphibole andesites and dacites (dome fields, ignimbrites, other pyroclastic rocks, etc.) with some pyroxene andesites. This unit is located in the Frailes Caldera (Cunningham *et al.* 1990). It contains some sedimentary, marine intercalations, especially abundant at the top of the sequence. Radiometric ages for this unit range from 10.8 to 12.2 Ma (Di Battistini *et al.* 1987).

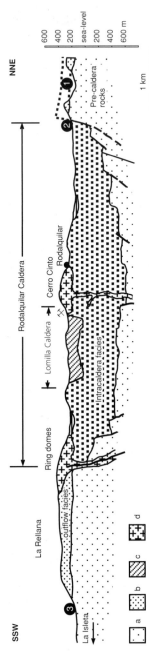

Fig. 3.3.3. NNE–SSW section through the Rodalquilar Caldera. The equivalent position of Stops 1–3 are shown: (a), pre-caldera rocks; (b), Cinto ignimbrite; (c), Lázaras ignimbrite; (d), lava domes. (Modified after Arribas, 1993.)

The rocks of the Frailes Caldera were partially covered by basaltic and pyroxene andesites of the Frailes volcano, dated at 8.5–8.6 Ma (Di Battistini *et al.* 1987). Other late pyroxene andesite volcanoes are located at Los Lobos hill and Mesa Roldán. These isolated volcanoes represent the most recent volcanic activity in the area and their geochemical composition is the least evolved in the Cabo de Gata sector (Toscani *et al.* 1990).

Peraluminous volcanics of south-east Spain (Excursion 3.3b, Stop 2)

The Hoyazo (or Joyazo) de Níjar structure belongs to a suite of cordierite-bearing calc-alkaline to shoshonitic volcanics that occur as small volcanic centres in several localities in south-east Spain (Fig. 3.3.1). The suite is distinguished by its strong peraluminous character, which seems to indicate a crustal anatectic origin for this group of rocks. One of the most remarkable features of this volcanic group is the abundance of inclusions of high-grade metamorphic rocks and, in some of the volcanic centres, enclaves of igneous rocks of basic–intermediate composition. There are significant variations in composition among the different centres.

In the Hoyazo volcano (Zeck, 1970, 1992; Cesare *et al.* 1997) the rocks are high-K calc-alkaline dacites, very rich in metamorphic rocks and euhedral garnet and abundant igneous inclusions. This volcanic centre appears in the Almería Basin, close to the calc-alkaline rocks of the Cabo de Gata Group.

Most of this magmatic activity appears directly linked to the late Neogene tectonic activity of the Eastern Betics, characterized by the existence of a large transcurrent zone, generated in response to indentation-related effects of the Aguilas Arc. According to Silva *et al.* (1993), the extensional collapse of the Betic orogen would have been disrupted by the indentation of the Aguilas Arc, producing a change from extensional to compressional tectonics. The Aguilas Arc is a tectonic block that has moved northwards, in relation to an approximately north–south compressive regime from the Middle Miocene to the Pleistocene, producing complex peripheral deformation resulting from a mechanism of horizontal indentation. It is bounded by sinistral strike-slip faults (NE–SW; Palomares Fault Zone) and dextral strike-slip to reverse faults (east–west; Almenara and las Moreras Fault Zones) whose kinematics has varied according to rotations of the regional shortening direction. These faults have also controlled the evolution of the sedimentary basins and the emission of magmatic products during the Neogene.

In the outcrops in the Vera Basin, and some volcanic structures near Mazarrón, the rocks are shoshonitic latites (called 'dellenítes' in older studies, e.g. Fúster, 1956) that show hybrid mineralogy and geochemical features interpreted as products of magma mixing between lamproitic and dacitic magmas (Venturelli *et al.* 1991b). In addition, occasional lamproite centres are also present (see below). These outcrops are related to the Palomares Fault Zone, the western margin of the Aguilas block.

According to K/Ar radiometric and biostratigraphic data (Bellon *et al.* 1983; Munksgaard, 1984), the peraluminous volcanic rocks erupted in the Tortonian–Messinian (11–6.6 Ma), and consequently are mostly coeval with the later stages of the calc-alkaline volcanism of the nearby calc-alkaline Cabo de Gata Group, and with the formation of the Murcia and Vera lamproites.

Ultrapotassic volcanism (lamproites) (Excursion 3.3b, Stop 3)

The Murcia and Almería lamproites were used by Niggli (1923) as the type locality for lamproites. The Spanish lamproites appear as small volcanic centres dispersed in a wide region from the Vera Basin (Almería, Betic Internal Zone) to Cancarix (Albacete, External Zone). Radiometric ages range from 8.5 to 6 Ma (Nobel *et al.* 1981; Bellon *et al.* 1983), but the stratigraphic position of some lava flows seems to indicate a Messinian age.

The ultrapotassic rocks of southern Spain comprise both vitrophyric and holocrystalline varieties. Most of the outcrops are nearly vertical pipes (*c.* 1 km) with brecciated borders, although some dykes and some lava flows can also be recognized. As a consequence of their petrographical diversity, they have received local names. Thus, they have been classified as *jumillites*, *fortunites*, *cancalites* and *verites*. Petrographically, they are characterized by an abundance of phlogopite, associated with other phases such as Mg-rich olivine, diopside, sanidine, potassic richterite, leucite and analcite. Some rare varieties contain hypersthene. Other minerals in these rocks are Cr- and Ti-spinel, apatite (abundant and commercially exploited in some locations), ilmenite, armalcolite–pseudobrookite (Fe–Mg titanates), roedderite, dalyite, and late-magmatic carbonate associated with warwickite and haematite. Diamond, a common occurrence in other lamproite provinces, has not been found here.

Geochemically, the rocks have very high MgO, Ni and Cr contents, as well as K_2O (up to 12%), P_2O_5, Ba, Sr, Th, U and Zr (Venturelli *et al.* 1984; Foley *et al.* 1987; Foley & Venturelli, 1989; Mitchell & Bergman, 1991). There is a wide range in SiO_2 contents (45–72%), from undersaturated rocks (jumillites and cancalites), through weakly undersaturated, saturated, to highly saturated (verites). Al_2O_3 and CaO contents are very low. Na_2O is highly variable, owing to secondary alteration processes (natrolitization). Rare earth element (REE) patterns are highly fractionated. Sr isotope initial ratios are extremely high (0.7148–0.7213) for magmas of mantle origin, and $^{143}Nd/^{144}Nd$ values are low. Pb isotopes indicate a crustal signature (Nelson *et al.* 1986). $\delta^{18}O$ values are also very high (López-Ruiz & Wasserman, 1991). According to Toscani *et al.* (1995), most of the lamproites are peralkaline (e.g. Jumilla, Calasparra and Cancarix) or subaluminous (Fortuna and Aljorra), whereas others (Zeneta, Vera and Mazarrón), although they retain a lamproitic affinity, cannot be considered lamproites *sensu stricto* as they show evidence of interaction with magmas of different composition or crustal rocks.

The genesis of the south-east Spain lamproites has been described in several papers (e.g. Fúster *et al.* 1967; López-Ruiz & Rodríguez-Badiola, 1980; Venturelli *et al.* 1984, 1988, 1991a; Nelson *et al.* 1986; Wagner & Velde, 1986; Foley *et al.* 1987; Mitchell & Bergman, 1991). The most widely accepted models propose a multistage origin by partial melting of a phlogopite and apatite-bearing, lithospheric mantle at moderate to low pressures (Venturelli *et al.* 1988). The mantle source was previously depleted and later metasomatically enriched in water and incompatible elements. Isotopic studies (e.g. Nelson *et al.* 1986) indicate that the metasomatizing component has the characteristics of sediments derived from the continental crust. The high Mg, Ni and Cr contents indicate that they are nearly primary magmas, although some interaction with the continental crust is evident in the most silica-rich facies. In some locations, mingling of lamproite magmas and Hoyazo-like dacite magmas is evident (Venturelli *et al.* 1991b).

3.3a Excursion: The calc-alkaline volcanics of the Cabo de Gata (half day)

JUAN M. FERNÁNDEZ-SOLER

Introduction

This excursion is dedicated to observations of the calc-alkaline volcanic rocks of the Cabo de Gata area, travelling approximately north to south (Fig. 3.3.4). Stops 1 and 2 are devoted to the study of facies associated with the northern margin of the Rodalquilar Caldera in the central sector of the chain, in places where they are not too affected by hydrothermal alteration. Stop 3 provides an overview of the Frailes volcano and the Lower Volcanic Group, and Stops 4–6 concentrate on different aspects of the volcanic rocks of the Lower Volcanic Group. Stop 6 is optional, depending on available time. The Quaternary geology and geomorphology of this area are dealt with in Excursion 6.5.

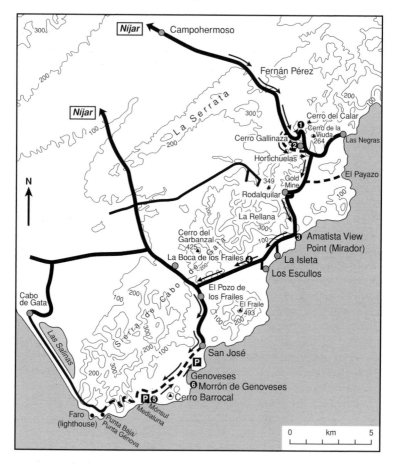

Fig. 3.3.4. Location map of the field-trip stops for Excursion 3.3a.

Directions to Stop 1

To reach Stop 1, travel along the E-15/N340 highway from Almería to Níjar, and turn off at the junction at KM 482, taking the main road towards Campohermoso, Fernán Pérez and Las Negras. Before arriving at Fernán Pérez (KM 10.3), the road crosses the Carboneras Fault System at the northern margin of La Serrata hills, where an olistostromic red breccia, made up of large volcanic clasts embedded in a sedimentary matrix, is clearly observed (Bordet, 1985). Past Fernán Pérez, the road goes through the Miocene carbonate sediments that cover the volcanic sequence of Cabo de Gata; some volcanic rocks are visible at isolated points along the road. At KM 14.9 the volcanic sequence is cut by the downhill section of the road. Stop 1 (Cerro Gallinaza) is located at KM 15.6 where the volcanic series is exposed in the slopes of Cerro de la Viuda (to the east) and Cerro del Calar (to the north). Walk to 865 827 indicated by the stop number on Fig. 3.3.4.

Stop 1: Cerro Gallinaza and the volcanic successions in the Cerro de la Viuda (861 825; Fernán Peréz 1 : 25 000)

At this point, near the village of Hortichuelas, the road to Las Negras cuts through a 11-Ma-old dome of amphibole dacite (Gallinaza dome, 313 m) that is generally quite weathered and covered to the NW by the Upper Miocene carbonate sediments. Looking north and NE to the opposite slope (Cerro de la Viuda, 264 m) a succession of different volcanic units comprising most of the characteristic lithologies in the central sector of the Cabo de Gata chain can be seen (Figs 3.3.5 and 3.3.6). They are as follows (from bottom to top).

1 A lower level of cream-coloured, friable, rhyolite–ignimbrite flows. In general these rhyolites are weakly or not welded, but in proximity to the road they are observed to have been compacted and folded to near vertical by the effect of the Gallinaza dome. The rocks contain plagioclase, biotite and quartz in a vitreous matrix. Good exposures, with a well-developed eutaxitic texture, are found immediately below the road.

2 Another level of grey- or cream-coloured hornblende-bearing ignimbrites and volcaniclastic rocks associated with layers of bioclastic calcarenites (visible along the ravine). These rocks contain an increasing amount of dark pyroxene andesite pebbles towards the top, giving the overall darker appearance.

3 A series of breccia layers comprising basaltic and pyroxene andesites make up most of the visible sequence (the 'Black Breccia', of Bordet, 1985). The two black layers correspond to coarse, dense, lithic breccias, whereas the intermediate, lighter layer is more fine-grained and rich in andesitic pumice lapilli. Towards the SE the black breccias grade to massive facies. These rocks contain plagioclase, augite and hypersthene as the main minerals and are some of the most mafic and least-evolved petrological materials in the Cabo de Gata volcanic area. Radiometric dating of these rocks has given contradictory values between 9.4 and 7.7 Ma.

4 A white layer of rhyodacitic ignimbrites, very rich in pumice and crystals of biotite and quartz. This layer is more extensive on the opposite slope of the hill and is characterized by the abundant presence of holocrystalline, tonalitic pebbles, formed by the accumulation of crystals in the edges of the magma chamber. This layer could be the lower part of the outflow facies of the Cinto ignimbrite (Fig. 3.3.5).

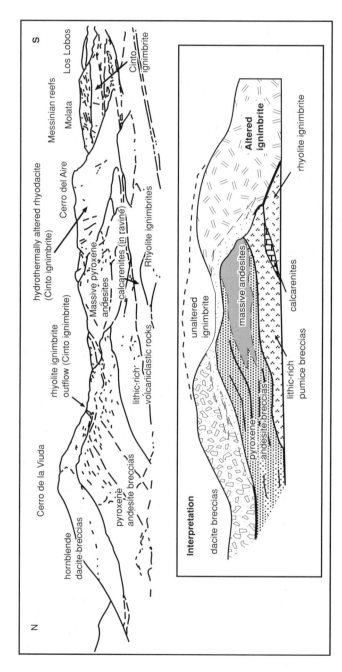

Fig. 3.3.5. Panoramic view and interpretation of the Cerro de la Viuda section, north of Hortichuelas (Stop 1). The view shows the overlap of the pyroclastic units of the Rodalquilar Caldera onto pyroxene andesite breccias and lavas, and other pyroclastic units. Upper Miocene carbonate covers the Cinto ignimbrites (Molatilla and Molata hills, looking SE).

Fig. 3.3.6. Volcanic sequence (Stop 1) in the Cerro de la Viuda slope. Black layers are pyroxene andesite breccias.

5 Lithic- and pumice-rich pyroclastic rocks with an amphibole dacitic composition (containing plagioclase, green hornblende, some pyroxene or cummingtonite, quartz and glass). This layer could also be related to the Cinto ignimbrite. Looking SE, the Cerro del Aire hill comprises highly altered dacites also belonging to the Cinto ignimbrite.

6 Several layers of lithic breccias formed as a consequence of the extrusion of a large group of dacitic domes that appear in various parts of the Las Negras area (e.g. the Calar and Bolinas domes, visible to the north from here).

This succession is covered by Upper Miocene marine sediments, observable both towards the north (Cortijo de Bornos) and towards the SE (Molata and Molatilla hills).

Transfer to Stop 2

Go back along the road towards Campohermoso for about 1.1 km. At KM 14.7, take an old, narrow road to the SW. This is the abandoned road to Las Negras, which is suitable only for cars or small minibuses. Go some 750 m along the road to reach Stop 2, situated on a steep slope. By coach it is possible to advance some 200 m along the road, and then walk the necessary distance to reach Stop 2.

Stop 2: Intracaldera facies in the northern border of the Rodalquilar Caldera
(853 819, Fernán Peréz 1 : 25 000)

At this stop the sequence comprising the fill of the Rodalquilar Caldera
(intracaldera facies) is exposed (Fig. 3.3.7). The most probable boundary
for this huge volcanic structure is a clearly visible fault along the old abandoned road to Las Negras, 1.5 km SE of this point. The succession of rocks
from the Cinto ignimbrite, scarcely altered at this point, comprise white,
pumiceous, dacitic and rhyodacitic ignimbrites and coignimbrite breccias

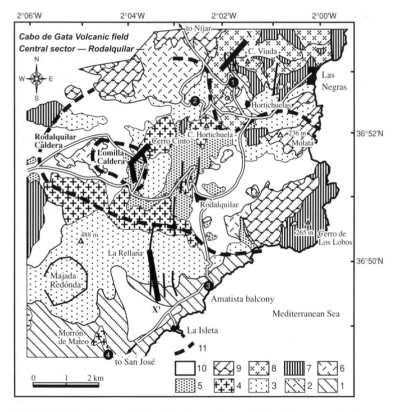

Fig. 3.3.7. Geological map of the Rodalquilar area (central Cabo de Gata zone) showing
Stops 1–4 and the situation of the Rodalquilar Caldera complex. 1, Pre-caldera rocks
(Frailes zone); 2, pre-caldera rhyolite ignimbrites (Hortichuelas), Rodalquilar Caldera
complex; 3, Cinto ignimbrite; 4, lava domes; 5, Lázaras ignimbrite; 6, hornblende andesites;
7, pyroxene andesites; 8, amphibole dacites; 9, Upper Miocene carbonate sediments;
10, Quaternary deposits; 11, caldera margins. Figure 3.3.3 shows an approximate
NNE–SSW section of this area (X–X′). (Modified after Arribas, 1993.)

containing abundant pebbles from other lithologies (amphibolite dacites, pyroxene andesites and hydrothermally altered rocks, etc.). These lithic fragments derive from the collapse of the substratum materials at the margin of the basin during the sinking of the caldera. In many cases these clasts almost obscure the true lithology of the ignimbrite. The ignimbrites are composed essentially of very friable fragments of pumice; which are easily weathered to form 'taffoni'.

The ignimbrite is more than 200 m thick at this point, in contrast with the few tens of metres observed in the ignimbrites at Stop 1. The beds dip some 15° towards the caldera at this point, but have been tilted 20–30° in the opposite sense near the centre of the caldera during the formation of a resurgent dome.

Associated with the caldera collapse was the extrusion of domes along the ring fracture, best represented on the southern edge of the caldera (near Rodalquilar). Looking south, the Hortichuela rhyolite dome can be seen, which was emplaced in the northern margin of the Rodalquilar Caldera.

It is also possible to see some andesite breccias overlying the Cinto ignimbrites. They lack pumice and are composed of more homogeneous and denser fragments of amphibole andesites that are darker, somewhat altered, and contain large amphiboles. These breccias also formed in relation to the later extrusion of domes, in various phases. Following the road towards the west, a massive, unaltered facies of one of these later domes (9.0 Ma; Rytuba *et al.* 1990) is encountered. A deeper intrusion of similar andesitic magma was probably the cause of the hydrothermal alteration and mineralization at Rodalquilar.

Transfer to Stop 3

From Stop 2, go back to the main road or proceed 2.5 km along the old road towards Hortichuelas, where it crosses the main road once again. The faulted margin of the caldera and the highly altered pre-caldera rocks are visible along the road. After reaching the main road, turn right and proceed 150 m to a junction and take the road that goes south to Rodalquilar and San José. Travel along this road for about 5 km. The road goes through the Rodalquilar Basin, which is part of the Rodalquilar Caldera complex. The resurgent dome of the Rodalquilar Caldera appears to the west, on the foreground beyond Rodalquilar village. The rocks are highly affected by hydrothermal alteration, and some mine workings are visible. Stop 3 is located on a view point balcony ('mirador') on top of Amatista hill. This spot is located along the road from Rodalquilar to San José. After leaving Rodalquilar, the road climbs to a hill crest and then rapidly descends to the Amatista Mirador.

***Stop 3: Mirador (view point) de la Amatista (856 762; El Pozo de los
Frailes 1 : 25 000)***

The top of Amatista hill provides an excellent overview of the organization
of the southern part of the volcanic chain of Cabo de Gata (Fig. 3.3.8). The
outflow facies of the Cinto ignimbrite comprises most of the La Rellana
plateau to the NW, which extends from this point to the west. There are
several subhorizontal individual flow units, easily discerned in the scarps of

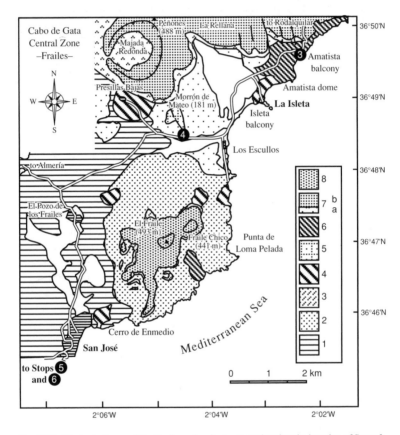

Fig. 3.3.8. Geological map of the Frailes volcanic structure showing the location of Stops 3
and 4. Lower Volcanic Group: 1, older rocks (mainly altered pyroxene andesites and hybrid
andesites); 2, amphibole andesites (Frailes Caldera, as proposed by Cunningham
et al. 1990); 3, main sedimentary and tuffaceous units; 4, large hornblende andesite
domes (Enmedio, Presillas, Isleta); 5, Morrón de Mateo lava dome; 6, pyroxene andesite
domes (Amatista, Tomate). Upper Volcanic Group: 7, Cinto ignimbrites (Rodalquilar
Caldera), (a) lower units and (b) upper units; 8, basaltic and pyroxene andesites of
Los Frailes (≈ 8 Ma). (Modified after Fernández-Soler, 1992.)

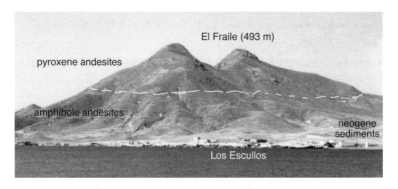

Fig. 3.3.9. Panoramic view of the Frailes volcano (493 m) looking south from La Isleta near (Stop 3). The contact between the lower amphibole andesites (10–12 Ma, Frailes Caldera) and the upper pyroxene and basaltic andesites (≈ 8 Ma) is shown. Sedimentary and some pyroclastic levels are present between both units. The volcanosedimentary unit extends more to the north (lower right of the image), where it is commonly exploited as bentonites.

the La Rellana plateau. Looking NE along the coast, the Lobos volcano, covered by uppermost Tortonian–Messinian carbonates, is made up of pyroxene andesites (7.9–8.4 Ma).

Several north–south-trending faults have displaced the tuffs. To the south, the Frailes volcano (Fig. 3.3.9; 493 m) is composed of an upper unit of breccias and lava flow of pyroxene andesites that erupted 8.6–7.9 Ma ago (Di Battistini *et al.* 1987), ending with the extrusion of two summit domes. Below this unit, the Lower Volcanic Group comprises a sequence of amphibole andesite breccias and tuffs (10.8–12.2 Ma) that fills most of the Frailes Caldera (Cunningham *et al.* 1990). The top of this unit is marked by a lighter coloured zone rich in marine sedimentary layers (sands and calcarenites) and tuffaceous, pyroclastic rocks, some of which will be visited at Stop 4. This pyroclastic sedimentary unit is intruded by some later andesite domes (e.g. Amatista dome and Isleta del Moro dome).

Transfer to Stop 4

Stop 4 is located along the road from Rodalquilar to San José, 3.4 km to the SW of Stop 3, where a large quarry and dump are found. Some other smaller quarries are close by. The route from Stop 3 crosses a large Quaternary alluvial fan near La Isleta and, to the left at Los Escullos, note the large Quaternary cemented dunes. Details of the Quaternary geology and geomorphology of this area are described in Excursion 6.5, Stop 3.

Stop 4: Bentonite deposits at Morrón de Mateo (828 741; El Pozo de los Frailes 1 : 25 000)

The top of the Lower Volcanic Group in the San José area (Frailes hill) is marked by a pyroclastic layer associated with sedimentary (shallow-marine or coastal) sediments. This tuffaceous unit contains several pyroclastic flow units (ignimbrites) and other pyroclastic facies. In the Morrón de Mateo quarries, near the locality of Los Escullos, the pyroclastic rocks were produced by hydromagmatic eruptions, resulting in the formation of alternating layers of pumice lapilli and crystal-rich ash, which is sandwiched between a bioclastic limestone and a calcarenite. The sedimentary layers are not directly observed in this quarry, but are present in other small quarries nearby. The whole sequence is cut by a reddish, amphibole–biotite dacite dome (Morrón de Mateo dome, 161 m).

The pyroclastic sequence was formed by cyclic layers of hydromagmatic deposits (10–20 cm thick) alternating with pumice-rich layers (*c.* 50 cm thick). The crystal-rich, sandy deposits are composed of lithic and crystal-rich ash, with fragments of amphibole andesites, broken amphibole and plagioclase crystals and altered glass. The predominant bedforms are flat-parallel and sheet-like, although at some points cross-bedding has been recognized. The vent was located to the west, as indicated by the polarity of impact sags produced by the ballistic fall of large blocks, and by the thinning of the individual layers towards the east. At the nearby coast to the east (Los Escullos), the crystal-rich layers are 1–2 cm thick and 'fallout' pumice layers are absent.

The pyroclastic sequence in this area is intensely altered to bentonites. This is one of the numerous bentonite quarries dispersed in the Cabo de Gata volcanics. Some of them (e.g. Los Trancos deposit) are very large and have an extremely pure and easily extractable bentonite (> 92% smectite). In Morrón de Mateo, the heterogeneity of the pyroclastic rocks produced a less pure bentonite, richer in the pumice-lapilli layers (approximately 70% smectite). The origin of these bentonites has been studied by numerous investigators. In general, their formation is related to the alteration of glass and plagioclase by hydrothermal fluids of meteoric origin. This alteration took place at temperatures lower than 70°C in Cabo de Gata (Leone *et al.* 1983). In Morrón de Mateo, the intrusion of the dacite dome seems to be responsible for the heating and circulation of low-*T* fluids of marine and meteoric origin. The alteration took place in several stages, recognized on the basis of stable isotope studies (Delgado-Huertas, 1993).

Transfer to Stop 5

To reach Stop 5, go along the road 4.7 km to the west, up to a junction, and take the road to San José. Frailes volcano lies to the south and Garbanzal volcano is to the north. After the junction, go for 3.4 km to the entrance of San José, where there is a sign to the Monsul and Genoveses beaches ('playas') to the right. Take this road and immediately turn right onto another some 50 m beyond. This road then circles San José and provides an excellent overview of the Frailes volcano. The paved road ends about 1 km later, on a crest from which a gravel road goes south to the Monsul and Genoveses beaches. The road is unpaved but usually well maintained and suitable for coaches. Go down the road *c.* 2 km to a junction where there is a road on the right that goes to Monsul (Stop 5), and another road that goes left to Genoveses (Stop 6). Proceed *c.* 2 km to Monsul beach, where there is an official parking area. Stop 5 observations are at Monsul beach and at Media Luna beach, situated 400 m westwards. Aspects of the Quaternary geology and geomorphology of this area are described in Excursion 6.5, Stop 2.

Stop 5: *Monsul and Media Luna area (763 656; Morrón de Genoveses 1 : 25 000)*

The Monsul beach and the Barrocal hill (Fig. 3.3.10) are made up of breccias (agglomerates) linked to massive (coherent) dome-like or dyke-like bodies of andesitic composition. The massive andesite shows well-developed fan-like columnar jointing. The contact between the massive and the volcaniclastic materials is clearly transitional. The massive zones give way to an autoclastic breccia without any fine-grained matrix. Domes or feeder dykes like those at Monsul beach are widely dispersed in this area, and their location is clearly related to crossed fracture systems running NE–SW and NW–SE.

Away from the massive zones, the breccia or agglomerate is more than 50 m thick, and consists of angular fragments of andesitic composition, ranging in size from 0.1 to 0.6 m, included in a light-coloured matrix of ash-lapilli size. It consists of crystal fragments (pyroxene and plagioclase) and glassy shards. The lithic fragments show jigsaw fitting and curviplanar edges. Some very large blocks (> 1 m) are visible also exhibiting columnar jointing. The breccia has no internal organization, apart from a faint colour-banding visible at Media Luna beach. In nearby areas, intervals of cross-bedded layered sequences are common, gradually passing from the blocky breccia to a whitish, ashy rock interpreted as an acidic hyaloclastite. The association of hyaloclastites, layered deposits and remnants of andesite domes (Figs 3.3.11 and 3.3.12) is typical of the submarine eruption of massive domes of acidic lavas (Pichler, 1965; McPhie *et al.* 1993).

Both the massive andesite and andesite blocks are porphyritic or

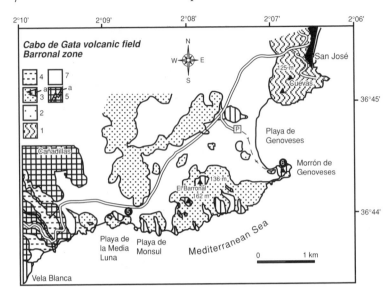

Fig. 3.3.10. Simplified geological map of the SE sector of the Cabo de Gata chain (Borronal area) and location of Stops 5 and 6. 1, Undifferentiated andesites (San José); 2, rhyolite ignimbrites (Morrón de Genoveses); 3, pyroxene andesite breccias, (a) massive bodies (dykes, lava flows); 4, andesite ignimbrites; 5, amphibole andesite lava flows, (a) vertical dyke of Cañadillas; 7, Quaternary sediments.

Fig. 3.3.11. Field appearance of the hyaloclastic Monsul breccias (Stop 5), showing jigsaw fit of dense, heterometric lithic clasts with curviplanar edges. The white lobes correspond to fine-grained portions, which laterally grade to distal hyaloclastites and resedimented hyaloclastites. A vertical dyke with signs of filling from upwards is also visible. Monsul beach (visible height is about 5 m).

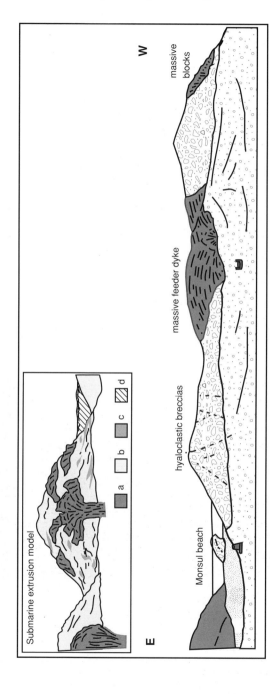

Fig. 3.3.12. Panoramic view of the Monsul area from the interpretation centre of the Cabo de Gata Natural Park (Stop 5), and proposed interpretation as an extrusion of intermediate–acid lavas in a submarine environment (see McPhie *et al.* 1993). (a) Massive feeder dykes and lobes; (b) hyaloclastic breccias (agglomerates); (c) fine hyaloclastic lobes; (d) later andesite ignimbrites ponded between coalescing domes.

vitrophyric rocks, bearing plagioclase and two-pyroxene (augite and hypersthene) phenocrysts. The crystallization temperature, calculated by Fe–Ti oxide thermometry, is 910–920°C for log (fO_2) = –11.3. K–Ar dating of rocks from this unit span from 12 to 10.8 Ma (Di Battistini *et al.* 1987).

Several pyroclastic deposits of pumice-flows (ignimbrites) are well exposed along the coast to the west of Media Luna beach, comprising large fragments of pumice (sometimes banded pumice) in a fine-grained, glassy matrix. In some places the pumice is flattened and welded, forming the classic *'fiamme'*. Occasional layers of pyroxene-andesite blocks in the ignimbrite can also be seen. This ignimbrite is covered by a massive, reddish, columnar-jointed unit (Cañadilla lava flow). At other points the top of the ignimbrite is marked by a sandy layer about 1 m thick, most probably formed by remobilization of loose fragments from the underlying ignimbrite. The pumice fragments of the ignimbrite are also of two-pyroxene composition, similar to the domes and the breccia blocks. Banded-pumice in ignimbrite deposits is commonly associated with eruptive processes caused by magma mixing in shallow magma chambers.

Transfer to Stop 6

Morrón de Genoveses is to the ENE, close to Genoveses beach. From Stop 5, go back towards San José for *c.* 2 km to the junction mentioned above, and then follow the signs to the Genoveses beach parking area, which is about 300 m away. From this point, walk for about 800 m to Morrón de Genoveses (Fig. 3.3.10). As this path is somewhat long, make this stop only if there is enough time available.

Stop 6: Morrón de Genoveses (790 665; Morrón de Genoveses 1 : 25 000)

In this zone, the formation of Monsul pyroxene andesites is underlain by a white pumice-and-ash pyroclastic flow (ignimbrite), which is a lateral equivalent of the Vela Blanca basal ignimbrite (see Excursion 3.3b, Stop 1). The black, massive pyroxene andesites lie unconformably over this unit, and the contact is marked by a yellowish, opaline zone of rubefaction. The welded ignimbrite, of which the upper 25–40 m are exposed, contains glassy shards, crystals, abundant pumice fragments and scarce black lithic fragments (pyroxene andesites). The composition is rhyolitic (rhyodacitic), with quartz, plagioclase and biotite. The ignimbrite is covered by finely stratified 'pyroclastic surge' deposits, made up of smaller, platy pumice fragments with a planar-laminar or wavy structure. The composition of the fragments is similar to the ignimbrite pumice. Over the surge deposits, several thinner layers of 'pyroclastic flows' are present, showing inverse

(coarsening-upwards) size-grading of pumice fragments, and normal grading of lithic fragments. These layers are intercalated with surge layers rich in cross-bedding and ripple structures.
Return to the main road network by retracing your route through San José.

3.3b Excursion: The volcanics of the Almería and Vera Basins (half day)

JUAN M. FERNÁNDEZ-SOLER

Introduction

This half-day excursion covers the entire range of volcanic rock types in the Almería Province. Stop 1 examines the calc-alkaline suite, Stop 2 the peraluminous and Stop 3 the ultrapotassic rocks. The trip follows up observations made in Excursion 3.3a and starts at the Cabo de Gata lighthouse (Fig. 3.3.13). Stops 2 and 3 will be dedicated to observations of peraluminous volcanic rocks in the Hoyazo de Níjar (Stop 2) and rocks with lamproitic affinities in the Vera Basin (Stop 3).

Directions to Stop 1

Stop 1 is located immediately east of the Cabo de Gata lighthouse, on the Punta Baja point (also called Punta Génova). To reach it from the end of Excursion 3.3a at San José, take the road from San José to Níjar to leave the Sierra de Cabo de Gata. At KM 8.2 take the road that goes west to Almería. Take another road 7.5 km later that goes to Cabo de Gata village, 5 km away. If driving from Almería take the N332 east from Almería and follow the signs for the Cabo de Gata. If arriving from the north leave the N340/E-15 at the Cabo de Gata signs. After arriving at Cabo de Gata, take a road that goes south, towards the Cabo de Gata saltworks ('salinas') and lighthouse ('faro'), some 7.5 km distant. This road goes between Cabo de Gata beaches and a large lagoon. Behind the lagoon, the rocks of the Lower Volcanic Group, highly affected by hydrothermal alteration, are exposed. After passing the lagoon, the road climbs the volcanic rocks and then descends to the lighthouse. The road is narrow at some points and not suitable for large buses. Some 200 m before reaching the lighthouse, take another road to the left (east) and go along it for another 500 m to a path that descends to a small beach. Walk down to the Punta Baja promontory, where some old cobble quarries are found. The Quaternary geology and geomorphology of this area are described in Excursion 6.5, Stop 1.

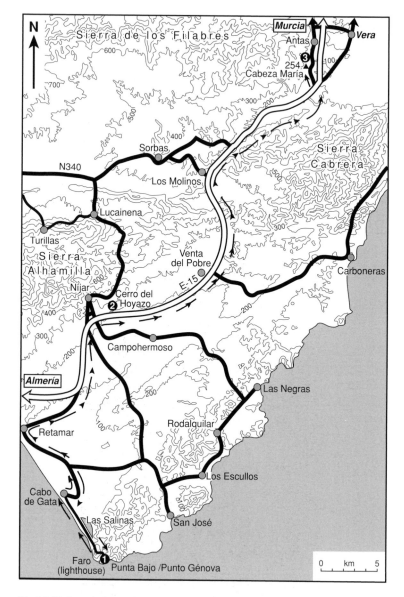

Fig. 3.3.13. Location map of the field trip stops for Excursion 3.3b.

Stop 1: Dome bodies around the Cabo de Gata lighthouse (729 645; Cabo de Gata 1 : 50 000)

The Cabo de Gata area, near the lighthouse, is made up of a dome complex of andesitic–dacitic rocks. These andesites have phenocrysts of plagioclase, amphibole and biotite, in an altered, devitrified groundmass. The rocks in this area have undergone extreme deuteric alteration, and so the mineral phases are transformed into sericite, alunite, carbonates, silica and iron oxides. Silica frequently fills the cracks in the rocks (the name 'Gata' derives from the abundance of agate fillings around this zone).

The complex consists of two principal domes, easily discerned by the disposition of igneous flow structures. The most prominent and spectacular structure to be observed in these rocks is the columnar jointing (Fig. 3.3.14), which developed because of the effect of the slow cooling of the andesite magma perpendicular to the walls of the dome. The complex disposition

Fig. 3.3.14. Fan-like columnar jointing in the Punta Baja hornblende andesites (Stop 1).

of columnar joints, with the frequent formation of fan-like formations, indicates that the magma was extruded in multiple stages, typical of acidic domes. The columnar jointing has been exploited for the production of cobbles. Other remarkable field aspects are the igneous-flow layering and banding, which are particularly well exposed at the dome boundaries, where flow folds can also be observed in some locations.

Looking eastwards, the coast shows the sequence of the Lower Volcanic Group in this area. The base is formed of whitish pumice flows or ignimbrites (Vela Blanca ignimbrites), highly altered, although their biotite–rhyodacite nature is still recognizable. This is the lowermost unit in this area of the Cabo de Gata chain. These white tuffs are covered by greyish tuffs (pyroclastic flows, surges and reworked rocks), which are found in the upper part of the road that ascends to the Vela Blanca tower. These tuffs are unconformably overlain by several lava flows of amphibole dacite composition that form prominent scarps in the slope, interlayered with light-coloured ignimbrite layers. The whole sequence is cut by the Vela Blanca dome, made up of altered, black pyroxene andesites. Several faults affect all the units.

Transfer to Stop 2

Stop 2 is close to Níjar and can be reached from the E-15 motorway from Almería to Murcia. From Stop 1, go back to the Cabo de Gata village, and then head towards Almería for about 12 km, up to Retamar village. At Retamar, turn right to reach the old road to Almería, and then proceed towards Níjar (eastwards) for 3.7 km to reach the E-15 motorway. Continue about 14 km more and leave the motorway at the second Níjar exit at KM 482. Immediately after the junction, take a small, unpaved service road to the right and travel along it for 1.2 km up to some workings where garnet is extracted from alluvial sediments. Then walk along a ravine for 500 m to reach the crater and outcrops of the volcanic rocks below the Messinian reefs described in Excursion 4.3.

Stop 2: Peraluminous volcanism in the Cerro del Hoyazo, Níjar (740 907; Campo Hermoso 1 : 25 000)

The Cerro del Hoyazo (or Joyazo) is a hill (335 m) 3 km to the east of Níjar on the southern side of the Sierra Alhamilla. It consists of the remnants of a dacite exogenous dome that was covered by Messinian reef carbonate deposits (visited in Excursion 4.3). The carbonate rocks and underlying dome have been eroded, producing a crater morphology within which the volcanic rocks crop out. The volcanic rocks consist mainly of autoclastic, disorganized breccias in a central, more massive zone, possibly corresponding to the feeding point, contoured by scarcely more organized flow-foot breccias. Sediments, biostratigraphically dated as Tortonian, appear under these breccias. Another outcrop of more altered dacites appears badly exposed *c.* 2 km east of the Hoyazo crater. A detailed description of the rocks is given in Zeck (1970, 1992) and Munksgaard (1984). The age of the Hoyazo dacites is 6.2 Ma (Turner *et al.* 1999).

In hand sample, the volcanic rock of the Hoyazo de Níjar is a highly porphyritic dacite with a large amount of glassy matrix (about 50–60%). It is black to light grey, showing abundant flakes of biotite, red garnet crystals (up to 10 mm), plagioclase and cordierite, and engulfed quartz. There are common, dispersed lumps (up to 10–15 cm) of shattered quartz.

Plagioclase and cordierite (up to 3–5 mm in size) each form about 10% of the rock. The plagioclase shows a variety of textural features: low-amplitude oscillatory zoning; sieve textures; inclusions of sillimanite, biotite or spinel.

Magmatic cordierite appears as bluish, euhedral crystals (10–100 μm), showing a well-developed sector twinning, and is usually zoned. Euhedral microcrystals of cordierite are also abundant in the groundmass of the dacite. Another type of cordierite is anhedral, sometimes embayed, and extremely rich in fibrolite needles, which in some cases are absent or less abundant in the rim of the anhedra. This cordierite seems to be of restitic origin.

Garnet constitutes about 1% of the rock. It is usually euhedral, sometimes rounded, and commonly 3–4 mm in size, but may exceptionally reach 10 mm. Garnets have long been exploited here for the production of abrasives.

Orthopyroxene and green hornblende are disseminated throughout the rock. These minerals are optically and compositionally similar to the orthopyroxene, and hornblende crystals present in the abundant mafic igneous enclaves. Other accessory mineral phases are spinels, graphite, sillimanite, apatite, zircon and monazite.

Inclusions of magmatic and metamorphic origin are very abundant in the dacite lava (1–2% are larger than 1 cm; Zeck, 1970). The main types of inclusions (*enclaves*) are the following.

1 *Garnet–biotite–sillimanite–plagioclase rocks (restite gneisses).* These are very dark inclusions, up to 0.5 m in size, showing a well-developed metamorphic foliation. They are very rich in lepidoblastic, dark brown biotite and lenses of fibrolite, with garnet (up to 1 cm) very frequently in the core of these lenses. The proportion of plagioclase is variable but usually small. A variable proportion of glass is present. In some cases these rocks pass gradationally to a rock very rich in cordierite, which forms very large crystals containing lenses of biotite, and large euhedral crystals of garnet, and some glass, but sillimanite is reduced to dispersed streaks in the cordierite.

2 *Quartz–cordierite rocks.* Globulous crystals of quartz included in large crystals of spongy cordierite. A variable proportion of plagioclase, biotite and sometimes garnet is present.

3 *Spinel–cordierite-rich rocks.* Gneisses and schists with biotite–sillimanite–staurolite–cordierite–sanidine–plagioclase and occasional garnet or
spinel. These rocks are foliated showing the superposition of two paragene-
ses. The first one corresponds to the amphibolite facies (garnet, staurolite,
biotite, sillimanite, plagioclase) and the second is marked by generalized
blastesis in high-temperature conditions (cordierite, spinel, sanidine, corundum, neocrystallization of biotite, sillimanite and plagioclase).

4 *Metamorphic inclusions.* Scarce fragments of amphibolites, fine-grained
schists and quartzites.

5 *Glass-poor mafic inclusions (dioritic).* Their mineralogy is quite variable. In general, they are formed by interlocking, euhedral–subhedral
crystals of biotite, orthopyroxene, hornblende, clinopyroxene and/or
cummingtonite, with abundant anhedral–subhedral plagioclase, some
quartz, and accessory amounts of apatite, ilmenite and zircon. Glass is
scarce or absent. Some of these enclaves contain Al-rich material consisting mainly of poikilitic cordierite, biotite lumps, plagioclases, sillimanite
and corundum.

6 *Basic microlithic inclusions.* These enclaves are very abundant. They are
more fine-grained, rounded, up to 50 cm, and usually have chilled margins.
These rocks are composed of sheet-like plagioclase, hornblende, orthopyroxene, biotite and sometimes clinopyroxene, in a large amount of vesicu-
lar diktytaxitic glass. Some inclusions contain large, rounded, corroded
quartz crystals surrounded by a reaction rim of glass and some needles of
orthopyroxene or amphibole. The mafic inclusions contain portions of the
biotite–sillimanite gneiss and related materials similar to those included in
the dacitic lava, with evident signs of reaction with the mafic rock.

Since the works of Zeck (1970, 1992), an origin through crustal anatexis
for the Hoyazo dacites and most of the peraluminous province (anatectic province) has been suggested. The Al-rich enclaves (biotite–sillimanite gneisses)
correspond to the restitic component for *c.* 30% melting of semipelitic
rocks. Garnet euhedra, part of the cordierite, biotite and plagioclase phenocrysts are also restitic phases. Some anhedral restitic cordierite is overgrown by euhedral magmatic cordierite. Melting conditions are estimated
at *c.* 850°C and 5–7 kbar (Cesare *et al.* 1997).

The participation in the anatectic event of mantle magmas of different
origins remains controversial. There is clear evidence of the interaction of
anatectic lavas with ultrapotassic rocks in Vera and Mazarrón, whereas in
the Hoyazo and Mar Menor (near Cartagena) mafic inputs seem to be calc-
alkaline. The relation of the mafic enclaves with the enclosing lavas is a
subject of debate. Zeck (1970) proposes that these enclaves correspond to
blobs of mafic magma injected in the lower part of a crustal magma chamber, providing the thermal energy necessary to promote the melting of a

pelitic source. However, Molin (1980) suggests that the mafic enclaves are the products of partial melting of pre-existing diorites; no new mafic magma is, in his opinion, emplaced during the anatectic event.

Transfer to Stop 3

Return to the N340/E-15. Travel east, towards Murcia and Vera for about 40 km, to the exit to Antas. The Cabezo María (243 m), a black hill with a small church on top, will be quite visible. Proceed towards Antas for about 2 km to where the road is closest to the volcano. At this point an unpaved track, suitable only for high-clearance cars, ascends to the very edge of the volcanic rock. Alternatively, a short walk of 300 m from the road will also reach the volcanic rocks and surrounding sediments.

Stop 3: Ultrapotassic rocks from the Vera Basin: the Cabezo María neck (948 192; Vera 1 : 50000)

The southernmost outcrops of ultrapotassic rocks are in the Neogene Vera Basin. The Cabezo María outcrop (Fúster & De Pedro, 1953) is located near the fault zone that separates the metamorphic area of the Sierra de los Filabres from the Tertiary sediments of the Vera Basin (Fig. 3.3.15). The Cabezo María outcrop is a neck interpreted as the remainder of an eroded volcanic centre, as several lava-flow outcrops appear in diverse places from the Cabezo María to the coast (Barragán, 1986–87). K-Ar ages of 8.6 Ma (Nobel *et al.* 1981) are inconsistent with biostratigraphical observations, as these lava flows, which locally display pillow-lava features, are intercalated among Messinian marls and turbidites (Barragán, 1986–87). The neck shows, at the contact with the marly limestone sediments, an intrusive brecci-ated border zone, where angular fragments of magmatic material are mixed with sediments. In this zone, which disappears at the neck core, the volcanic fragments are highly glassy and vesiculated. This breccia is thus a *pepperite*, formed as a consequence of the intrusion of hot magma into wet sediment.

The rock, which was termed *verite* by Fúster & De Pedro (1953), is very dark and contains olivine (> Fo_{90-95}) and phlogopite phenocrysts in a glassy groundmass (> 70%) with tiny phlogopite flakes. Other phases present in the groundmass of some samples are diopside, sanidine and leucite. The most common accessory mineral is apatite. Some larger, xenocrystic olivine contains frequent spinel inclusions. The rock is rich in vesicles filled with carbonate and/or tridimite. In some places, especially near the contact with the sediments, the rock is very altered to clay minerals and alunite-like material.

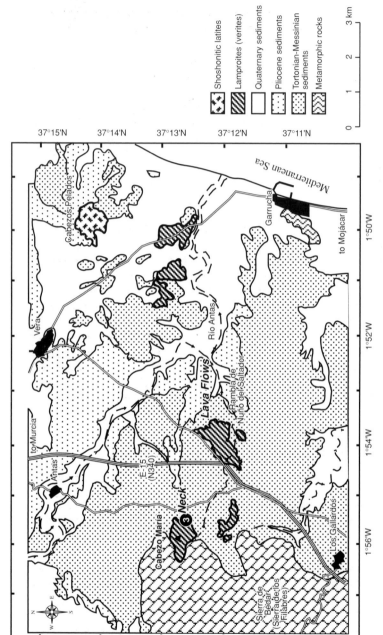

Fig. 3.3.15. Simplified geological sketch of the southern Vera Basin showing the location of volcanic outcrops (lamproites and shoshonitic latites), and the location of Stop 3.

3.4 Tectonics and sedimentation: the evolving turbidite systems of the Tabernas Basin

PETER HAUGHTON

Introduction

The oldest sedimentary fill of the Tabernas Basin is represented by ?Serravallian conglomerates which record a subaerial to marine transition, reflecting initial flooding of the basin. Most of the sedimentary fill is represented by Tortonian–Messinian deep-water successions (subwave base, up to 800 m water depth according to Kleverlaan 1989c). The Tabernas Basin (and the Sorbas Basin into which it passes eastwards) had an elongate trough-like geometry at the time, *c.* 10 km wide, and several tens of kilometres long, extending in an east–west direction (Montenat *et al.* 1987a,b; Ott D'Estevou & Montenat, 1990; Fig. 1.1.1). Sediment stripped from surrounding basement uplifts and narrow shelves was redistributed, primarily by sediment gravity flows, to deposit a variety of submarine fans, fault-controlled slope aprons and laterally extensive sheet turbidites (Kleverlaan, 1989a; Dabrio, 1990; Haughton, 1994).

The area to be visited lies at the western end of the Tabernas Basin, to the north and west of the Sierra Alhamilla basement block (Fig. 3.4.1). Here the upper Tortonian–Messinian deep-water deposits crop out in the core of an open anticline (Fig. 3.4.2) which plunges gently to the west. Other aspects of the Tabernas Basin are examined in Excursions 6.3 (Quaternary succession) and 6.4 (modern badland morphology). The geology of the Filabres (from which much of the sediment in the Tabernas Basin was sourced) is examined in Excursion 3.2.

Local succession

A composite, stratigraphical section for the deep-water sediments of the area is shown in Fig. 3.4.3. The overall package (up to the Gordo megabed) is *c.* 700 m thick, but the total thickness is complicated by offset and 'growth' of the succession on account of syn-depositional faulting (see below). Five informal stratigraphical divisions are recognized (labelled A to E, Fig. 3.4.3). The different units preserve turbidites and associated deposits with quite different geometries and distributions, and contain beds with vertical profiles which record a wide range of flow behaviour. It is argued below that the succession records an upward change from relatively high gradient slopes with axial bypass (units A and B), through to flow containment in structurally controlled enclosed depressions or mini-basins (unit C) followed by healing of the basin floor topography (unit D)

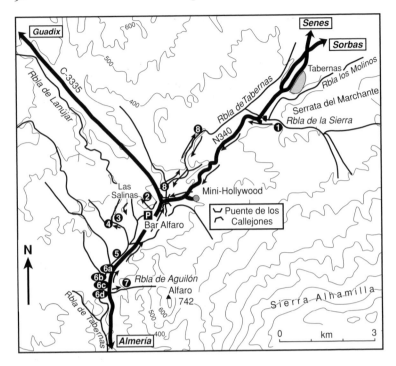

Fig. 3.4.1. Location map of the field-trip stops for Excursion 3.4.

with containment restricted to rare, large-volume collapse events (unit E, the Gordo megabed) which outran the scale of the basin. The interval is placed entirely in the Late Tortonian by Kleverlaan (1989a), but Cronin (1994) has indicated that the upper part of the section may extend into the Messinian on the basis of preliminary coccolith determinations (by A.T.S. Ramsay).

3.4   Excursion: The turbidites of the Tabernas Basin (one day)

PETER HAUGHTON

Introduction

The aims of this excursion are first to examine the continental to marine transition of the older basin-fill and then to concentrate on the development and evolution of the deep-water deposits. These turbidites and related

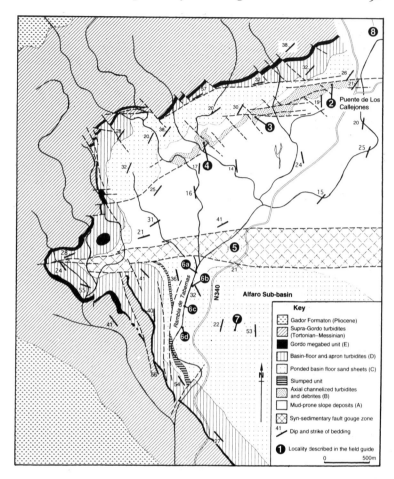

Fig. 3.4.2. Detailed geological map of the area highlighted in Fig. 3.4.1 annotated with the localities described in the text. (Modified after Haughton, 2000.)

sediment gravity flow deposits record an upward evolution in style which can be related to a fundamental change in basin geometry reflecting syn-depositional tectonism.

This field excursion starts with the non-marine fanglomerates and coarse-grained fan-deltas related to the early development and flooding of the basin (Dabrio, 1990) exposed in the Rambla Sierra. The excursion then continues up the succession from the core of the anticline in Fig. 3.4.2, as far as the Gordo megabed, a probable seismite which forms a distinctive basin-wide marker (Kleverlaan, 1987) draping the earlier succession. In

Unit

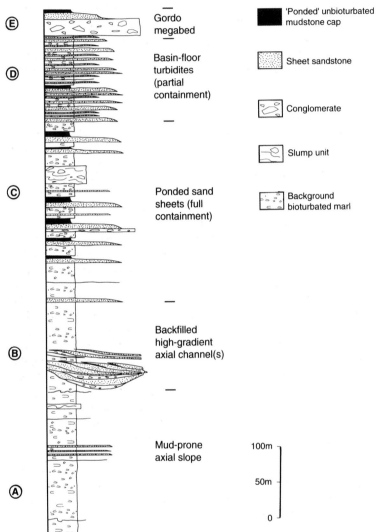

Fig. 3.4.3. Summary section for the succession found beneath the Gordo megabed in the area shown in Fig. 3.4.2.

the area of the basin to be visited, the oldest deep-water deposits seen are mud-prone marine marls which accumulated following the flooding of the trough (Middle–Late Tortonian according to Iaccarino *et al.* 1975, although Early Tortonian flooding was implied by Kleverlaan, 1989c).

Please note that Stops 2–7 may have restricted access because of the construction of the new Granada–Almería Autovia along the route of the N324.

Directions to Stop 1

The first stop is reached by way of the N340, from either Tabernas or Rioja. From either direction, the location of the metamorphic blocks flanking the basin are apparent, forming the mountainous ground to the north (the Sierra de los Filabres, with the observatory on top) and to the south (the Sierra Alhamilla). During the Tortonian, the Sierra de los Filabres block appears to have been the dominant source area with coarse-grained conglomerates interfingering southwards, forming an apron of underwater fan-deltas banked against the northern margin of the basin. These rocks form the prominent ridge (the Serrata del Marchante) running eastwards from Tabernas and Stop 1. The conglomerates can be followed as far east as Lucainena, where they disappear beneath a local unconformity separating the latest Tortonian, Messinian and younger fill of the Sorbas Basin. Although the Sierra Alhamilla block was evidently less important as a sediment source during the Tortonian, there is evidence that it was a positive area in that the sediment gravity flow deposits thin and pinch out as the block is approached (Haughton, 1994). At the time, the block would have had a carapace of lower grade Alpujárride lithologies which were largely absent from the Sierra de los Filabres, and there is evidence for local reworking of these rocks together with contemporary carbonate grains in the Tortonian basin-fill. The Alhamilla block was therefore probably a shallow water sill, perhaps with low-relief islands which were cored by basement.

The other prominent topographic feature is the conical hill (Alfaro) lying immediately to the west of the Sierra Alhamilla. This is composed of sheet turbidites belonging to unit C. The western dip slope of Alfaro coincides with a major slump sheet exposed at the base of the mountain in the Rambla de Tabernas (Stop 6, this excursion).

The excursion starts 1 km SW of Tabernas village near the road bridge of the N340 over a rambla (530 000). Just south of the bridge the road takes a sharp bend to the NW. Descend into the rambla on the east side of the road, using the abandoned service road. Once in the rambla, head upstream (east). Where the river forks, take the right-hand tributary. Follow this rambla for *c.* 400 m to a point just before a sharp bend to the right in the rambla floor. Park here. Beware of falling rocks from steep canyon sides.

Stop 1: Rambla de Sierra, terrestrial to marine transition
(538 996; Tabernas 1 : 50 000)

The canyon of the Rambla de la Sierra exposes a succession of steeply dipping ?Serravallian–Tortonian sediments on the southern flank of a faulted anticline SE of Tabernas. The succession represents the transition from terrestrial to marine conditions during the early development of the Tabernas Basin. Resting unconformably on these rocks are horizontal Quaternary fluvial conglomerates described in more detail in Excursion 6.3, Stop 5.

The red ?Serravallian conglomerates are composed dominantly of material derived from the Sierra de los Filabres (amphibole mica schist and rare augen gneiss). This provenance, combined with rare palaeocurrent data from imbricated clasts in the red conglomerates, suggests derivation from the NE. The red conglomerates range from massive, disorganized to weakly bedded, better organized bodies.

Walk upstream through the southern limb of the anticline. On the left (east) side of the rambla some beds can be observed containing subrounded clasts (up to 3 m long) and associated with imbricated conglomerates indicating stream flow. Further upstream, some 10 m below the transition to grey/green conglomerates, the beds show inverse grading with a *c.* 10 cm thick sand/granule-rich base with a sheared fabric passing abruptly into a clast-supported conglomerate (individual clasts up to *c.* 1 m) which contains push fabrics (Fig. 3.4.4). These features suggest deposition from a debris-flow. The calibre and angularity of the material suggest the section is proximal to the source area.

Further up the section along the rambla floor grey/green conglomerates are exposed. The general organization of the conglomerates increases, although beds may still be massive and up to 2 m thick. Sandstones become progressively more common up the section. Associated with this transition, a combination of both *in situ* and reworked *megabalanus* are found. These, together with trace fossils, oysters and rare echinoderm fragments, indicate progressively deepening marine conditions. The macrofossil assemblages indicate that this deepening was pulsed (Doyle *et al.* 1997). These observations suggest that mass flows were entering marine conditions from a steep and rocky shoreline. The pulsed nature of the progressive deepening probably reflects the interaction between sedimentation and tectonics (Doyle *et al.* 1997). The sequence fines upwards into deeper water marls.

Transfer to Stop 2

Return to the N340 and head west to the junction with the C-3335 (to Guadix) at Puente de Los Callejones, parking close to the petrol station/café bar Alfaro. The first two stops are visited on foot from this point; if

Fig. 3.4.4. Detail of sheared base of inversely graded debris flow, Rambla de la Sierra, Tabernas. (Photo A. Mather.)

time is limited, Stop 3 can be omitted as it involves a 1.5-km walk to the west from Stop 2.

Stop 2: 'Solitary channel complex' west of Puente de Los Callejones (493 974; Tabernas 1 : 50 000)

The aim of Stop 2 is to introduce the nature of the lower marl-prone section of the basin-fill (unit A), and to consider the origin of a curious ribbon-like body of sandstone and conglomerate beds enclosed within it (unit B). The pair of prominent hills behind the petrol station are capped by a bundle of variably amalgamated sandstones and conglomerates. Stop 2 is located in the saddle between the two hills. From afar, note the marl-prone nature of the lower hillslopes which lack any interbedded sandstones, and the sharp surface running across the hillslope juxtaposing this and the overlying sand-prone section. The sandstones can be seen to run further to the west, capping the ridge which extends in this direction. The sand-prone package is also exposed in the main road just to the north of the road junction. The section seen in this vista is thus a longitudinal section through a narrow (up to *c.* 200 m wide) belt of anomalously coarse-grained lithologies (in the context of the surrounding succession), extending at least 3 km westwards from here (where they dip below the level of exposure) with a linear ENE–WSW orientation (Fig. 3.4.2).

Climb up to the saddle between the two hills, noting the 'background' lithology *en route*. This is best seen in well-washed exposures in the gully running down from the saddle. It is composed of micaceous sandy marls which are very heavily bioturbated, preserving few indications of primary sedimentary structures. This is typical of the lowermost marl succession (unit A, Fig. 3.4.3) which is widely exposed over the badlands to the south of the sandstone-capped ridge.

The coarse-grained package exposed in the saddle is disrupted by local faulting but details of the bedding and sedimentary structures are well displayed. The succession is dominated by sandstones and conglomerates, but is visibly more mud-prone with isolated sandstone beds in the southernmost 'marginal' position seen in the slopes to the east. Beds are generally 0.15–0.8 m thick, and range from fine- to coarse-grained, highly lithic sandstones (with obvious flake-like dark schistose grains, often forming a strong grain fabric), through to matrix- and clast-supported conglomerates. Clasts in the latter are predominantly of dark-coloured schist, with subordinate paler psammite, angular vein quartz and bioclasts, in addition to local rip-up clasts. Coarse-grained, more amalgamated bed bundles alternate with slightly more mud-prone, less amalgamated, thinner bedded sections at a scale of several metres. Textures are all highly immature and suggest limited transport. Beds are sharp-based, locally discordant (particularly the conglomerates) and many show normal grading with evidence for bioturbation extending downwards from bed tops. Sole marks are difficult to see. The coarse-grained beds tend to be separated by thin orange-coloured silt and marl horizons which again are intensely bioturbated. A typical graphic log from this locality is shown in Fig. 3.4.5. A significant feature of many of the sandstone beds from the upper part of the section is the occurrence of large-scale cross-stratification (foresets up to 30 cm high). The cross-bedding is restricted to bed tops, is sometimes laterally discontinuous, and is partially overprinted by burrows extending from the bed top. These indicate unidirectional and steady currents, transporting sand-grade material from west to east approximately parallel to the strike of the sand-prone ribbon, which here is striking NW–SE on account of local structural rotation.

The coarser grained beds record deposition from sediment gravity flows with a range of rheologies. The matrix-supported conglomerates were emplaced as debris-flows, whereas the structureless but graded sandstones and some of the better sorted clast-supported conglomerates are interpreted as high-concentration turbidity current deposits. Locally flows were steady and dilute enough for dunes to become established and to migrate downslope towards the east, implying significant bypass. An analogy can be drawn with the coarse-grained cross-stratified facies of Mutti & Normark (1987). Flows were sufficiently episodic for fine-grained background deposits to accumulate between events and these were only locally removed by erosion to produce amalgamated bed bundles and rip-up conglomerates. The provenance of the materials suggests a source in the Nevado–Filábride basement complex (to the north or west).

The long, ribbon-like outcrop, abrupt contact with the underlying marls which are devoid of sandstone beds, and evidence for the high energy of the flows (transporting large metre-scale blocks out onto an area of the basin floor previously associated with marl deposition) suggests a channelized origin for the coarse-grained deposits.

Before leaving Stop 2, note the view to the south which reveals the Sierra Alhamilla basement block and the unit C sheet turbidites which cap Alfaro. Note also to the SE, in the middle distance, a series of sandstone-capped hills running to the south of Mini-Hollywood. These again rest on sand-starved bioturbated marls with an erosional contact and may represent another channel system, this time tapping into a mixed siliciclastic–carbonate sediment source. Is the channel we have been looking at isolated or part of a family of such conduits?

Transfer to Stop 3

Stop 3 is reached on foot from Stop 2 by walking northwards out of the saddle (note the marls capping the sandstones and conglomerates seen at Stop 2) and bearing west, following the contour of the slope. After 0.5 km, descend to a vantage point on a spur overlooking a wide, flat-based, north–south-orientated dry valley. This valley is an abandoned meander of the Rambla de Lanújar, the incised river to the west (see Excursion 6.4, Stop 2b). From here, it is possible to get an overview of the local stratigraphy. Running northwards, along the eastern side of the valley, the succession overlying the channelized package is seen, albeit thinned by faulting (Fig. 3.4.2). Pale-coloured marls with thick, isolated sandstone sheets (unit C) sit stratigraphically above the channelized package (unit B) and starved marls (unit A). These pass upwards into a more sand-prone mixed

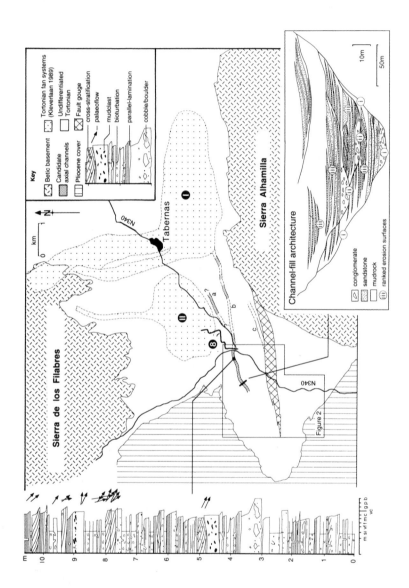

Key

Betic basement | Tortonian fan systems (Kleverlaan 1989)
Candidate axial channels | Undifferentiated Tortonian
Pliocene cover | Fault gouge

cross-stratification
palaeoflow
mudclast
bioturbation
parallel-lamination
cobble/boulder

Sierra de los Filabres

Sierra Alhamilla

Tabernas

N340

N

km

Channel-fill architecture

conglomerate
sandstone
mudrock
ranked erosion surfaces

10m

50m

Figure 2

N340

m si vf f mc l g p b
vc

succession of interbedded sandstones and marls, again with sheet-like geometries (unit D). The local succession is capped by the Gordo megabed (unit E) which is distinguished by its characteristic dark grey/black colour in distant outcrops (a function of the blackschist blocks concentrated in the lower section of this unit). From the map (Fig. 3.4.2), it can be seen that these upper units (D, E and to lesser extent C) can be traced continuously around the open anticline lying to the south. Unit C has a less continuous expression with local thinning and thickening which we will examine later. The key point at this stage is that we move stratigraphically upwards from turbidites and related deposits which are confined within an erosional channel to those which are laterally much more extensive and sheet-like.

Stop 3 is located on the western side of the dry valley and entails a walk of some 0.8 km along a prominent fault-controlled knickpoint on the valley floor (Fig. 3.4.2). The western extension of the channel-fill seen at Stop 2 is extensively faulted on the intervening low ground, and Quaternary–Recent travertines coat many of the surfaces which descend to the south off the perched valley floor (see Excursion 6.4, Stop 2b).

Stop 3: 'Solitary channel complex' north of Las Salinas (482 979; Tabernas 1 : 50 000)

The aim of this stop is to examine further some aspects of the unit B channel complex, specifically some of the thinner bedded fill and the stratal patterns which include common inclined bedding; the latter is difficult to discern in the strike-parallel sections seen at Stop 2.

The ribbon of coarser grained channel-fill deposits again caps a prominent ridge running westwards from this locality but it is displaced on a series of faults which bring it down to valley floor level at this point. A NW–SE-trending fault at the base of the escarpment downthrows the whole complex to the NE, and a second east–west-trending fault which is exposed on the western valley floor juxtaposes the lower channel-fill against what is interpreted to be a thin-bedded upper fill section to the south.

The thin-bedded section to the south is composed of several bundles of

---

Fig. 3.4.5. Sketch map and graphic log illustrating the character and context of the ribbon of marl-enveloped sandstones and conglomerates seen at Stops 2–4. Also shown are the transverse feeder-lobe systems (the sandy system I and muddy system II) identified by Kleverlaan (1989a) which sit at approximately this stratigraphic level. Alternative eastward extensions of the channel system and possible additional channel-fills are labelled a, b and c. See text for further discussion. The inset cross-section is a composite showing the main features of the incision-fill close to its western exposure limit looking downslope to the east. The location of the rambla section to be visited at Stop 8 is also indicated. (Modified after Haughton, 2000.)

lenticular, sharp-based sandstone beds, *c.* 0.10–0.25 m thick, interbedded with bioturbated marls. These are arranged in low-angled, shingled to sigmoidal packages (seen here in sections at a high angle to the inferred palaeoflow). The sandstones are normally graded with structureless bases and parallel laminated, partially bioturbated tops, features consistent with deposition from relatively low-concentration turbidity currents. The lenticular and low-angle inclined bed geometries and stratal bundling suggest a stack of small, possibly meandering channel-fills in which sand was episodically deposited on inclined channel banks (lateral accretion). Similar facies overlie and occur marginal to the sandier lower channel-fill further to the west. The thin-bedded deposits are interpreted here as nested channels sitting within a larger-scale incision, and not as extrachannel levee deposits. Previous interpretations (Cronin, 1995) have drawn attention to the possibility of levee building and westerly bedform migration at this point.

Faulted exposures of the lower sand-prone fill to the north resemble those seen at Stop 2, and further examples of cross-bedding indicating transport to the east can be seen.

The western continuation of the ridge of sandstones and conglomerates can be viewed in longitudinal section from a shoulder of high ground reached by climbing the slope immediately west of the thin-bedded sandstones. The sharp erosional base to the sandy lower fill can be seen, as can further bundles of low-angle inclined beds. Although these apparently dip mainly to the WSW, these are not indicative of palaeoflow and are interpreted as oblique sections through laterally inclined bed bundles sweeping out of the plane of section (see also Clark & Pickering, 1996).

Transfer to Stop 4

Return to the vehicle(s) eastwards across the knickpoint, descending down across the main travertine amphitheatre to pick up a track leading back to the petrol station. Stop 4 (Fig. 3.4.2) is again reached on foot, although depending on rambla conditions, it may be possible to drive some distance towards the exposure. Drive SW *c.* 1.5 km almost to the bridge across the Rambla de Tabernas, leaving the road by the exit track on the right *c.* 200 m before the bridge. Travel westwards *c.* 150 m along the rambla, turning northwards and following the tributary (Rambla de Lanújar), parking the vehicle(s) as appropriate and proceeding on foot.

Stop 4: The 'solitary channel' in cross-section, Rambla de Lanújar
(479 967; Tabernas 1 : 50 000)

The exposures on the walk in are composed of heavily bioturbated marls belonging to unit A. Occasional thin and bioturbated sandstone ribs are seen (centimetre-scale), and locally the crude fissility within the marls is disrupted and contorted, probably as a result of slump remobilization. About 1 km NW of the Rambla de Tabernas, the rambla is incised through the westward continuation of the channelized system examined earlier at Stops 2 and 3. The value of this section is that it is transverse to the trend of the sand ribbon and provides a detailed picture of the architecture of the incision-fill.

The exposure is best first examined in the outer (eastern) flank of an incised meander loop. Initially identify the three main types of bounding surfaces (see inset in Fig. 3.4.5).

1 A first-order erosion surface which 'contains' all the coarser-grained beds and which can be seen high on the southern part of the exposure, dipping gently to the north. This is offset (to below the rambla bed) by a NW–SE-trending fault which downthrows to the NE (look for the offset conglomerate unit dragged up into the fault zone). This fault is probably related to the NW–SE-trending fault encountered at Stop 3. Note the onlap of sandstone beds against the first-order incision surface. The depth of incision on this surface is of the order of 35–40 m, with a V-shaped profile expanding upwards to over 150 m wide (giving a narrow aspect ratio of *c.* 1/4 or 1/5).

2 Second-order surfaces, of which the one with the overlying plug of striking boulder conglomerate is the most prominent. Although this may be mistaken for the base of the main incision, closer examination shows that the conglomerate rests on sandstone beneath it. This conglomerate plug is a cohesive debris-flow deposit, preserving basal shear and lobe-frontal compression structures in the clast fabric. These are related to a palaeoflow direction from west to east. The first-order surface is below the rambla bed here. The surface beneath the conglomerate plug can be traced upwards to the north truncating earlier sand packages and clearly represents a major reactivation or flushing out of the partly infilled incision (to a depth of > 15 m). Such surfaces probably span the width of the original incision and may locally modify the first-order surface.

3 Third-order bounding surfaces separate bundles of lenticular sandstone beds, many of which again have low-angle inclined geometries, dipping to the north (at a high angle to overall ribbon trend). Inclined beds imply lateral migration of flow pathways by several tens of metres, with some examples showing a change from initial lateral migration to final aggradational infilling of shallow (several metre deep) channel relief. The implication is that small and mobile channels were active inside the large-scale incision as it was infilled, and that periods of aggradation of the incision-fill alternated with periods when it was partially flushed out.

Additional features of the incision-fill may be seen on the inside of the meander loop, where there are deep, steep-sided scour-and-fill sandstone plugs, several further conglomerate bodies resting on second-order erosion surfaces, convolution of laminations at the tops of some sandstone beds, local concentrations of bioclastic material (reworked oysters) and good flute structures on the base of a thin, isolated, sandstone bed (above the knickpoint in the stream). The last of these features are consistent with the eastward palaeoflow deduced from the cross-bedding seen earlier, and with other flute and ripple-lamination observations made elsewhere along the incision-fill outcrop. The succession overlying the sandstones is marl-prone, although bundles of thin-bedded lenticular sandstones similar to that seen at Stop 3 occur high on the cliffs to the west of the meander loop and seem to represent mud-prone channels still within the incision-fill. It is possible to climb up to some of these thin-bedded sandstones and see ripple-laminations in the bed tops again indicating flow towards the east. Sheet sandstones of a different character appear in the back of the high cliffs to the north and record the incoming of unit C. Before moving southwards and up the stratigraphy to discuss this switch in style in more detail, it is worthwhile summarizing the key aspects deduced for the lower part of the succession and placing it in a wider basinal context.

Summary of 'solitary channel complex' setting

The western segment of the channel complex has a relatively linear map expression (Fig. 3.4.2), despite evidence for abundant laterally inclined beds within the fill. Less continuous exposure to the east of the N340 obscures the eastward course of the system; Cronin (1995) followed Kleverlaan (1989a) in suggesting a NE–SW route to link with the northern entry point ('a', Fig. 3.4.5). However, there are appropriate exposures of analogous sandstones directly to the east and the system may extend in this direction ('b', Fig. 3.4.5) to give a lateral extent of at least 8 km (in a direction parallel to the axis of the basin). To the west, the incision-fill dips below the level of exposure, although at this point it is finer grained and includes a thicker central mud-prone unit than seen elsewhere along its length.

Taken together, the internal structure, evidence for eastward-directed palaeoflow, ENE–WSW orientation, and low aspect ratio are difficult to reconcile with a NE transverse source for the incision-fill within an 'outer fan' setting (cf. Kleverlaan, 1989a; Cronin, 1995). As the system is wholly enclosed in bioturbated marls in which there is local evidence for slump remobilization, the isolated conglomerates and sandstones are interpreted as the fill to a slope incision which opened out on to the basin floor in an easterly direction. This area of the basin was thus part of an eastward-facing slope at the time, to which sediment was probably delivered by drainage systems exploiting the dominant east–west structural grain to the west (Fig. 3.4.6a). Fission-track data (Johnson *et al.* 1997) establish that uplift of the basement proceeded from east to west (with uplift in the west continuing through into the Tortonian) and hence this is consistent with easterly sediment transport off the more recently uplifted basement highs and a regional gradient from west to east.

The low aspect ratio, lack of obvious overspill, width, depth and setting (in a narrow trough-like basin) resemble the proximal parts of modern delta-fed, high-gradient (0.5–1) fjord-bottom channels (Prior *et al.* 1986; Zeng *et al.* 1991). The depth of incision and lack of sand outside the conduit implies significant bypassing. Thus axial supply was probably more important than has hitherto been recognized on the basin floor to the east, where the emphasis has tended to stress the transverse supply systems (Kleverlaan, 1989a). The punctuated infilling and change from a relatively straight course during incision to more sinuous embedded channels during filling (to explain the dominance of inclined bedding) may record the onset of gradient reduction (the overlying succession requires a flat and enclosed basin floor setting). A reduced gradient would have allowed the incision to back-fill (over at least 8 km) and could account for the upward and westward transition to an increasingly fine-grained style of fill (Fig. 3.4.6b).

Transfer to Stop 5

Return to the vehicle(s) along the rambla floor. The next exposures to be visited lie to the south (Figs 3.4.1 and 3.4.2) along the Rambla de Tabernas and can be accessed via the rambla on foot (leaving a vehicle at the end of the traverse at Stop 6) or by moving a high-clearance vehicle progressively down the rambla. Alternatively, access can be gained on foot from suitable points on the old main road.

Stop 5: A syn-depositional fault gouge zone in the Rambla de Tabernas (481 957; Tabernas 1 : 50 000)

Working south from the road bridge along the Rambla de Tabernas, typical exposures of the sand-starved bioturbated marls occur in which bedding is defined by a crude fissility and subtle alternations in texture and ichnofabric. Dips are moderate to the NW. This is part of the marl-prone unit A section seen earlier surrounding the incision-fill unit. About 300 m down the river bed the character of the succession changes. The best exposures are in the gullied sections beneath the road and in the roadside cuts. On the western side of the rambla, the most notable change is in the

(a) Axial slope incision and bypass

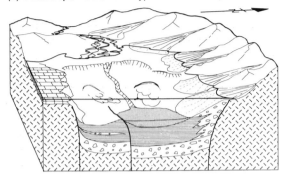

(b) Slope collapse and channel back-filling

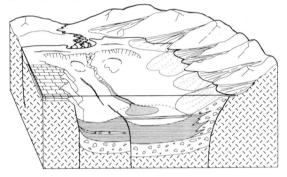

(c) Flow ponding in structural depressions

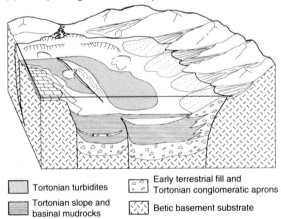

Tortonian turbidites

Tortonian slope and basinal mudrocks

Early terrestrial fill and Tortonian conglomeratic aprons

Betic basement substrate

Fig. 3.4.6. Schematic illustration of the inferred depositional evolution from axial slope incision and bypass (a) through slope collapse and partial back-filling of the incision (b) to the subsequent development of an irregular basin floor with deeps flanked by propagating faults (c). (After Haughton, 2000.)

weathering characteristics and in the unusual concentrations of vein calcite fragments strewn on the degraded marl surface. Where clean exposures are examined along the gullied slopes to the east, coherent bedding which was originally obvious is no longer visible. Instead, the marls have taken on a strong scaly or sheared fabric, with many subvertical shear planes separating panels in which contorted bedding and lozenges of different lithology (including sand) are juxtaposed. The principal vertical shear planes have an east–west orientation and striae on the scaly surfaces are predominantly subhorizontal. Calcite veins have fibres again suggesting subhorizontal oblique opening of cracks. Vein breakage suggests a complex history of veining and further deformation. Occasionally, larger blocks of coherent sandstone appear to have been entrained in the zone. One of these can be examined forming a prominent knickpoint in a tributary merging with the main rambla from the west. When walked out, the width of the zone of sheared rock is *c.* 300–400 m and it is interpreted as an important fault zone.

Several additional aspects of this gouge zone can be seen from a vantage point at road level looking westwards, armed with the detailed geological map shown in Fig. 3.4.2. To the left (SW) a thick section of laterally extensive sandstones interbedded with marls can be seen in the high bluff to the west of the rambla. This is the unit C sheet system, which is particularly well developed south of the gouge zone. Note the progressive break-up of the lowermost sheet sandstone exposed immediately west of the rambla floor as it is traced into the southern margin of the deformation zone. Looking due west, the gouge zone runs along a prominent break in slope, narrowing westwards. Sheet sandstones cannot be observed in the low ground immediately north of the gouge zone, but several wedging sandstone sheets can be seen pinching out northwards by onlap against the underlying marls in the distant hillslopes. Note that this is the full expression of unit C at this point, implying significant lateral thickness variations at this level. In contrast, units D and E can be traced across the westward extension of the zone (with continuity of individual beds and overall thicknesses) and only a minor vertical (10 m) brittle displacement. The implication is that this was an active zone of deformation during deposition prior to sealing by units D and E; it appears to have had a significant impact on the distribution of unit C. The broad width of the gouge zone could reflect the propagation of a basement fault through a weakly lithified, unconsolidated sediment cover to produce a melange-like unit.

Transfer to Stop 6

Stop 6 (Fig. 3.4.2) entails a traverse down the Rambla de Tabernas, working southwards from the southern margin of the gouge zone to examine a section through the unit C sheet sandstones exposed on either side of the rambla. Continue to walk southwards down the rambla until the first intact sandstone sheet appears.

Stop 6: Rambla de Tabernas traverse through contained sheet turbidites (from 478 954; Tabernas 1 : 50 000 to 476 948; Almería 1 : 50 000 or Gador 1 : 25 000)

The aim of this stop is to examine the curious structure of the sandstone sheets which typify unit C. The evidence, discussed below, suggests that this is a function of flow containment in the deep bathymetry which developed in association with the propagation of syn-depositional faults (cf. Stop 5).

A composite log through the lower part of unit C is shown in Fig. 3.4.7. The most easily accessible sections sit beneath a major slump horizon (*c.* 30 m thick, Figs 3.4.2 and 3.4.7) which traverses the steep western slopes forming the backdrop to the rambla (identified as the interval conspicuously lacking sandstone sheets). This slump unit is thought to control the dip slope running eastwards towards the summit of Alfaro, as the sheet sandstones there all correlate with those sitting beneath the slump horizon in the main rambla section. The succession intersects the rambla floor at a low angle (dips are to the WSW) and progressively higher sheets are generally encountered to the south. However, extensive faulting (mainly north–south, NNW–SSE normal faults downthrowing to the east, only some of which are indicated in Fig. 3.4.2) mean that it is difficult to relate the successions either side of the rambla, and the section shown in Fig. 3.4.7 has been assembled by carefully tracing and correlating sections through the faulted outcrops. This is possible because: (i) individual beds extend for at least 2 km laterally through different fault blocks; and (ii) individual beds and successions of beds have distinctive attributes—either compositional variations, distinctive thickness patterns or recurrence intervals, and different internal structures.

Perusal of the log shown in Fig. 3.4.7 highlights a number of key features of this part of the succession. First, very thick beds are common, in addition to thinner beds. Secondly, there is not the familiar log–normal distribution of bed thicknesses seen in many turbidite successions. Thicker beds appear to be over-represented, and thinner beds less frequent than encountered in many turbidite successions. Thicker beds reach a maximum of *c.* 11 m thick. Thirdly, all the thicker sandstone beds are isolated in monotonous sections of 'background' bioturbated sandy marls. However, each of the sandstone beds is overlain (either gradationally or abruptly) by an unbioturbated mudstone cap which scales with the thickness of the underlying sandstone bed.

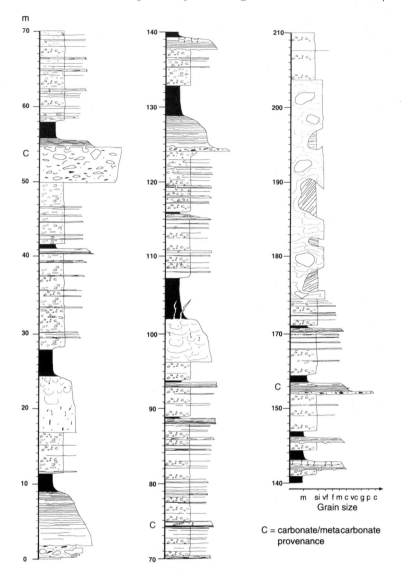

Fig. 3.4.7. Graphic log of the lower part of the unit C sheet system exposed in the Rambla de Tabernas and its tributaries at Stops 6 and 7. (Modified after Haughton, 2000.)

Walking south from the gouge zone along the rambla, the first pair of thick sandstone sheets are seen on the western bank (Stop 6a, 478 954). The lower partially faulted unit is *c.* 11 m thick and comprises a lower clast-rich and 'slurried' division overlain by a thick, graded, medium- to very fine-grained, internally laminated sandstone unit passing gradationally upwards into an unbioturbated mudstone. The overlying unit, again *c.* 11 m thick, is dominantly very fine-grained sandstone with dispersed, elongate mudclasts (defining a vertical fabric) and scattered basement pebbles, and a more clay-rich 'balled' top reflecting intense internal liquefaction. As a rule, the finer grained sandstones show more intense internal deformation. This point is borne out in the next obvious thick sandstone bed (Stop 6b, 478 952), this time seen on the east bank of the rambla, *c.* 200 m further downstream. This corresponds to the sandstone–mudstone couplet shown in the log at *c.* 100 m (Fig. 3.4.7). The sandstone is fine- to dominantly very fine-grained, highly micaceous with scattered carbonaceous flecks. It is 5.5 m thick and again is overlain by a mudstone cap (5 m thick), which is poorly exposed at this locality. Again the internal structure of the sandstone reflects intense liquefaction, with a swirled and 'balled' internal structure.

The same sandstone bed is exposed several hundred metres further along the rambla (Stop 6c, 476 950) in well-washed exposures actually in the rambla bed where the rambla dog-legs to the east, beneath a promontory extending from the western cliff face. The internal contortion and de-watering fabrics are well seen in this outcrop, as are vertical anastomosing clay-lined seams (a consolidation feature?). It is possible to walk across the 'linked' mudstone cap here (trace the unit across the rambla to see the full thickness of the couplet exposed on the eastern cliff face beneath the ruined building at road level). Thin sandstone dykes extend upwards through the mudstone cap from the underlying sandstone. From this point, it is also possible to observe the major slump sheet in the western cliff wall (approach these unstable cliffs with care). Large fragments of marl float within a chaotic, sheared matrix and the whole rests on undisturbed marls with thin sheet sandstones (showing that this is a layer sandwiched within the stratigraphy and not later deformation as seen at Stop 5). Note the fault which truncates the SE extension of the slump sheet and juxtaposes it against a distinctive bundle of five sand–mud couplets.

The five couplets just alluded to can be examined at ground level in exposures where the Rambla de Aguilón joins the Rambla Tabernas from the east (Stop 6d, 476 948; Fig. 3.4.1) just south of the derelict building at road level. The lower four beds are mainly siliciclastic sandstones, the upper-most bed is a calcarenite with a basal debrite unit. All five beds show a dominance of primary parallel laminations, variably disrupted by water escape structures. Good sole marks are seen on the bases of the beds at 138 and 142 m (Fig. 3.4.7) on the south side of the tributary. These indicate that the burrowed marls were scoured by flows moving from west to east before deposition of the sand began.

Transfer to Stop 7

Stop 7 is reached on foot from Stop 6 and involves walking *c.* 700 m up the Rambla de Aguilón. The aim of this stop is to look at the structure of some of the thinner sandstone sheets interleaved with the thicker sandstones examined in the main rambla section, and to examine a calcarenitic megabed with an unusual provenance. Follow the river bed eastwards from its junction with the Rambla de Tabernas, passing under the storm drain beneath the road. Alternatively, climb up the overgrown track leading up to the road (north of the tributary mouth) and examine the lowermost and thickest of the five sandstone sheets just seen exposed to the east of the road in an escarpment facing the lay-by. The feature to note here is the deformation of the substrate seen beneath the graded sandstone, with the marls and thin-bedded sandstones deformed in a zone immediately beneath the sandstone, and close to the top of the exposure, pushed up to form a ruck onto which the sandstone sheet onlaps.

Stop 7: Sheet deposits of the Rambla de Aguilón (483 947; Almería 1 : 50 000 or Gador 1 : 25 000)

The thick sandstone sheet immediately to the east of the road is the same unit seen earlier in the main rambla floor section (Stops 6b and 6c). It is characterized by a 'balled' or 'pillowed' internal structure. Two thinner sandstone beds occur immediately beneath it, directly to the east of the tunnel beneath the road. Pause and examine the lower of these. It shows an unusual vertical profile, having a lower graded division with parallel and ripple laminations, immediately overlain by a structureless sandstone division with sigmoidal dewatering sheets cutting through it (Fig. 3.4.8). Ripples appear to be propagating in a NW direction. This same structure is seen where this bed is traced laterally over at least 1 km and it cannot be explained as two separate events because the two disparate parts of the bed are always in contact and many of the thinner isolated sandstone beds within the succession have vertical profiles of this type, i.e. a lower structured division and an upper structureless (dewatered and/or sheared) division.

Following the rambla upstream, it dog-legs north (be particularly wary of loose blocks overhead here) and then returns to the east. Sheet turbidite sandstones (both siliciclastic and calcarenitic) dip into the rambla bed and are interbedded with thick bioturbated marl sections. Note the succession is repeated by faulting. Well-washed exposures allow details of the thinner sandstone sheets to be observed. Note that all the beds seen here are present

Fig. 3.4.8. Unusual vertical structure of thin-bedded 'contained' turbidite sheets at Stop 7.

in the Rambla de Tabernas exposures and in other ramblas to the south, demonstrating their lateral continuity. Many of the beds reveal a similar bipartite structure to the bed examined earlier, demonstrating that this is a recurring phenomenon in the thin-bedded sandstones of this system. Bioturbation extends into the top of some of the sandstone beds, with spectacular traces preserved on bedding plane exposures. About 200 m from the dog-leg, a distinctive megabed unit dips down into the rambla. This is a tripartite bed with a lower debris-flow unit (8 m thick) composed of angular marl clasts floating in a marly bioclastic matrix with scattered, rounded, extrabasinal pebbles (dark grey dolomite and pale red psammites, the former with borings). A very different provenance is implied, in a shallow-water area with abundant contemporary carbonate grain production and reworking of Alpujárride basement lithologies (contrast the provenance with that of the unit B system). The debris-flow unit is overlain by a graded calcarenitic sandstone sheet which partly balls down into, and is incorporated as clasts in, the upper part of the debris-flow unit. This then fines up into a graded mudstone cap which lacks the bioturbation and is texturally more uniform than the normal background bioturbated sandy marls.

Unit C sheet system: a synthesis

Drawing together the observations from Stops 6 and 7, unit C can be seen to comprise a stack of laterally extensive sheet-like sandstones of variable composition (siliciclastic, calcarenitic and mixed), the thicker of which have prominent mudstone caps. The latter imply that the muddy portion of the flows which introduced the sandstones were now unable to escape downslope but were 'contained' by basin topography (Pickering & Hiscott, 1985; Marjanac, 1987). The smaller volume flows ran out normally, decelerating on the basin floor, and then collapsing rapidly as the flow met counter-slopes off which they reflected. The larger volume and more energetic flows probably sloshed to and fro within the containing bathymetry, inducing extensive liquefaction in the finer grained sandstone sheets. Alternatively, the latter may have been prone to post-depositional liquefaction as a result of cyclical and dynamic loading. Some of the liquefaction appears to have post-dated the fallout of the mudstone caps, judging by the sandstone dykes at Stop 6c. The sealing mudstone caps would have prevented efficient drainage of pore waters during early consolidation.

The depositional setting envisaged for unit C is clearly markedly different from that established in the underlying unit where the flows were bypassing and incising an axial slope. The geometry of the basin must have changed substantially in the intervening period. The infilling of the incision may have presaged the change in basin configuration, as a result of a reduction in gradient. Faults then propagated through the basin-fill to break the surface, opening up localized depressions (mini-basins) into which flows were diverted and from which they could not escape (Fig. 3.4.6c); hence the upward transition to ponded flows. The ponded sandstone sheets wrap over the top of the channel-fill, but pinch out against a muddy high immediately north of the inferred zone of syn-sedimentary shearing. The sheet system is greatly expanded south of this fault zone and the latter is thus inferred to have been active during deposition, producing a 'sump' to the south which collected rare (seismically triggered?) flows from westerly (and possibly southerly or SW) sources. The major slump sheet intercalated with the sheet turbidites may record collapse of a poorly lithified, submarine fault scarp, or even venting of mud volcanoes sitting above the active fault system (cf. Cronin *et al.* 1997).

Transfer to Stop 8

To complete the succession up to the Gordo megabed, the best exposures are found by returning to the Puente de Los Callejones, parking the vehicle opposite the garage, and dropping down into and walking up the Rambla de Tabernas. Alternatively, if time is short, aspects of the upper stratigraphical units can also be seen alongside the N340 to the south of Stop 6 where the trace of the Gordo megabed crosses the road (Fig. 3.4.2). A feature of the upper succession is that there is little differentiation of exposures north and south of the inferred syn-sedimentary fault zone across which it drapes. The implication is that the irregular basin floor topography with mini-basins had been healed by this time, and that deposition was now taking place within an enlarged basin.

Stop 8: Gordo megabed, Rambla de Tabernas (499 979–510 995; Tabernas 1 : 50 000)

The Rambla de Tabernas north of Puente de Los Callejones reveals a section through the succession overlying the incision-fill examined at Stops 2–4 (Fig. 3.4.2). Just north of the bridge, the roadside exposures of the unit B incision-contained turbidites are seen, dipping to the north at $c.$ 30°. For the first 0.5 km, exposure in the western rambla bank is limited to tectonized examples of thick-bedded sheet sandstones equivalent to those seen in unit C further to the south. Better exposures are seen when the main rambla adopts an east–west orientation, with high cliffs on the southern side of the rambla showing several thick-bedded sandstone sheets with complex internal structures, again with unbioturbated mudstone caps. These appear to be thinning northwards when traced along the exposure, and are not encountered when this stratigraphic level is seen again further to the north. These sit structurally in the same position as the contained sheet sandstones further to the south and are thought to again reflect flow trapping in small fault-controlled 'sumps' on the basin floor. There appears to be less thin-bedded sand in the succession intervening between the thick sandstone sheets compared with that at Stops 5 and 6, and the smaller volume flows may have been abstracted by deeper bathymetry lying to the south.

Following the rambla to the east (the valley loops northwards and then eastwards again), high cliffs on the western and northern side of the valley begin to reveal the nature of the deposits overlying the contained sandstone sheets. At the top of the cliffs, a thick and chaotic unit can be discerned in which there are boulders (of metamorphic basement) and large rafts (many metres across) of intrabasinal material. This is the basal division of the Gordo megabed (Kleverlaan, 1987), a laterally extensive debrite bed (up to 50 m thick) capped by a sandstone sheet grading up into a mudstone cap; neither of the upper divisions is seen at this point. Kleverlaan (1987) has traced this unit, interpreted as a possible seismite, for $c.$ 13 km laterally forming a basin-wide marker (see Fig. 3.4.2), and capping the stratigraphic slice considered here.

The base to the debrite unit is abrupt and it sits on a bundle of relatively thinly bedded sandstones with laterally extensive geometries. These descend to the rambla floor level and can be examined in strike-parallel exposures running eastwards along the northern margin of the valley for several hundred metres. This bundle of mostly thin-bedded sandstones is designated as unit D and it can be traced southwards around the open anticline, having a similar geometry to the overlying megabed (Fig. 3.4.2). The sandstone beds are graded, generally poorly amalgamated and have a schistose provenance with obvious blackschist flakes in the coarse-grained bed bases. Significantly, most sandstones lack the mudstone caps seen in unit C (although there are several thicker 'key' beds with mudstone caps) and the thinner beds do not have the distinctive internal characteristics of the thin beds seen at Stop 7. The overall impression is one of a 'normal' bed thickness frequency distribution (perhaps reflecting a different source mechanism—slope oversteepening rather than seismic triggering?), thinner sandstone beds showing vertical sequences of structures more akin to 'classical' turbidites, greater amounts of terrigenous mud in the intervening fine-grained packages, and very diverse ripple orientations.

The palaeoflow of these sandstones is complex and subject to continuing study. The diverse ripple orientations probably relate to flow reflection, and the rare thicker sandstones still retaining mudstone caps are consistent with continuing flow containment, but perhaps over a larger length scale than the underlying mini-basins to explain the widespread distribution and coherence of the package. Seen in a broader context, there is no marked change in bed character and overall sand content when this unit is traced out laterally to the south and west. A laterally restricted lobe origin for part of this package (as entertained by Kleverlaan, 1989a; his System II) is difficult to reconcile with the apparent lateral extent of the package, the absence of systematic lateral trends and the difficulty of defining an obvious lobe margin (see postulated position of lobe fringe, Fig. 3.4.5).

The stratigraphic package is completed by moving up the rambla ≈ 1 km where the Gordo megabed is accessible in large exposures in either valley wall. *En route*, the rambla valley swings abruptly to a northerly direction as it exploits a fault which juxtaposes the Gordo megabed to the east against supra-Gordo turbidites to the west. Continue up the valley until the large, westward-facing cliff exposure which reveals a huge rafted and partially deformed block of turbidites, slumps and background fines. This is the local expression of the basal debrite unit here, where it is dominated by material rucked up and transported (?) as a result of substrate deformation. Note the complex shear planes within the raft, and the overlying 'linked' sandstone–mudstone couplet which can be traced along the outcrop about halfway up the cliff section. Scattered dark schist blocks directly beneath the overlying couplet reveal the typical exotic component entrained in the debrite. The unit is dominated by such blocks a little further to the north. It is possible to examine all three divisions of the megabed on the west side of the rambla *c.* 200 m north of the large ripped-up raft. Here the irregular upward contact between the debrite unit (with many smaller contorted blocks of local slope/basin floor sediment admixed with exotic schist clasts) into sandstone is seen. Climbing the low slopes demonstrates a thick (5 m) graded sandstone sheet, the lower part of which preserves dewatering sheets in carbonate cemented nodules overlain by internally 'slurried' very fine-grained sandstone. Eventually, the sandstone grades imperceptibly upwards into a fine, unbioturbated mudstone cap. Looking back at the large deformed raft, the linked sandstone unit appears to thin over the underlying raft, and the latter may have stood proud of the surface on which the sand was deposited.

Return to the vehicle(s) either by walking back down the rambla, or meeting up with transport on the main road to the east, which can be reached up the track which descends into the valley just beyond the Gordo megabed exposure.

Synthesis

The scheme in Fig. 3.4.9 attempts to pull together aspects of the Tortonian–Messinian turbidite geology illustrated by the exposures visited

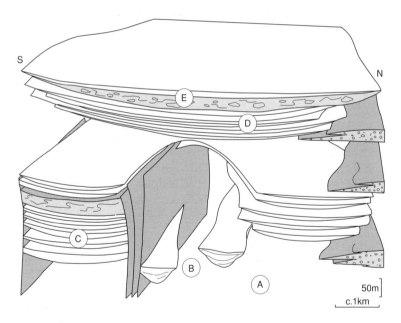

Fig. 3.4.9. Schematic illustration of the stacked turbidite systems encountered beneath the Gordo megabed to the west of the N340. Viewed towards the west. (Modified after Haughton, 2000.)

during this section of the field guide. Early sand starvation and deposition of marls gave way to incision and bypass with the establishment of one or more axial channels which fed sand eastwards along the trough. West–east flow is a persistent feature of the trough-fill further to the east (Haughton, 1994). Punctuated back-filling of the incisions took place during foundering of the axial gradient as basement faults propagated through the basin-fill, resulting in the opening of mini-basins which acted to trap and pond infrequent, seismically generated failure products. Localized sediment deposition infilled the structurally controlled sub-basins, healing the basin floor relief and resulting in a change in flow containment. Re-establishment of updip sediment feed allowed more frequent influx of turbidites, although occasional seismically generated events, including the Gordo megabed, were sufficiently large to still outrun the scale of the basin, which must have remained closed or at least silled. Note that despite evidence for significant intrabasinal relief, the basin remained well oxygenated throughout. Changing basin geometry was evidently a key control on the nature of the turbidite systems preserved in this basin.

3.5 Plio-Pleistocene regional deformation

ANNE E. MATHER & MARTIN STOKES

Introduction

After nappe emplacement in the Middle Miocene there was a significant phase of extensional tectonism associated with the creation of the Alborán Sea (Comas *et al.* 1992, 1996). From the Late Miocene onwards strike-slip deformation became dominant. The associated faulting defined the Neogene basins in the region, which occupy the regional left-lateral shear zone which is characteristic of the NE–SW-trending, left-lateral, Trans-Alborán shear zone (Larouziére *et al.* 1988). This shear zone stretches from Alicante in the NE to Almería, and offshore into the Atlantic to the SW. Faults within this region are associated with regular earthquake activity (University of Granada Geophysics web page, http://www.ugr.es/iag/iagpds.html) and localized hot springs in the Sierra Alhamilla and Sierra Gádor (Weijermars, 1991). Temperature of the spring waters has been shown to rise following earthquake events (Tapia Garrido, 1980) associated with the faults (the Baños de Alhamilla thermal spring temperature rose from 42 to 53°C after an earthquake in 1865). Currently inactive faults affect the local hydrology, with gypsum-rich and carbonate-rich meteoric water springs leading to active travertine deposits in the Tabernas Basin (Mather & Stokes, 1999).

Regional fault pattern

The distribution of the regional fault patterns (Fig. 3.5.1) is broadly similar to simple shear models (Mather & Westhead, 1993). Applying this regionally, NNE–SSW faults such as the Palomares Fault would be the equivalent of synthetic shears, ENE–WSW faults such as the Carboneras Fault

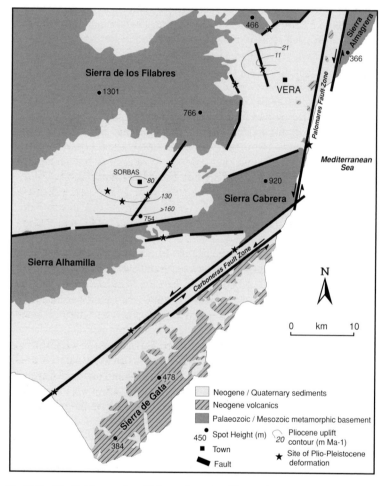

Fig. 3.5.1. Plio-Pleistocene sites of deformation and uplift rates from the Sorbas and Vera basins calculated in metres Ma^{-1} from the present elevation of Pliocene and Pleistocene marine units (Mather, 1991; Stokes, 1997).

represent the master faults and WNW–ESE reverse faults (such as the Cabrera–Alhamilla Northern Boundary Fault) reflect north–south compression related to the overall bulk simple shear (Mather & Westhead, 1993). The Carboneras Fault changes from NE–SW orientation (south of Níjar), to NNW–SSW north of Carboneras before becoming the north–south Palomares Fault which parallels the coast north of Mojácar. This creates a restraining bend which has uplifted the Sierra Cabrera (Keller *et al.* 1995). The changes in orientation, combined with changes in dip direction of the fault, have led to spatially complex deformation styles associated with the dominantly NNW–SSE compression (Hall, 1983; Mather & Westhead, 1993; Keller *et al.* 1995; Stokes, 1997).

Deformation rates

The two most active fault zones in the region today are the NNE–SSW Palomares Fault Zone and NE–SW Carboneras Fault Zone. Estimates of displacement over the Carboneras and Palomares Fault Zones since the Burdigalian (some 16 Ma) are for some 35–40 km laterally and 5–6 km vertically (Hall, 1983). Quaternary lateral slip rates have been estimated for the Palomares Fault in the region of 2 mm yr^{-1} (Weijermars, 1987) and 1 mm yr^{-1} for the Carboneras Fault (Weijermars 1991), rates which are comparable with segments of the San Andreas Fault of California. Work by Bell *et al.* (1997) suggests the Carboneras Fault underwent dominantly left-lateral strike-slip (0.2–0.3 mm yr^{-1}) from the Early to Middle Quaternary and dominantly vertical slip over the last 100 ka (estimated at 0.05–0.1 mm yr^{-1}). Geological evidence suggests that slip along the faults has been temporally and spatially non-uniform, and has switched from one strand of the fault zone to another through time (Keller *et al.* 1995), partially explaining the apparent disparity in displacement rates obtained by various workers.

Away from the Carboneras and Palomares Fault Zones, net uplift rates from the Early Pliocene to the present have been calculated to be in excess of 160 m Ma^{-1} for the main axes of the Sierras Alhamilla and Cabrera (Mather, 1991; Fig. 3.5.1), but are typically much lower for the basins (80 m Ma^{-1} for the Sorbas Basin, 11–21 m Ma^{-1} for the Vera Basin; Mather, 1991; Stokes, 1997).

Most earthquakes in the region appear to be shallow, originating at less than 60 km, and related to fault movement in the crust (Udías *et al.* 1976), although continuing research suggests that the Palomares Fault Zone may be a much deeper feature (E. Rutter, personal communication, 1997). Analysis of recent earthquake data with regional east–west tensional stresses suggests that dominantly normal rather than lateral displacements are occurring today on the NE–SW orientated fault systems (Buforn *et al.* 1988). Some researchers (Bell *et al.* 1997) suggest that the coastline between Carboneras and Cabo de Gata has been uplifted some 2 m since the 15th century (1475), perhaps in response to the larger historic earthquakes such as those which destroyed Almería (1487, 1522, 1659 and 1804; Udías *et al.* 1976) and Vera (1518 and 1865; Bousquet, 1979).

Deformation styles

The faulting activity in the Almería Province has affected the basins throughout their evolution. The deformation covers a range of types including: (i) syn-depositional soft sediment deformation (folding, sediment intrusion, remobilized sediment) which is fairly large scale and would require coseismic activity (Mather & Westhead, 1993); (ii) syn-depositional uncon-formities (including progressive unconformities); (iii) syn-depositional faulting; and (iv) post-depositional brittle faulting. In addition, Quaternary landforms, such as river terraces, have been offset by faulting, and back-tilted, and the evolution of modern river systems has been influenced by the orientation of master faults such as the Carboneras Fault Zone (Bell *et al.* 1997). The range of deformation will be examined in the following excur-sions, concentrating on the regional faults well known to the literature (Palomares and Carboneras Fault Zones) in Excursion 3.5a. Excursion 3.5b will explore the lesser studied Sierra Cabrera Northern Boundary Fault and Plio-Pleistocene basinal lineaments of the Sorbas Basin.

3.5a Excursion: The Carboneras and Palomares Fault Zones (one day)

ANNE MATHER & MARTIN STOKES

Introduction

This full-day excursion examines the region's two master fault zones: the Carboneras Fault Zone (Stops 2–6) and the Palomares Fault Zone (Stop 1). The stops (Fig. 3.5.2) have been selected to cover sections though the fault zones (Stops 2 and 3), and their impact on the associated Quatern-ary fluvial and coastal features.

Directions to Stop 1

Take the Al-151 to Mojácar old town. To the NW of the town follow the signs for the football ground, which lead to the town's main car park. From here follow the road which climbs up to the main plaza.

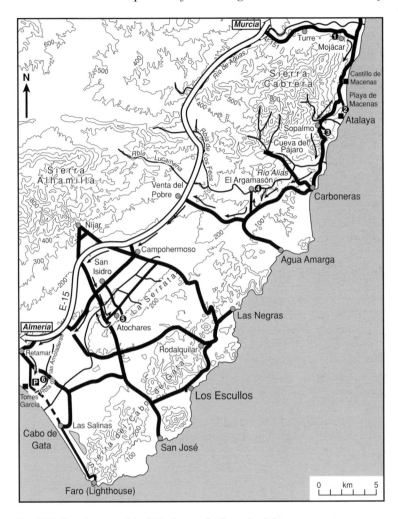

Fig. 3.5.2. Location map of the field-trip stops for Excursion 3.5a.

Stop 1: Mojácar plaza (022 113; Mojácar 1 : 25 000)

This stop provides a view across the southern margin of the Vera Basin. Look north from the main plaza. The line of the Palomares Fault Zone is clearly delimited by: (i) the straight coastline; (ii) the detached mountain range of the Sierra Almagrera which was attached to the Sierra Cabrera in the Mio-Pliocene (Weijermars, 1987; Stokes, 1997); and (iii) a series of pop-ups associated with the fault zone, mainly forming residual hills of purple Triassic basement in the middle distance. Most faults in the region are restricted to the crust; however, recent research suggests that the Palomares Fault may penetrate as far as the lithosphere–asthenosphere boundary (E. Rutter, personal communication, 1997). The fault zone was important in providing a source area for Pliocene fan-deltas around Vera (visible on a clear day to the NW; see Excursion 5.2). The prominent area on which Garrucha town sits is associated with a range of raised beaches (documented up to 30 m above sea level, see Excursion 6.5, Stop 7; and 80 m above Mojácar town, Stokes, 1997).

To the west, the mountain ranges which delimit the western and northern margins of the basin can be observed—the Sierras Bédar and Almagro. Note the *c.* east–west-striking, northward tilting, vertically bedded Tortonian–Messinian (Azagador Member) beds which abut the Sierra Cabrera just north of the view point. These reflect the active uplift of the Sierra Cabrera from the Late Tortonian onwards.

Transfer to Stop 2

From the parking area retrace the road to the junction with the Al-151. Turn right towards the playa. At the junction for Mojácar town (main entrance) turn left towards the coast. At the coast turn right (you should now be driving along the coastal road with the sea on your left). The blackschists through which you are driving form part of the Nevado–Filábride schist of the Sierra Cabrera. Note also the Pleistocene cemented raised beach levels (one of which the road is built upon) associated with red soils to your right, and cropping out beneath the road to your left and right. These are described in Excursion 6.5, Stop 6.

Drive through Mojácar Playa area, and proceed out of town until you see the Castillo de Macenas (signposted) on your left. Coaches are advised to park here. If travelling by car continue driving along the coastal track to the SE, across the river bed of the Rambla de Macenas, and along the coast for about 1.5 km. Park in one of the lay-bys to your right or in the area in front of Macenas Tower. Walk up to Macenas Tower.

Stop 2: Macenas Tower (Atalaya) (023 023; Castillo de Macenas 1 : 25 000)

This facilitates a brief examination of a range of Quaternary raised beach deposits which record the relative sea-level fluctuations driven by regional uplift and eustatic sea-level changes (Excursion 6.5, Stop 5). The dominant basement is composed of black Nevado–Filábride schist of the Sierra Cabrera, buff ?Oligocene carbonates; brecciated Triassic metacarbonates and ?Oligocene conglomerates. These form a series of slivers associated with the Carboneras Fault. Slickenlines can be observed along the track which indicate predominantly left-lateral movement with varied directions indicating the complexity of this zone (the interaction between the Palomares and Carboneras Fault systems). The fault does not appear to deform the upper Pleistocene raised beach sequence (explored in Excursion 6.5, Stop 5) at this locality, although there was evidence for deformation by faulting with small (0.5 m) lateral displacements in higher beach levels north of the Río Aguas, parallel to the Palomares Fault, but now destroyed by construction. There has also been a *c.* 5 m vertical displacement recorded between upper Pleistocene (*c.* 100 ka) marine terrace sequences north and south of the trace of the Carboneras Fault Zone (Bell *et al.* 1997).

Transfer to Stop 3

Retrace the track back to the junction with the road. Turn left to head southwards towards Sopalmo and Carboneras, driving parallel to the Rambla Macenas which drains along the strike of the Carboneras Fault Zone. At Sopalmo, immediately after the pedestrian crossing, turn left and take the track, signposted for the beach, that descends into the Rambla Granatilla (coaches will have to be left on the road). It is advisable to check the state of this track before progressing as it is prone to damage from flash flooding on a regular basis. Follow the gradient down, turning left at the bridge (main tributary). After 1 km from where you left the road you will observe brightly coloured badlands to your left. This is Stop 3.

Stop 3: Sopalmo (000 011; El Agua del Medio 1 : 25 000)

This stop offers a second chance to observe a section of the Carboneras Fault Zone (Fig. 3.5.3). The Carboneras Fault Zone is some 40 km long and 1.4–2.5 km wide, with the narrowest portions being located in the vicinity of Stop 3. Its northern portion has been subdivided into the Polopos, Sopalmo and Colorados Fault Zones by Keller *et al.* (1995). The faults are steeply dipping to the NNW and SSE and in plan form create an anastomosing pattern. The fault zone you are now looking at forms part of the Colorados Fault Zone, exposed as a series of multicoloured badlands with individual units 4–5 m wide, containing a range of cataclastic and mylonitic features. Material incorporated in the fault zone includes purple marls, volcanic rocks and mica schist. Careful observation can reveal a number of kinematic indicators for the deformation. Note the steep gradient of the modern river channel. This reflects, in part, tectonic uplift along the fault, and has been significant in controlling the drainage evolution of the river systems (see Excursion 6.5, Stop 4).

Chapter 3

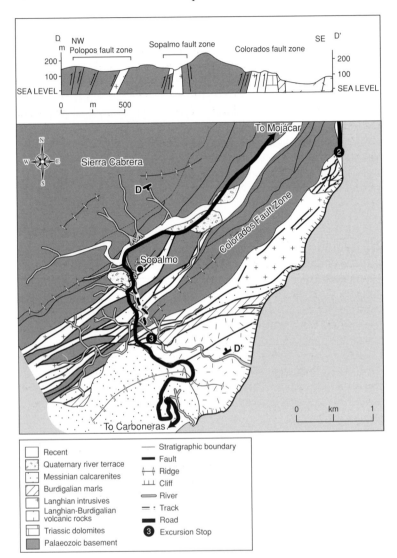

Fig. 3.5.3. Geological map and cross-section of the Carboneras Fault Zone at Sopalmo (Stop 3) showing location of Stops 2 and 3. (Modified after Keller *et al.* 1995.)

Transfer to Stop 4

Retrace the track back to the road junction at Sopalmo. Turn left towards Carboneras. Proceed through Carboneras, following the signs for Almería. After the turning for Agua Amarga, take the turning on the right for El Argamasón. The road to El Argamasón traverses yellow, Pliocene, shallow marine sediments (Cuevas Formation) and grey sands and gravels belonging to Pliocene fan-delta sediments. At the bridge take the old road/track to the right which descends into the river bed. Note the volcanic rocks to the left in the facing cliff which are juxtaposed by faulting. As the road climbs up, find somewhere to park. Take the track (on foot) which cuts below the hill to your right (east). Traverse to the top of the hill.

Stop 4: El Argamasón (917 949; Carboneras 1 : 25 000)

This stop continues the examination of the deformation associated with the Carboneras Fault System. To the north is a view of the southern margin of the Sierra Cabrera and Azagador Member. In the middle distance Messinian gypsum and marls crop out, overlain by the yellow, Pliocene, Cuevas Formation sediments. The hill on which you are standing is composed of gypsum, faulted along the fault zone.

To the east, brightly coloured badlands pick out the line of the Carboneras Fault Zone. To the south, the rectilinear nature of the river course can be seen (Fig. 3.5.4) as the river traverses the Carboneras Fault Zone. The ridge immediately in front is dominated by large cross-beds—part of a Gilbert-type fan-delta, detached from its source area.

Pleistocene fluvial terrace gravels can be seen draped and deformed over the fault zone to the SW, suggesting pre-lithification, dominant vertical fault activity. These gravels form part of an abandoned meander loop of the river, which can be traced behind the hill in front of you. It has been filled with colluvial sediments post-abandonment which show extensive gully development.

Transfer to Stop 5

Retrace your route back to the main road. Turn right and continue to the west. At Venta del Pobre turn onto the motorway (toll-free) towards Almería and continue as far as the second Níjar exit (signed for Níjar, San Isidro). From the motorway ramp turn left (now signed for San José) and follow this road towards La Serrata ridge. The road descends the distal slopes of the large Níjar Quaternary alluvial fan, past San Isidro to the rambla at the base of the fan. Note the lack of any significant incision.

Chapter 3

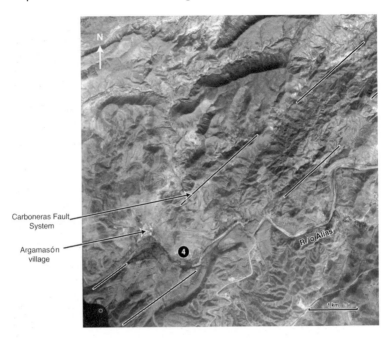

Fig. 3.5.4. Aerial photograph of the Carboneras Fault Zone and offset river system (Stop 4).

Cross the rambla. The road now starts to climb the distal slopes of the smaller and steeper La Serrata alluvial fans. About 1 km after the rambla, take the small road to the left to Atochares. Turn right into the village. If in coaches, park them here and walk. If in cars/minibuses, proceed past the football ground to the north, until after 300 m you join a track which heads east. Follow this up the hill to the TV mast. Park by the mast. Follow the ridge some 200 m to the SW.

Stop 5: Sierra Serrata pressure ridge (756 819; Fernán Pérez 1 : 25 000)

From the vantage point of La Serrata you can observe the 90° inflexion of the Barranco de Fuentecilla along the Carboneras Fault which delimits the north side of the serrata to the SW. The stream parallels the Carboneras Fault for some 200 m before exiting through the pressure ridge (on which you are standing) onto the alluvial fans in the north. This offset has been estimated as Early to Middle Quaternary in age (Bell *et al.* 1997). There is no clear evidence for any vertical slip within the later Quaternary sequence (last 100 ka) at this locality.

Fig. 3.5.5. Aerial photograph image of La Serrata showing shear structure and drainage (Stop 5).

If these streams are examined along the Sierra Serrata it appears that streams on the north of the ridge are deflected to the west (Harvey, 1990; Boorsma, 1992; Bell *et al.* 1997), whereas streams to the south of the sierra are deflected to the east. This highlights the structure of a small shear zone (Fig. 3.5.5). What is unclear is how much of this deflection of the streams is by active tectonics and how much simply expresses a passive response to the physical attributes of the local shear zone.

Transfer to Stop 6

Return to the E-15 as indicated in Fig. 3.5.2 and continue towards Almería until Retamar. From Retamar drive to the coast and head SE at the shoreline. Note the coastline displays three prominent steps. These correspond to subparallel traces to the Carboneras Fault which extend offshore. At the barrier next to the Torres García park and walk *c.* 500 m upstream in the Rambla de las Amoladeras. This is Stop 6.

Stop 6: Rambla de las Amoladeras (643 756; Cabo de Gata 1 : 50 000)

Individual vertical displacements of 3 m can be observed in Pleistocene (Tyrhennian II *c.* oxygen isotope Stage 5; Bell *et al.* 1997) marine terraces. Cumulative vertical offsets across the subparallel traces are of the order of 5–10 m (Goy & Zazo, 1986). No measurable lateral offsets are evident. Note that the main line of the Carboneras Fault follows the Rambla de Morales to the SE, but is less well exposed.

3.5b Excursion: The Sierra Cabrera northern boundary fault and adjacent areas (half day)

ANNE E. MATHER

Introduction

This half-day excursion explores the northern boundary fault of the Sierras Alhamilla and Cabrera (Stop 1) and then examines deformation of Plio-Pleistocene fluvial sediments along a left-lateral, NE–SW basin fault in the Sorbas Basin (Stop 2). There then follows an examination of an east–west orientated structural lineament which generated extensive soft-sediment deformation in the Plio-Pleistocene deposits of the central Sorbas Basin (Stop 3). It is possible to extend the theme of this excursion to a full day by incorporating the deformation at Urra in the itinerary, details of which are given in Excursion 6.2, Stop 5.

Directions to Stop 1

Take the Junction 504 Sorbas turn-off from the E-15 motorway and head south. The road passes through a series of road-cuttings through Tortonian turbidites, which have been tilted and faulted, towards the basin centre as a result of the uplift of the Sierra Alhamilla along the boundary fault (a basin-bounding reverse fault). After 2.2 km, at Peñas Negras, turn left towards Gafarillos. Proceed along this road, which takes you through some of the earliest (?Serravallian) red, continental basin-fill for 2 km to a rubbish dump. Park here and descend to the stream and head upstream. Take the barranco which parallels the narrow ridge in front of you. Follow this for *c.* 100 m.

Stop 1: Gafarillos (874 014; Polopos & El Agua del Medio 1 : 25 000)

You should now be facing a fault wall which depicts part of the northern boundary fault zone (NBF) of the Sierra Alhamilla. The fault is trending

85° to the south. Slickenfibres pitch 16° towards the east. Note the tension gashes in the Tortonian fault-wall sediments and the brecciated Triassic metacarbonate and mica schist in the fault zone which crop out in the river bed. This fault zone delimits the southern margin of the Sorbas Basin and was operative during the uplift of the Sierra Alhamilla during and after the Tortonian.

Transfer to Stop 2

Retrace the road back to the junction at Peñas Negras. Turn right towards Sorbas (do not rejoin the motorway). Continue along this road (the Al-104) to the junction with the old N340. At the junction turn left. Proceed past the town of Sorbas (to your right) which rests on Sorbas Member sediments, explored in Excursion 2.1, Stop 5. After passing the town look for KM 495. Just past here on the left is a house, with a track between two buildings signposted (on the right) for La Cumbre. Take this track (unsuitable for coaches). If in a coach omit Stop 2. This track is not advisable in any vehicle after rain.

Proceed along the track, following the signs for La Cumbre. After 3 km you will arrive at a weak track to your left. The main track descends to the right along the top of a steep cliff. Park by the track at 769 043. Follow the track on foot to the ridge top, follow the ridge along to the east, traversing three small cols. You are at Stop 2.

Stop 2: La Cumbre (771 041; Polopos 1 : 25 000)

Looking south from the ridge you can see the white carbonates of the Messinian Sorbas Member juxtaposed by faulting against red Plio/Pleistocene sands (Góchar Formation). The Barranco de las Contreras has exploited the fault system to develop rapidly and capture the former south–north draining cols over which you have traversed (Mather, 2000b).

To the east (771 041; point 2a in Figs 3.5.6 and 3.5.8) is a spectacular example of a progressive unconformity (Fig. 3.5.7) within Pliocene Góchar Formation sands and conglomerates. Note the changes in thickness in the conglomerate suggesting the syn-sedimentary nature of the deformation. At least three unconformities can be identified. This deformation was important in influencing the alluvial architecture of the Plio/Pleistocene sediments (Mather, 1991; 1993a; 2000a).

Follow the track to the NE to point 2b on Figs. 3.5.6b and 3.5.8, located at the end of the ridge. The facing cliff to the north shows considerable syn-sedimentary thickening and a number of angular unconformities, the most pronounced at the base of the section. If traced further to the east, these sediments are tilted (dips up to 70°) and dragged along a left-lateral

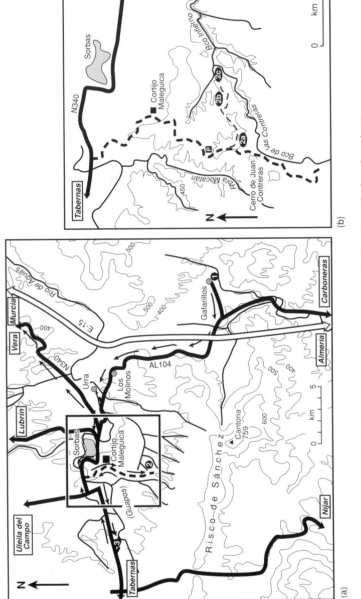

Fig. 3.5.6. (a) Location map of the field-trip stops for Excursion 3.5b. (b) Detail of Stop 2 covering boxed area in (a).

Fig. 3.5.7. Progressive unconformity indicated by key beds (a–d) from the Pliocene Góchar Formation (Stop 2a). View is 150 m across

strike-slip fault (Infierno–Marchalico structural lineament; Fig. 3.5.8). Sections through the fault zone (including fault gouge) are located in the Rambla de los Contreras from point 2c in Figs 3.5.6b and 3.5.8 and upstream. This lineament was probably responsible for the Late Pleistocene deformation at Urra (805 060; Mather 1991; Mather *et al.* 1991; Excursion 6.2, Stop 5) which affects D_2 terraces estimated to be of Late Pleistocene age (based on soils and landform development, Harvey *et al.* 1995) or possibly even Holocene (based on U/Th dating of pedogenic calcretes; Kelly *et al.* 2000). The same alignment (NE–SW) to many portions of the river systems can be identified on aerial photographs indicating the significance of structural control (Fig. 3.5.9). It also passes through an offset in the Cuesta Encantada–Risco de Sánchez (Fig 3.5.9) and faults Tortonian sediments and basement close to 739 003 at the southern margin of the basin, with sinistral displacements of *c.* 250 m (D. Hodgson, personal communication, 2000), but over a much narrower zone than that observed in the more poorly lithified and heterogeneous Plio-Pleistocene lithologies observed at La Cumbre.

Transfer to Stop 3

Return to the main road and at the road junction turn left towards Tabernas. Before the road changes to three lanes (after 1.4 km) take the abandoned road (former route of the N340) to the left. If in a coach, park

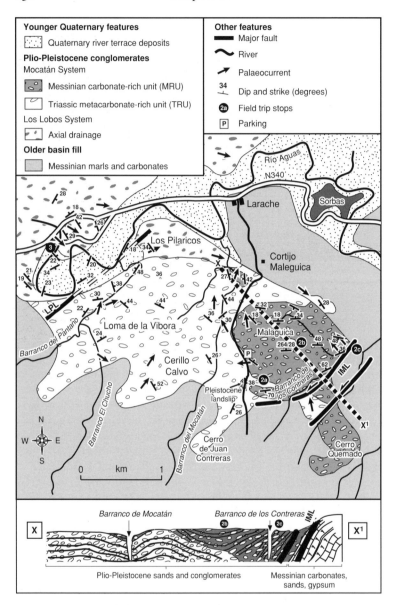

Fig. 3.5.8. Simplified geological map of the area visited by Stops 2 and 3. (Modified after Mather, 2000a.)

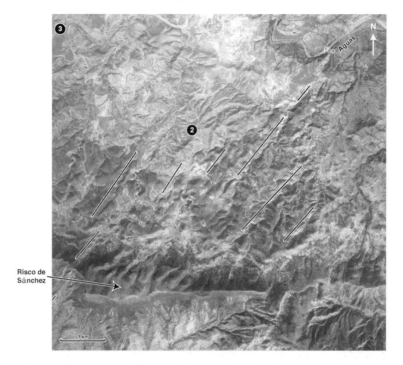

Fig. 3.5.9. NE–SW drainage alignments in the Río Aguas and its tributaries around La Cumbre (Stop 2). (Circled numbers are excursion stops.)

here. Proceed along the old road, through the cutting and over the old bridge. At about 900 m from this junction take the track on your left immediately beyond the old road bridge and descend to the river bed. Park here.

Stop 3: Abandoned road (747 058; Sorbas 1 : 25 000)

Examine the facing river cliff which exposes Plio/Pleisocene conglomerates of the Góchar Formation. Note the well-cemented beds which sag into the centre of the section, and the vertical weathered recesses (three of them) to the left and right of this feature (Fig. 3.5.10).

Closer inspection reveals that the weathered recesses are metre-wide clastic dykes of sands and silts which have been intruded into the overlying conglomerates (the latter are described in more detail in Excursion 5.4, Stop 1). Locally the conglomerates have been completely liquefied and lost their coherence. In other parts of the sections the sediments have been partially liquefied, sufficient to allow clasts to be reorientated parallel to

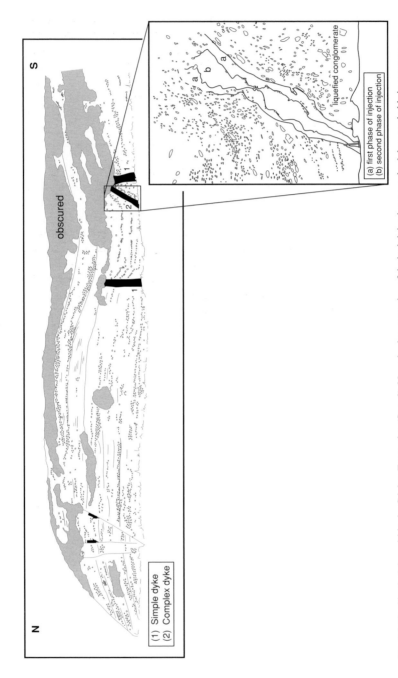

Fig. 3.5.10. Clastic dykes within the Góchar Formation (Stop 3). Note the downwarping of the beds lying between the dykes, probably in response to sediment withdrawal at depth. (Modified after Mather, 1991.)

the dykes. The overlying beds sag into the sediment withdrawal hollow between the dykes (Fig. 3.5.10). Soft sediment deformation features of this scale are indicative of major seismic activity (Mather & Westhead, 1993). Many other intrusive structures (pillars and liquefaction pockets) and examples of folded and remobilized soft sediment deformation have been observed in this area (Mather, 1991; Mather & Westhead, 1993). These features pick out a NE–SW structural lineament (the Los Pilaricos lineament, located approximately in Fig. 3.5.8) which was active in the Plio-Pleistocene (Mather & Westhead, 1993; Mather, 2000a).

This area of the basin is also affected by abundant syn-sedimentary folding and associated unconformities which have generated cumulative thickness variations of up to 130 m within the succession (Mather, 2000a). The deformation forms a series of closed synclines with wavelengths of the order of 150 m and amplitudes (estimated from sedimentary thicknesses in the synclines) of 10–60 m. One of these is evident in the new road cutting through the hill, with horizontal, pedogenically reddened horizons overlying folded sediments which thicken into a syncline (Mather, 1991; Mather & Westhead, 1993). This can be viewed by climbing back to the abandoned road, turning right and taking the track at 748 059 to the top of the hill, overlooking the newer road cutting of the N340.

Chapter 4
Shallow Marine Sedimentation

JOSÉ M. MARTÍN & JUAN C. BRAGA

4.1 Introduction

The first evidence of late Neogene, shallow-water, marine carbonate sedimentation in the Almería area is from the Agua Amarga Basin, where temperate carbonates of Early Tortonian age crop out extensively and are made up of bryozoans, bivalves, echinoids, barnacles, benthic Foraminifera, coralline algae, brachiopods and solitary corals (Brachert *et al.* 1996; Braga *et al.* 1996b; Martín *et al.* 1996; Betzler *et al.* 1997).

In the Sorbas Basin, shallow-water, marine carbonate sedimentation started in the Late Tortonian. At that time, a narrow carbonate platform developed along the northern margin of the basin (rimming the Sierra de los Filabres; Braga & Martín, 1997; Martín *et al.* 1999). Small, scattered, patch reefs occurred all across the platform. In these reefs the most representative corals are *Porites*, *Tarbellastraea* and *Siderastrea*. They have a similar structure to other reefs of the same age occurring in other Neogene basins of the Betic Cordillera, such as the Almanzora Corridor (Martín *et al.* 1989), the Fortuna Basin (Santisteban, 1981; Santisteban & Taberner, 1988) and the Granada Basin (Braga *et al.* 1990).

Shallow-water, marine sedimentation continued during the latest Tortonian and the Messinian in most of the Almería basins, with the deposition of the Azagador Member temperate water carbonates, the Cantera Member reef units, the gypsum unit (Yesares Member) and the Sorbas Member (see Section 2.1). The 'Messinian' deposits are arranged in two depositional sequences (Braga & Martín, 1992; Martín & Braga, 1994), which can be correlated with the TB 3.3 and TB 3.4 cycles of Haq *et al.* (1987) (Fig. 4.1.1). This stratigraphic record is believed to be the result of eustatic sea-level changes combined with local tectonic activity (Martín & Braga, 1996). Temperate carbonates (Azagador Member: unit 1 of Fig. 4.1.1) developed during the cold phase of the lower (pre-evaporitic) cycle. Tropical carbonates with coral reefs (Cantera Member: units 2 and 3 of Fig. 4.1.1) formed as sea-level rose and seawater temperature rose. These reefs grew in a very marginal position, close to the limits of the tropical reef belt (Martín & Braga, 1994). Oxygen isotope studies carried out on planktonic Foraminifera tests from the marls laterally equivalent to

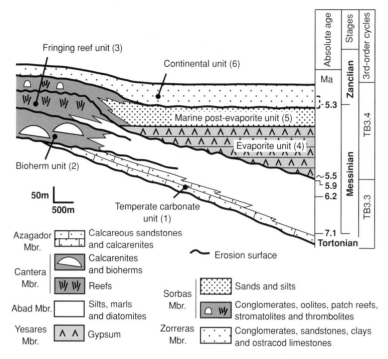

Fig. 4.1.1. Latest Tortonian–Early Pliocene stratigraphy of the Sorbas Basin (Modified after Martín & Braga, 1994; absolute dates after Berggren *et al.* 1995; lithostratigraphic names after Ruegg, 1964.)

these units yield an average sea-surface palaeotemperature of 16°C ± 2 for the Azagador Member carbonates which rose to 24°C ± 2 by the time the bioherms developed and fell to 21°C ± 2 during fringing-reef development (Sánchez-Almazo *et al.* 1997; Martín *et al.* 1999).

Climatic conditions alternated between temperate and tropical several times during the Late Neogene in the area (Martín & Braga, 1994; Brachert *et al.* 1996; Esteban *et al.* 1996; Martín *et al.* 1999) (Fig. 4.1.2). The first temperate episode corresponds to the Early Tortonian, which was succeeded by a tropical phase in the Late Tortonian. The next temperate span was Latest Tortonian to Early Messinian. Tropical conditions prevailed again for most of the Messinian, except when the Messinian evaporites were precipitated and the average sea-surface palaeotemperature was 20°C ± 2 (estimated from the isotopic analysis of the tests of some of the planktonic Foraminifera in the silts and marls intercalated with the gypsum). Temperature rose to 21°C ± 2 in the latest Messinian, during the deposition of the Sorbas Member, which is again tropical (Sánchez-Almazo *et al.* 1997;

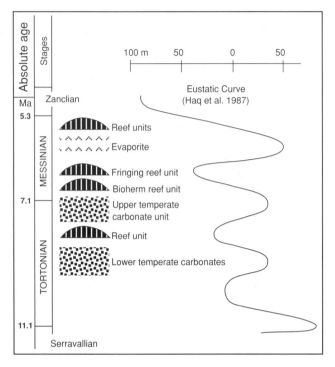

Fig. 4.1.2. Temporal distribution of temperate and tropical (reef) carbonates in southern Spain during the Neogene (modified after Brachert *et al.* 1996; Braga *et al.* 1996b), and their correlation with the eustatic curve of Haq *et al.* (1987). Temperate carbonates were deposited during low sea-levels of third-order cycles whereas coral reefs grew during high sea-levels. Note that Messinian evaporites formed during the cold phase (low sea-level stand) of a third-order cycle.

Martín *et al.* 1999). Temperate conditions were established from the Pliocene onwards.

The Mediterranean area has been especially sensitive to climatic oscillations. The present-day limit between the temperate and tropical climatic zones is just south of the Mediterranean Sea. Climatic variations induced by the low-order Neogene eustatic cycles (probably connected to major glaciations: Shackleton, 1984; Kendall & Lerche, 1988; Williams, 1988; Müller *et al.* 1991) imposed significant changes in the latitudinal position and amplitude of the temperate and tropical climatic belts, with displacement of the temperate–tropical boundary to the south during the cold stages and to the north during the warm ones. Consequently, in cold periods the Mediterranean was a temperate sea, as it is now, becoming tropical during the warm episodes (Martín & Braga, 1994).

These fluctuations between temperate and tropical climates were prob-ably accompanied by significant changes in water-current patterns within the western Mediterranean. During temperate periods, cold Atlantic surface-water entered the Mediterranean and temperate shelf facies developed along the coast in the western Mediterranean. Storm-induced mixing of water precluded stratification. During tropical periods, warmer surface water spread into the western Mediterranean from the east over cold, bottom-related Atlantic waters, resulting in thermal stratification and promoting the development of coral reefs. Deposition of basinal euxinic facies and diatomites lateral to reefs reflects bottom restriction, together with local upwelling (Martín & Braga, 1994; Martín *et al.* 1997).

Evaporite (gypsum) deposits of the Neogene Almería basins are thought to be connected to the 'Messinian Salinity Crisis'. Mediterranean deep-sea desiccation (Hsü *et al.* 1977) resulted in the deposition of more than a thou-sand metres of salt on its floor (Rouchy & Saint-Martín, 1992). At the time these basin-centre evaporites formed, marginal Almería basins, such as the Sorbas Basin, were emergent and subject to erosion. This resulted in a major unconformity affecting the top of the fringing reef unit and the top of the laterally equivalent Abad Member marls (Fig. 4.1.3; Dabrio *et al.*

Fig. 4.1.3. The cliff-forming Yesares Member gypsum filling and onlapping a 30 m deep depression in the underlying Abad Member marls at Molinos del Río de Aguas. The lowermost gypsum horizon pinches out laterally, delineating the margin of the north–south-trending palaeovalley. Note the canyon cut through the gypsum by the modern Río Aguas (see Chapter 6).

1985; Riding *et al.* 1991a, 1999; Martín & Braga, 1994). The relief related to this unconformity is, in the case of the Sorbas Basin, at least 240 m (Riding *et al.* 1998).

Deposition of the gypsum in the Almería area took place in small, perched satellite basins, temporarily isolated from the main Mediterranean because of local uplift (Riding *et al.* 1998). Gypsum deposits of the Sorbas Basin formed during reflooding of the Mediterranean, after drawdown (Riding *et al.* 1999) (Fig. 4.1.4). Sulphur (δ^{34}S values ranging between + 23.5‰

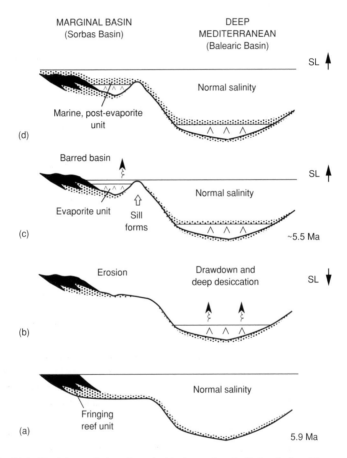

Fig. 4.1.4. Messinian evolution of marginal basins, such as the Sorbas Basin, with respect to the deep Mediterranean (a–d). Note that in this model evaporite deposition in the marginal basins (c) took place after drawdown and main desiccation (b) as the basins became silled and barred by uplift. Marine reflooding (d) is believed to have been completed before the end of the Messinian. (From Martín *et al.* 1999; modified after Riding *et al.* 1998.)

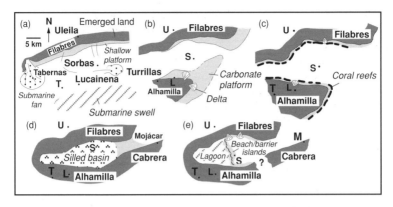

Fig. 4.1.5. Late Miocene palaeogeographical evolution of the Sorbas Basin (Martín *et al.* 1999; modified after Braga & Martín, 1997). During the Late Tortonian the basin was completely open to the south. Small carbonate platforms with local coral reefs developed along its northern margin (a). A huge submarine fan formed in the NW part of the basin (Kleverlaan, 1989a,b; see Chapter 3.4). A smaller fan was located to the east, in the middle of the basin (Haughton, 1994). The present-day southern relief (Sierra Alhamilla) started to emerge during the latest Tortonian (Martín & Braga, 1996). Temperate shelf carbonates were deposited at that time and during the earliest Messinian along both margins of the basin (b). When the Messinian reefs formed (c), the Sorbas Basin was a narrow, east–west-trending corridor. During deposition of the gypsum unit the basin was open to the east (d), with a sill (between Sierra de Filabres and Sierra Cabrera) restricting its connection to the open sea. During deposition of the last Messinian unit the basin was open to the east (Dabrio & Polo, 1995) and presumably to the south as well (e).

and 22.0‰) and strontium ($^{87}Sr/^{86}Sr$ values ranging between 0.70893 and 0.70895 $\pm$ 10^{-5}) isotope data are both clearly within the range for Messinian marine waters (Claypool *et al.* 1980; Hodell *et al.* 1989; Müller & Mueller, 1991) and indicate a marine origin for this gypsum (Playà *et al.* 1997). Most of it formed subaqueously, crystallizing directly at the bottom of the brine.

The Sorbas Basin shows a complex palaeogeographical evolution from the Late Tortonian onwards, reflecting its progressive restriction and isolation (Fig. 4.1.5). At the time of gypsum deposition the Sorbas Basin was a narrow, east–west-trending bay with a topographic sill placed at its eastern margin (Braga & Martín, 1997; Martín *et al.* 1999).

The last marine Messinian unit, which overlies the evaporites, includes large microbial carbonate domes (stromatolites and thrombolites). They occur in a variety of environments, such as fan-deltas (Martín *et al.* 1993; Braga *et al.* 1995), beaches (Braga & Martín, 2000) and oolitic shoals (Riding *et al.* 1991b; Braga *et al.* 1995). Within the domes an impoverished, normal-marine biota of corals (*Porites*), coralline algae, serpulids, bivalves and encrusting foraminiferas can be found (Martín *et al.* 1993; Braga *et al.*

1995), Its presence rules out the possibility of abnormal salinities, suggested by several authors (Esteban, 1979–80; Rouchy & Saint-Martín, 1992) to account for the occurrence and proliferation of these structures. The most plausible explanation is that microbes acted as opportunistic biota and, for a time, during the initial stages of marine recolonization of the Mediterranean Sea (after drawdown and deposition of the evaporites) outcompeted other organisms, settling and growing successfully in most of the available environments (Martín & Braga, 1994).

It should be noted that the last Messinian reefs survived the 'Salinity Crisis'. They occur in this uppermost Messinian unit, together with the giant microbial domes.

The basin silts and marls, laterally equivalent to the coastal and shallow-marine facies with microbial carbonates, have small, unreworked planktonic Foraminifera (Riding *et al.* 1998), together with marine ostracods (Nachite, 1993) and calcareous nanoplankton. This unit, placed immediately on top of the evaporites, represents the final stages and completion of the marine reflooding, which is undoubtedly Messinian in age. A comparable situation is found in the western Mediterranean. In the cores drilled in that area, the Messinian evaporites appear directly overlain by Messinian marls which contain small, planktonic foraminiferas (DSDP site 372: Hsü *et al.* 1977, 1978; ODP site 975: Comas *et al.* 1996). The existence of this Messinian normal-marine unit on top of the evaporites argues against the long-held view of early Pliocene (Zanclian) reflooding (Hsü *et al.* 1977; Cita, 1991).

4.2 Excursion: Temperate water carbonates of the Agua Amarga Basin (one day)

JOSÉ M. MARTÍN & JUAN C. BRAGA

Introduction

The Agua Amarga Basin is a small, subordinate, east–west elongated depression at the northern limit of the Neogene, volcanic Sierra de Cabo de Gata range (Fig. 4.2.1). Although this basin was temporarily connected to the Almería Basin to the west, it has an independent sedimentary evolution resulting in a unique stratigraphic record.

Temperate carbonates crop out extensively in the central part of the Agua Amarga Basin (Fig. 4.2.2). They occur on top of volcanic rocks that yield a radiometric age of 9.6 Ma (Bellon *et al.* 1983; Montenat, 1990). They consist of calcarenites and calcirudites with abundant fragments of bryozoans and bivalves together with smaller amounts of echinoids, barnacles, benthic foraminiferas, coralline algae, brachiopods and solitary

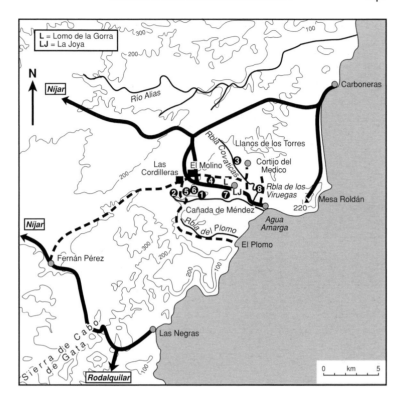

Fig. 4.2.1. Location map of the field-trip stops for Excursion 4.2.

corals. Two main sequences can be distinguished (Fig. 4.2.3): the lower one consists of beach/barrier deposits of Early Tortonian age (Braga *et al.* 1994; Betzler *et al.* 1997); whereas the upper one is made up of coastal to platform deposits of Latest Tortonian to Early Messinian age (Braga *et al.* 1994; Martín *et al.* 1996). These two sequences can be correlated with the Early Tortonian and the Latest Tortonian to Early Messinian temperate episodes represented in other basins of the Betic Cordillera (Martín & Braga, 1994; Brachert *et al.* 1996; Fig. 4.1.2). An important emergent event, marked by a well-developed red soil and, locally, by a karst surface, occurs between the two sequences (Braga *et al.* 1994, 1996b).

To the south and NE of the Agua Amarga Basin, volcanic rocks around 8.1–8.7 Ma old (Bellon *et al.* 1983; Fernández-Soler, 1992) appear intercalated between the two temperate carbonate units. In places they incorporate huge (up to several hundred metres long and several tens of

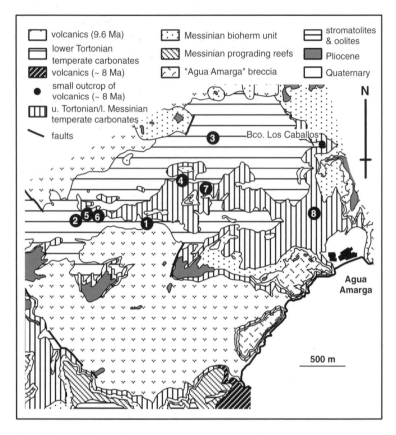

Fig. 4.2.2. Detailed geological map of the Agua Amarga Basin and northern Sierra de San Pedro, with location of Stops 1–8. (From Braga *et al.* 1996b; modified after Martín *et al.* 1996.)

metres thick), irregular bodies of the underlying Lower Tortonian, temperate carbonates (Braga *et al.* 1996b). In both these areas reefs of Messinian age lie unconformably upon the upper temperate carbonates and/or the upper Tortonian volcanics, and they are in turn unconformably overlain by locally brecciated Upper Messinian oolitic and stromatolitic limestones (Addicot *et al.* 1977; Dabrio & Martín, 1978; Esteban & Giner, 1980; Van de Poel *et al.* 1984; Riding *et al.* 1991a). In Mesa de Roldán this latter unit includes small *Porites* patch reefs (Riding *et al.* 1991a). A thin unit (up to 5 m thick) of Pliocene, calcareous conglomerate/sand beach deposits (Martín *et al.* 1994) completes the Neogene stratigraphy of the Agua Amarga Basin.

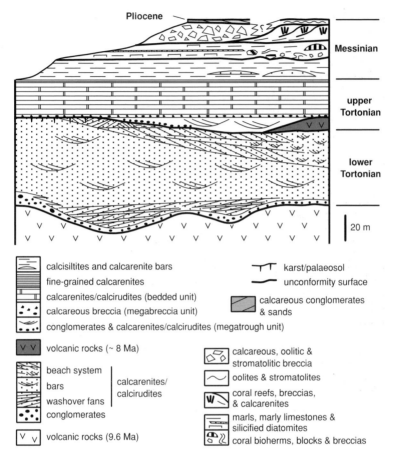

Fig. 4.2.3. Stratigraphy of the Agua Amarga Basin and northern Sierra de San Pedro. (From Braga *et al.* 1996b; modified after Martín *et al.* 1996.)

The Agua Amarga Basin probably developed as a small pull-apart basin related to the Carboneras strike-slip fault system (Braga *et al.* 1996b). In the northern margin of the basin, facies trends were arranged ENE–WSW during the two Tortonian temperate episodes (Martín *et al.* 1996; Betzler *et al.* 1997). During the Early Messinian, the palaeogeography was strongly modified because of intense tectonic activity and subsequent uplift and emersion of the NE and southern areas, around which Messinian reefs developed. Further modifications occurred during the Pliocene and Quaternary (Van de Poel *et al.* 1984).

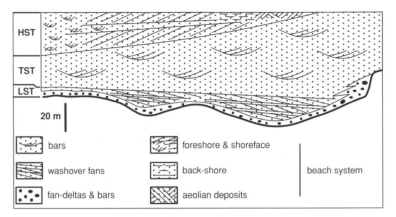

Fig. 4.2.4. Stratigraphy and sequence interpretation of the lower temperate carbonate unit of the Agua Amarga Basin. HST, highstand systems tract; LST, lowstand systems tract; TST, transgressive systems tract. (From Betzler *et al.* 1997.)

Lower temperate carbonate unit

This unit developed as an accretionary coastal prism. It is an unconformity-bounded depositional sequence (parasequence) with three system tracts, each characterized by distinct depositional systems (Betzler *et al.* 1997, 1998) (Fig. 4.2.4). The lower part of the lowstand systems tract consists of a succession of fan-delta deposits (up to 13.5 m thick) showing evidence for wave reworking. The upper part of the lowstand systems tract is formed by calciruditic/calcarenitic washover-fan deposits (up to 20 m thick) with sets of tabular cross-beds with lateral extensions of several hundred metres (Fig. 4.2.5). The transgressive systems tract of the sequence shows a more extensive encroachment onto the volcanic basement, and consists of trough cross-bedded calcirudites/calcarenites (Fig. 4.2.6) belonging to a complex of submarine bars and dunes which reached a maximum thickness of 60 m. The troughs indicate flow towards the east, with a subordinate westwards direction. Patchy red palaeosols and intraformational breccias indicate episodes of local emergence. At the northern margin of the basin, these bar deposits are attached to vertical/subvertical palaeocliffs of volcanic rocks. The highstand systems tract is a wedge increasing in thickness (up to 25 m) towards the SE. The wedge is subdivided into two packages of beach deposits with low-angle planar stratification, changing towards the SE to trough cross-bedded shoreface sediments. A silty clay and/or calcarenite deposit up to 7 m thick interfingers with the topmost foreshore sediments. These rocks probably represent back-shore lagoonal and aeolian deposits (Betzler *et al.* 1997, 1998).

Fig. 4.2.5. Washover-fan deposits of Stop 2 with well-developed sets of tabular cross-bedding separated by extensive, horizontal to low-angle, planar surfaces. Lower Tortonian units of the Agua Amarga Basin.

Fig. 4.2.6. Trough cross-bedded, shoal deposits of Stop 2. Lower Tortonian units of the Agua Amarga Basin.

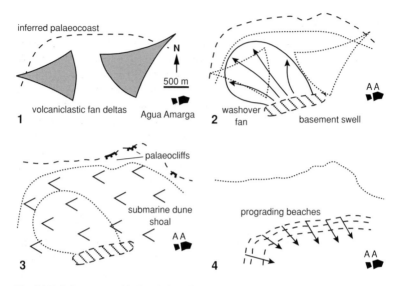

Fig. 4.2.7. Palaeogeographical evolution of the Agua Amarga Basin during deposition of the lower temperate carbonates. (From Betzler *et al.* 1997.)

Sedimentological evidence indicates that lowstand materials formed in a coastal area with local development of fan-deltas. Bars and barrier islands, probably promoted by substrate swells, sheltered small areas in the western half of the basin where washover fans developed. With ascending sea-level most of the basin was covered by a shoal area with locally emergent bars. Beach deposits, prograding to the SE, eventually filled the basin at the end of the sedimentary cycle (Betzler *et al.* 1997) (Fig. 4.2.7).

The occurrence of large benthic Foraminifera (*Heterostegina*, *Amphistegina*) shows that the sediments formed in a warm temperate setting with minimum surface water temperatures of approximately 17°C (Betzler *et al.* 1997). As a result of intense reworking of particles by waves and currents, biofacies differentiation is poor and characterized by small variations between some end members. Along the palaeocliff line, sediments are dominated by echinoderm debris and benthic foraminiferas. Facies rich in benthic Foraminifera occur in beach deposits. The rest of the deposits contain abundant bryozoans, but the assignment of a defined bryozoan growth-type to a specific depositional environment is not possible (Betzler *et al.* 1997, 1998).

Directions to Stop 1

Agua Amarga is a small, seaside village 7.5 km to the SW of Carboneras (Fig. 4.2.1). From Agua Amarga drive along the paved road to the WNW for *c.* 6 km to a junction near El Molino (windmill) farm. Turn left at the junction, follow the track to Fernán Pérez for *c.* 600 m, and turn left again in the direction of El Plomo beach. Coaches should be parked here. It is possible to continue by car following this new track for another 700 m to the Cañada de Méndez ravine. The outcrop can easily be reached from the track that runs east–west, parallel to the Cañada de Méndez ravine to El Plomo beach. This track is followed eastwards for around 500 m. The observations are to be made on a small hill to the north of the track.

Stop 1: *Cañada de Méndez section (912 894; Carboneras 1 : 25 000)*

Lower Tortonian temperate carbonates are well exposed along cliffs at the northern side of the western part of the Cañada de Méndez valley (Betzler *et al.* 1997, 1998) (Fig. 4.2.8). The temperate carbonates lie on top of volcanic rocks which are in places heavily altered to bentonites that are locally mined. The largest bentonite quarries are those of Los Trancos, 3.5 km to the west of the study outcrop, although there are some smaller bentonite quarries, now abandoned, only 1 km to the SW (see Excursion 3.3a, Stop 4).

At the base of the valley, the volcanic basement is erosionally overlain by volcaniclastics, followed by washover-fan deposits with characteristic planar cross-bedding. The washover deposits are separated from the overlying shoal deposits by a 70 cm thick volcaniclastic siltstone with reddish

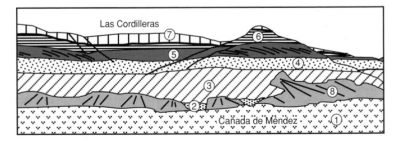

Fig. 4.2.8. Lower Tortonian temperate carbonates at Cañada de Méndez. 1, Volcanic basement; 2, volcaniclastic rocks; 3, washover-fan deposits; 4, shoal deposits; 5, beach deposits; 6, lagoonal and aeolian dune deposits; 7, upper temperate carbonates; 8, scree deposits (Recent).

fissures representing a palaeosol. Shoal deposits exhibit conspicuous trough cross-bedding. The local emergence of bars is indicated by the presence of a discontinuous, *in situ* brecciated limestone layer up to 25 cm thick. The section also provides an overview of the planar foresets of the fore-shore deposits of the beach system, which occur on top, and also allows observation of the back-shore silts. Two cycles of beach deposits can be recognized.

Transfer to Stop 2

Stop 2 is located in the same valley 1 km to the west of Stop 1, close to a large greenhouse (*c.* 0.5 km to the NW). The easiest way to reach this outcrop is to follow the river bed (on foot or by car) to the west up to the point where the dirt road to El Plomo beach from El Molino farmhouse crosses the Cañada de Méndez valley (Fig. 4.2.1). If travelling by car you must park the car here and walk westwards for some 100 m, following the ravine.

Stop 2: Westernmost part of Cañada de Méndez section (902 894–903 891; Carboneras 1 : 25 000)

At this locality there are some spectacular outcrops of the washover-fan and shoal deposits of the Lower Tortonian temperate carbonate unit (Figs 4.2.5 and 4.2.6). Washover-fan sediments exhibit well-developed sets of tabular cross-bedding separated by extensive, horizontal to low-angle, planar surfaces. Echinoid burrows along bedding planes are sometimes very conspicuous. Shoal deposits are trough cross-bedded. Individual troughs can be up to several hundred metres in length and a few metres thick.

Transfer to Stop 3

This stop is 0.5 km to the NW of Cortijo del Médico (Fig. 4.2.1) and is only suitable for those travelling by minibus or car. If travelling by coach go directly to Stop 4 (turn right along the Agua Amarga paved road for *c.* 1.5 km and park on the road to the SE of Los Murcias). To get to Cortijo del Médico follow a dirt road heading east from the Agua Amarga road just in front of the El Molino farmhouse. Cortijo del Médico is *c.* 3 km ENE of El Molino. The outcrops are in the area of Llanos de los Torres, specifically at Rambla de los Covaticas, 200 m to the NW of the Cortijo del Médico. To reach the first outcrop walk to the NNW along the rambla for *c.* 0.5 km, and then another 300 m to the west from this point to reach the second outcrop.

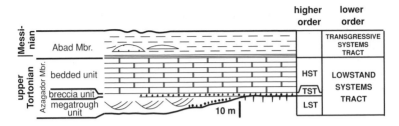

Fig. 4.2.9. Stratigraphy and sequence stratigraphical interpretation of the uppermost Tortonian–lower Messinian temperate carbonates (Stop 3). Symbols as in Fig. 4.2.3. HST, highstand systems tract; LST, lowstand systems tract; TST, transgressive systems tract. (From Martín *et al.* 1996.)

Stop 3: Cortijo del Médico section (933 914; Carboneras 1 : 25 000)

At this locality Lower Tortonian temperate carbonates lie at the base of a palaeocliff, which they abut, lying on top of the basement volcanic rocks. Volcanic boulders are embedded in the bioclastic sediments, and fissures in the volcanic rocks are infilled with bioclastic sediments.

Upper temperate carbonate unit

This unit, of latest Tortonian–early Messinian age, unconformably overlies the lower Tortonian temperate carbonates or the upper Tortonian volcanics. It consists of carbonates overlain by silts and silty marls. The carbonates correlate with the Azagador Member of the Sorbas Basin and form a parasequence which can be divided into three subsequences corresponding in turn to a higher-order lowstand, a transgressive and a highstand systems tract (Martín *et al.* 1996). The overlying silts and marls belong to the lower Abad Member (Fig. 4.2.9).

The lowstand systems tract forms a shorewards thinning wedge (maximum thickness of 10 m) of megatrough cross-bedded calcirudites which pass laterally into a volcanic conglomerate with boulders up to 1 m in diameter. The transgressive systems tract is represented by a well-cemented breccia 1–3 m thick, encroaching the lower Tortonian carbonates to the north. Clasts (up to 30 cm long) are volcanic pebbles and carbonate lithoclasts from the underlying temperate carbonates, together with abundant centimetre-sized bioclasts of bryozoans, oysters, pectinids, solitary corals, coralline algae, barnacles and gastropods. They are all embedded in a calcarenite/fine-grained (granule-sized) calcirudite bioclastic matrix. The highstand systems tract consists of a thick wedge (up to 25 m thick) of bioclastic, bryozoan/bivalve-dominated calcarenites/calcirudites.

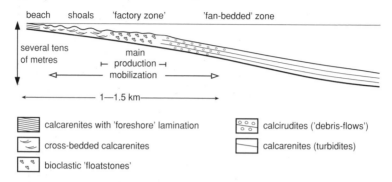

beach   shoals   'factory zone'   'fan-bedded' zone

several tens   main
of metres   ⊢ production ⊣
               ◁——————— mobilization ——————▷

◀———————— 1—1.5 km ————————▶

▦ calcarenites with 'foreshore' lamination   ▦ calcirudites ('debris-flows')

▱ cross-bedded calcarenites   ▦ calcarenites (turbidites)

▦ bioclastic 'floatstones'

Fig. 4.2.10. Depositional model for the Azagador Member carbonates in the Agua Amarga Basin. These temperate carbonates were deposited on a gentle ramp, in a series of different subenvironments. A beach system formed at the coast. In front of the beach, a shoal area developed, with well-sorted calcarenites/calcirudites exhibiting very conspicuous trough cross-bedding. Calcareous organisms lived prolifically in a narrow area ('factory zone') located immediately seawards of the shoals, and their skeletons accumulated to form bioclastic floatstones. Further offshore the so-called 'fan-bedded' zone occurs, consisting of gently dipping, sloping beds, thickening downslope, and composed of highly fragmented bioclastic floatstones/rudstones, with grain size diminishing very rapidly downslope. (From Martín *et al.* 1996.)

The depositional model for the highstand deposits (Martín *et al.* 1996) (Fig. 4.2.10) is that of a gentle ramp with prograding beaches and shoals in its higher parts. Seawards of the shoals was the 'factory zone', where most organisms lived and maximum carbonate production took place. From the 'factory zone' some of the skeletal carbonates were washed landwards by waves and/or currents during storms and incorporated into the shoals and beaches; others moved downslope along the ramp as mass flows, accumulating to form the 'fan-bedded zone'. Facies trends were roughly arranged in an east–west direction, with the coastline to the north of the basin (Martín *et al.* 1996) (Fig. 4.2.11).

The internal structure of the 'carbonate factory' deposits is more or less massive with occasional laterally discontinuous and irregular, convex-upwards stratification. Micritic lithologies prevail ('floatstones') with well-preserved delicate bioclasts. The absence of cross-bedding and the high matrix content, in association with the comparatively large size of bioclasts and lower degree of fragmentation, suggests that most particles are parautochthonous and were originally produced either within this zone or close by. The biogenic association is dominated by nodular and branching bryozoans and bivalves (Fig. 4.2.12), together with barnacles, benthic Foraminifera, echinoids, brachiopods, solitary corals and coralline algae (Martín *et al.* 1996; Brachert *et al.* 1998). A shallow setting for the 'factory zone', just below wave base (*c.* 10 m), is evident from lateral facies changes,

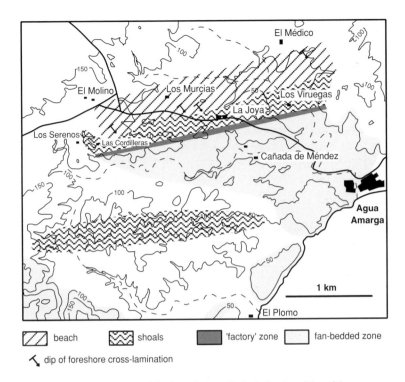

beach ▨ shoals ▨ 'factory' zone ▨ fan-bedded zone □

⊀ dip of foreshore cross-lamination

Fig. 4.2.11. Palaeogeography of the Agua Amarga Basin during deposition of the Azagador Member carbonates. (From Martín *et al.* 1996.)

Fig. 4.2.12. Floatstone from the 'factory' facies consisting of abundant nodular bryozoans (Stop 3), Azagador Member carbonates. Coin is 2.5 cm in diameter.

Chapter 4

the ramp topography, the presence of seagrass-dwelling foraminifers (*Cibicides, Cibicidella, Rupertina, Carpentaria balaniformis, Elphidium*) and the abundance of central growth cavities in bryozoan encrustation fabrics interpreted as representing traces of seagrass (*Posidonia*-type) (Betzler *et al.* 1998; Brachert *et al.* 1998).

The highstand systems tract deposits display a high-frequency cyclicity. The 'factory area' and fan-bedded sediments are intercalated with five well-defined, thick beds of calcarenites/fine-grained calcirudites (Figs 4.2.13 and 4.2.14). They show bar morphologies (single or amalgamated), or make up sand waves with very consistent tabular cross-bedding pointing landwards. These beds formed in a very shallow, wave/current-influenced, coastal environment. The bars and sand waves in the fan-bedded zone developed during lowstands of high-order, short precession cycles, whereas those located higher up in the ramp interbedded with the factory facies are related to transgressive stages. Prograding beaches, shoals, factory facies and fan-bedded layers developed during the highstands (Martín *et al.* 1996) (Fig. 4.2.15). Net skeletal production occurred mainly during the highstands.

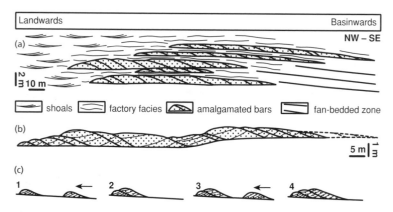

Fig. 4.2.13. (a) Field sketch of the Las Cordilleras area (Stop 6) where five thick calcarenite/calcirudite beds intercalate between coarse-grained floatstones ('factory' facies). The latter grade landwards into shoals and basinwards to fan-bedded layers. (b) Detail of one of the beds (the one in the middle) showing the internal structure, consisting of amalgamated bars giving a 'false', but very conspicuous cross-bedding to the SSE. Note that internal lamination within single bars points in the opposite direction (NNW). The higher elevation of the bed at the right-hand side is a result of the morphology of the immediately underlying bed. (c) Hypothesized genesis of calcarenite/calcirudite beds by landwards migration, superposition and amalgamation of successive bars. (From Martín *et al.* 1996.)

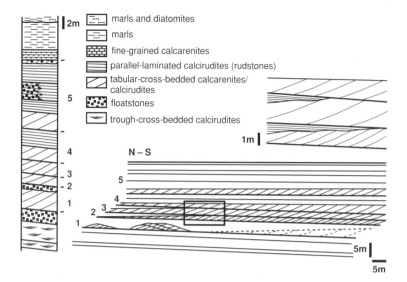

Fig. 4.2.14. Stratigraphic column from Rambla de los Viruegas (Stop 8) and schematic cross-section showing five distinct beds. Each of these exhibits tabular cross-bedding to the NNW and local bar morphologies (bar and sand-wave deposits), and is overlain by medium- to fine-grained calcirudite (rudstone) beds with distinctive parallel lamination, dipping gently to the south (fan-bedded). Note (inset) how the three central beds become amalgamated to the south. Floatstones immediately underlying unit 1 are considered to be distal equivalents of the 'breccia unit'. (From Martín *et al.* 1996.)

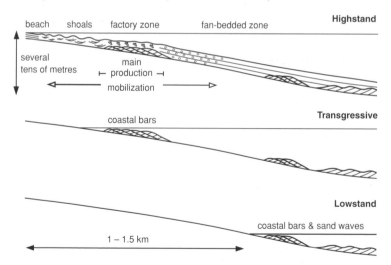

Fig. 4.2.15. High-frequency, internal cyclicity of the Azagador Member carbonates in the Agua Amarga Basin. In the lowstand stage, coastal bars and sand waves developed at a low point down the ramp. In the transgressive stage, they shifted to a higher position. At the highstand stage all these sediments were covered and buried by the highstand deposits described above (see Fig. 4.2.10). (From Martín *et al.* 1996.)

Transfer to Stop 4

From Cortijo del Médico return to the Agua Amarga paved road and drive eastwards for 1.5 km. The section is located on the flank of a low, flat-topped hill SE of Los Murcias. It is *c.* 100 m north of the road.

Stop 4: La Gorra (919 902; Carboneras 1 : 25 000)

Here the beach deposits of the highstand systems tract of the upper temperate carbonates are exposed. These deposits form a wedge pinching out towards the NW and thickening towards the SE. Foreshore beds up to 1.2 m thick show internal layering with well-developed, low-angle parallel lamination dipping to the SE. They overlie and pass laterally into trough cross-bedded shoreface deposits (Fig. 4.2.16).

Transfer to Stop 5

To get to the outcrop for Stop 5 drive back to El Molino farmhouse junction and follow the track to Fernán Pérez for *c.* 500 m. Coaches must be

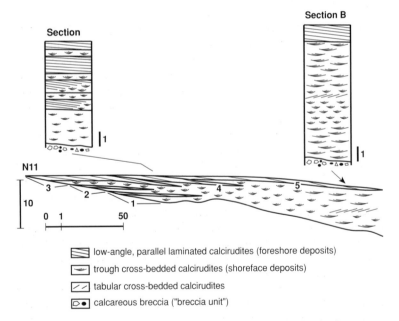

Fig. 4.2.16. Sketch of the La Gorra area (Stop 4) and selected sections showing five land-attached beach cycles numbered 1–5 each marked by foreshore beds prograding SE over shoreface deposits. (From Martín *et al.* 1996.)

parked here. If travelling by car proceed along the track to the left that takes you directly to Los Serenos houses. Stop 5 is a small cliff south of here.

Stop 5: Los Serenos section (905 895; Carboneras 1 : 25 000)

The outcrop provides an overview of the shoal deposits placed just in front of the shoreface sediments. Trough cross-bedded, bioclastic calcirudites predominate. The thickness of the unit and the dimensions of the sedimentary structures increase very rapidly in a north–south direction. Most cross-bedded structures are orientated (foreset azimuths) between N240° and N280° with gentle dips (maximum 10–12°).

Transfer to Stop 6

This section is situated at the western side of the low hill east of the Los Serenos houses, *c.* 200 m to the SE of them and can be reached only on foot.

Stop 6: Las Cordilleras section (907 895 and 908 894; Carboneras 1 : 25 000)

In this locality the 'carbonate factory' deposits (Fig. 4.2.12) alternate with cross-bedded bar deposits (Fig. 4.2.13). This alternation reflects a high-frequency internal cyclicity. Five well-defined, thick calcarenite/fine-grained calcirudite beds are intercalated between very coarse, pebble- to cobble-sized calcirudites (floatstones) with well-preserved remains of nodular and branching bryozoans, bivalves, barnacles, echinoids and coralline algae. Individual bioclasts may be up to 10 cm in size.

The calcarenite/calcirudite beds (up to 2 m thick) extend laterally in a north–south direction for some tens of metres before thinning out and disappearing (Fig. 4.2.13). In detail, they show a complicated internal structure. Tops and bottoms are sharp and irregular. They normally start as a small bar with well-defined morphology to which other bars are attached laterally and partly superimposed, forming as a whole a well-defined, tabular cross-bedding pointing seawards (SSE). Internally, most of these bars show poorly defined cross-lamination dipping to the NNW, clearly indicating that they moved landwards before merging and becoming attached to previous bars.

Floatstone deposits from the 'factory zone' show a broad (up to 0.5 m thick and 10 m across), poorly defined, gently undulating bedding. The large size and low degree of fragmentation and abrasion of their bioclasts indicates that most of these remains come from organisms that lived in this area or close by.

Transfer to Stop 7

Return to the El Molino farmhouse junction and drive eastwards for *c.* 2 km towards Agua Amarga. Park all vehicles by the road. The outcrop is by the road to Agua Amarga, *c.* 100 m to the south of La Joya.

Stop 7: La Joya section (927 899; Carboneras 1 : 25 000)

In the La Joya section, 2 km to the east of the Las Cordilleras section and within the same facies belt, carbonate factory facies alternate with NNW-dipping tabular cross-beds. These beds are equivalent to the composite bars described at Stop 6.

Transfer to Stop 8

Stop 8 is 1 km to the NW of Agua Amarga, in a 30 m high, NW-facing cliff by a dog-leg bend of the Rambla de los Viruegas. To get to the outcrop follow a track (unsuitable for coaches) branching from the Agua Amarga road, some 0.7 km to the west of the village and running parallel to the Rambla de los Viruegas. This track is followed for *c.* 0.5 km before reaching the locality. If travelling by coach, park where convenient and proceed on foot.

Stop 8: Los Viruegas section (944 897; Carboneras 1 : 25 000)

Here the bedded unit shows a fan-bedded (fan-array) disposition, with single beds increasing in thickness from north to south, and dipping gently to the SSE (Fig. 4.2.14). To the north, poorly sorted, coarse-grained calcirudites (floatstones) containing bioclasts up to 8 cm in size, are intercalated between, and grade laterally southwards into, coarse- (2–3 cm) to fine-grained (a few millimetres) calcirudites (rudstones) with well-developed, parallel lamination. These two types of sediments could be genetically related and represent small debris-flows and associated, finer-grained, turbidite deposits.

To the south, medium- (1 cm) to fine-grained (a few millimetres) calcirudites (rudstones) exhibiting very distinctive, parallel lamination predominate. These sediments are intercalated (and in some cases thin out and disappear) between five well-defined 2–3 m thick beds (Fig. 4.2.14). The latter beds consist of calcarenites/fine-grained calcirudites (rudstones) with highly abraded bioclasts, and with a very consistent internal tabular cross-bedding directed to the NNW. They represent migrating sand waves that pinched out and disappeared to the north, through an area where only small, isolated bars occurred. They increase very rapidly in thickness southwards and locally become amalgamated (Fig. 4.2.14).

4.3 Excursion: Tropical carbonates of Níjar (one day)

JOSÉ M. MARTÍN & JUAN C. BRAGA

Introduction

In the Neogene basins of Almería there are two important stages of Messinian reef development preceding evaporite deposition: (i) the bioherm stage; and (ii) the fringing-reef stage. The bioherms developed as isolated coral and algal (*Halimeda*) mounds on the platform and at the platform slopes, respectively (Martín *et al.* 1997). In the coral bioherms *Porites* is locally accompanied by abundant *Tarbellastraea*. At the base of the bioherms laminar coral growth is most common. These are followed by vertical stick-like growths, which are often surrounded by stromatolitic crusts. At the very top of the bioherms, coral heads up to 1 m in diameter can be found. The flank facies of the coral bioherms are bioclastic packstones, with abundant *Halimeda*, coralline alga, serpulid and bivalve remains (Riding *et al.* 1991a; Esteban *et al.* 1996) (Fig. 4.3.1). There are some good examples of *in situ* coral bioherms at the northern side of the Cerro de la Molata in the area of Las Negras in Cabo de Gata, formerly considered as reef blocks (Franseen, 1989; Franseen & Mankiewicz, 1991; Franseen & Goldstein, 1996).

The fringing-reef stage is the main episode of Messinian reef development. Reefs are localized around basin margins and topographic highs (Fig. 4.3.2), from which they prograded basinwards in a series of clinoform wedges (Braga & Martín, 1996a) (Fig. 4.3.3). Internally, each wedge consists proximally of reef, and distally of reef slope (Fig. 4.3.4). The reef includes three vertical divisions: (i) a thick lower pinnacle zone; (ii) a thicket zone; and (iii) a reef crest zone. Pinnacles are isolated, circular coral growths up to 15 m high found at the base of the reef, whereas the thicket and reef crest correspond to more continuous biostromal growths, each a few metres thick. The pinnacle and thicket zones are dominated by stick-like *Porites* colonies, connected by laminar bridges. Stromatolitic crusts occur around the corals (Fig. 4.3.5) and increase rapidly in importance upwards. The reef crest is mainly stromatolitic with minor, very thin, irregular and contorted laminar *Porites* growths (Riding *et al.* 1991a). In the reef slope three distinct parts are differentiated (Dabrio *et al.* 1981; Fig. 4.3.4):

1 reef-talus slope (uppermost) with coral breccias and blocks together with abundant *Halimeda*, bivalves, bryozoans, serpulids and some laminar *Porites*;

2 proximal slope (middle) characterized by bioclastic sand and gravel composed of fragments of corallines, together with rhodoliths and minor quantities of *Halimeda* and serpulid debris; and

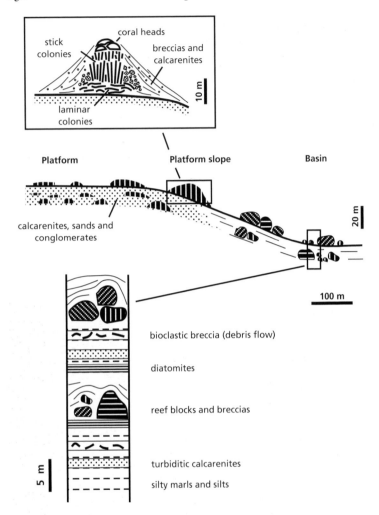

Fig. 4.3.1. Messinian coral bioherms. Sedimentary model and internal structure. (Modified after Esteban *et al.* 1996.)

3 distal slope (lowermost) consisting of bioclastic sands and silts with mostly unrecognizable skeletal fragments.

Most of the organisms inhabited upper slope environments and their remains accumulated more or less *in situ* (Martín & Braga, 1989). Mankiewicz (1988) reports *Halimeda* concentrations at particular levels on the slope, which she relates to periodic fluctuations in productivity caused by episodic upwelling.

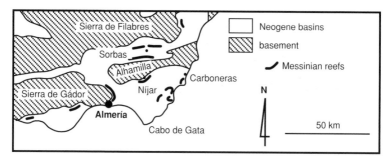

Fig. 4.3.2. Messinian coral–stromatolite fringing reefs in the Almería Province. (Modified after Dabrio *et al.* 1985.)

Fig. 4.3.3. Fringing reef front at Barranco de los Castaños, 1 km west of Cariatiz (Excursion 4.4, Stop 3). Reef-framework facies prograding on top of talus-slope breccias changing laterally to fore-reef slope calcirudites/calcarenites. Reef-slope facies dip steeply basinwards and wedge out distally. This panoramic view corresponds to the southernmost part of the Barranco de los Castaños section.

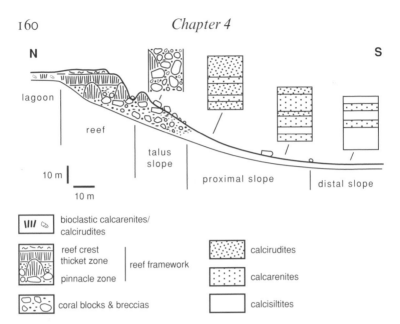

Fig. 4.3.4. Facies model of the Messinian fringing reefs. (From Braga & Martin, 1996a.)

Fig. 4.3.5. Laminated, domed stromatolitic crusts developed on the upper surface of a laminar coral bridge (dissolved) between vertical *Porites*, Cariatiz, pinnacle zone. Coin is 2.5 cm in diameter.

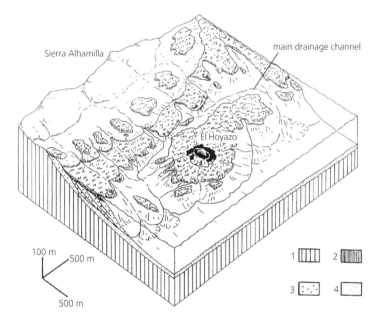

Fig. 4.3.6. Ideal reconstruction of the Níjar reef complex during the last episodes of reef progradation. At the reef front, reef growth was discontinuous and occurred as separated lobes with channels between. The drainage system of the fringing reef rimming the Sierra de Alhamilla interfered with the one developed around the volcanic cone of the El Hoyazo and two main drainage channels resulted. 1, Pre-reef substratum, 2, reef-core deposits; 3, reef-slope sediments; 4, basinal sediments. (From Dabrio *et al.* 1985.)

The general gross morphology of the fringing reefs is best preserved at Níjar, where the reef front appears as a series of lobes, laterally separated by sand channels (Dabrio *et al.* 1981, 1985) (Fig. 4.3.6). At this locality there are also good exposures of the reef slopes. The fore-reef-slope facies are volumetrically the most important in the reef complex. They are exposed as giant, decametre-scale cross-beds, with original dips of as much as 30° close to the reef core, gradually decreasing downslope to 3–5°. The thickness of the individual beds also diminishes downslope, where they interfinger with, and grade laterally into, the basinal facies (Dabrio *et al.* 1981). The reef deposits at Níjar are now completely dolomitized (Martín, 1980; Meyers *et al.* 1997).

Directions to Stop 1

Níjar is a village at the southern foot of the Sierra Alhamilla, 30 km to the NE of Almería and close to the motorway. From the motorway take the

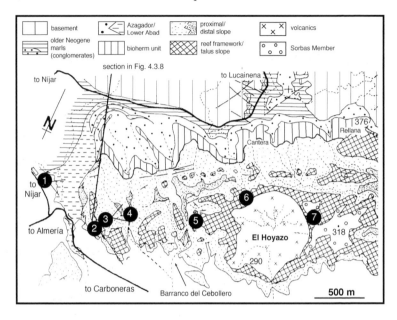

Fig. 4.3.7. Detailed geological map of the Níjar reef and itinerary with alternative stops and panoramic views. (From Martín & Braga, 1989; modified after Dabrio *et al.* 1981.)

exit for Níjar (first exit if travelling towards Almería, second exit if travelling from Almería). The first stop is a panoramic view of the western side of the Níjar reef. This panoramic view is from one of the corners of the village sports field, which is 100 m to the east of the cemetery, located at the SE entrance road to Níjar, *c.* 1 km to the SE of the village centre.

Stop 1: Panoramic view of the Níjar cross-section (715 905; Níjar 1 : 25 000)

The Níjar reef (Fig. 4.3.7) developed along the southern flank of the Sierra Alhamilla (Dabrio *et al.* 1981, 1985; Mankiewicz, 1987, 1996; Martín & Braga, 1989; Serrano, 1990; Jimenez & Braga, 1993). This sierra consists of Palaeozoic and Mesozoic rocks from the Internal Betic Zones. Langhian and Serravallian marls and marly limestones were unconformably deposited at this point, on top of the Triassic dolomite basement. They are, in turn, unconformably overlain by uppermost Tortonian–lower Messinian sandstones and conglomerates, which grade laterally into marls with fossiliferous turbiditic sandstone and conglomerate intercalations ('Marginal Terrigenous Complex' of Dabrio *et al.* 1981; Azagador Member in our interpretation). Upwards in this unit there is a gradual increase in the amount of skeletal particles, mainly fragments of bivalves, bryozoans, echinoids and corallines.

N–S

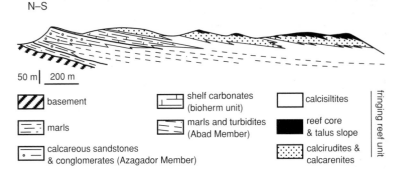

50 m | 200 m

⟋⟋⟋ basement

shelf carbonates (bioherm unit)

calcisiltites

⎓⎓ marls

marls and turbidites (Abad Member)

reef core & talus slope

calcareous sandstones & conglomerates (Azagador Member)

calcirudites & calcarenites

fringing reef unit

Fig. 4.3.8. The Níjar section (Stop 1). (From Jimenez & Braga, 1993; modified after Dabrio *et al.* 1981.)

Above these sediments lie platform carbonates with abundant rhodoliths, bivalves and bryozoans, and frequent small- to medium-scale sedimentary structures, such as channels and cross-bedding. These platform carbonates belong to the 'bioherm unit' in our interpretation. The reef complex proper developed on top of these carbonates (Fig. 4.3.8).

Three episodes of reef growth can be seen, apparently descending in step with contemporaneous relative sea-level fall (Fig. 4.3.8). The reef-core facies and the reef-talus-slope facies are indistinguishable from a distance. The proximal-slope facies are yellowish calcarenites exposed as giant cross-beds (Fig. 4.3.8) reflecting the basinwards progradation of the reef. The depositional dip of the layers progressively flattens out towards the bottom of the slope, where the calcarenites interfinger with the white, silty calcarenites of the distal-slope facies.

Transfer to Stop 2

From the cemetery drive to the SE for *c.* 400 m to a junction in front of a large pig farm, and turn left. Continue for *c.* 100 m, crossing the bridge at KM 31.9 of the old Níjar to Carboneras road. Park on the right shoulder. Return to the bridge and climb up the slope on the right, to the top of the hill. From here Stops 2–7 will be reached on foot. Coach passengers can be dropped here. The coach can continue along the road towards the motorway, and wait at the entrance to the unpaved service road on the left, before the bridge over the motorway.

Stop 2: Reef-complex facies at the southern end of the Níjar section (721 906; Níjar 1 : 25 000)

A complete section across all the different reef-complex facies can be observed in the upper third of the hill (Fig. 4.3.7). The distal slope deposits consist of very fine sand and silt-sized peloidal wackestones, with local fine-grained packstones. Thin wedges of proximal-slope facies pinch out downslope. Centimetre-scale planar lamination is frequent, probably reflecting the settling of suspended sediment. Most of the skeletal constituents are unidentifiable. Higher up in the proximal slope the dominant grain size is coarse to fine sand. Planar parallel laminae of centimetre- to decimetre-scale are the most obvious sedimentary structure, produced by size sorting of skeletal grains. Abundant burrowing disturbs this layering. Decimetre-thick, channelized, debris-flow deposits consisting of bioclastic calcirudites are locally present. *Halimeda*, coralline algae, bivalve and bryozoan remains are abundant. Poorly developed rhodoliths of *Mesophyllum*, *Sporolithon* and *Lithothamnion* are scattered in the sediment. The calcarenites/calcirudites of the proximal slope are overlain by the reef-talus slope coral breccias. *In situ Porites* growths occur on top of coral breccias and tilted *Porites* blocks. Bivalve coquinas and *Halimeda* calcarenites/calcirudites are also abundant in the reef-talus slope. *Halimeda* in particular seems to be mostly concentrated in discrete beds that extend laterally from the reef-core to the distal-slope facies.

The view from the reef at Níjar takes in many aspects of the geology and geomorphology of the Almería–Níjar basin. To the NW, immediately behind the town of Níjar, is the Sierra de Alhamilla, comprising Triassic dolomites of the Alpujárride nappe. To the SE, the low ridge in the middle distance, behind the plastic greenhouses, is La Serrata de Níjar, formed mainly of Cabo de Gata volcanics brought in along the Carboneras Fault Zone. In the far distance are the volcanics of the Sierra de Cabo de Gata. The low skyline to the SW is on a variety of Neogene rocks forming the basin-fill of the Almería Basin, bevelled by Early(?) Pleistocene pediment surfaces. The low hills below are formed mainly of Tortonian rocks poking through the Quaternary cover. The extensive surfaces extending from Níjar to La Serrata are Quaternary alluvial fans, large fans fed by the Sierra Alhamilla, and smaller fans in the middle distance fed by small catchments in La Serrata. The proximal fan surfaces are deeply incised by fan-head trenches. The older fan surfaces are capped by magnificent pedogenic calcretes, showing oolitic structures, laminar calcretes and multiple sequences of brecciation. These are excellently exposed near the roadside *c.* 300 m SE of the bridge east of the pig farm.

Transfer to Stop 3

This stop is located *c.* 200 m to the NE of the previous one.

Stop 3: Panoramic view of the western half of the Níjar reef
(722 907; Níjar 1 : 25 000)

From this point there is a complete view of the western half of the Níjar reef. The reef complex prograded to the SE in most of the Níjar area. This situation was complicated somewhat by the local presence of a seawards volcano (Fig. 4.3.6), the volcanic cone of El Hoyazo, that was colonized by corals in an atoll-like fashion. The volcanics are dealt with in more detail in Excursion 3.3b, Stop 2. Fore-reef slopes radiated from this atoll and some of them descended in an opposite direction to that of the major fore-reef slope which prograded from the shelf. This development produced complicated drainage patterns in the reef complex (Fig. 4.3.6), including two main drainage channels and a radial system of elongated reef buttresses around El Hoyazo. The present-day topography closely reflects the original reef-complex morphology. Buttresses (lobes) and channels from the final episodes of reef growth stand out clearly. In the centre of the panoramic view a major Messinian drainage channel, resulting from the convergence of the drainage systems of the reefs fringing the Sierra de Alhamilla and from those surrounding the El Hoyazo volcano, can be seen.

Transfer to Stop 4

Stop 4 is located on the same hill as Stop 3, some 200 m downslope to the SE.

Stop 4: Halfway along the walk from Níjar to the El Hoyazo
(723 906; Níjar 1 : 25 000)

White laminated calcisiltites of the distal slope with intense horizontal burrowing can be observed here.

Transfer to Stop 5

Walk to the NE for *c.* 0.5 km until you can see the Barranco del Cebollero and then walk towards it. Follow the rambla bed to the south for *c.* 100 m and climb up the hill to the left. Stop 5 is in a small ravine merging from the east with the Barranco del Cebollero. It is a steep climb on the way up to the El Hoyazo, once you cross the Barranco del Cebollero.

Stop 5: Reef complex in the eastern side of the Barranco del Cebollero
(732 909; Campohermoso 1 : 25 000)

Stop 5 shows the transition from the reef-slope sediments into the reef-core framework. At the bottom of the section proximal-slope, coralline algae calcarenites occur. Upwards in the section, *Halimeda* and bivalve coquinas become more abundant, mixed with *Porites* breccias and blocks (reef-talus-slope facies). At the top *in situ Porites* colonies appear. This same succession of facies can be observed laterally. Corals are bored by *Clionia* and *Lithophaga* and coated by microbial, micritic crusts. Calcarenitic internal sediment fills in the residual interskeletal gaps. From the top of the cliff there is an excellent view of the lobe and channel system of the reef.

Transfer to Stop 6

Complete the walk to the top of the hill, then continue to the NE for 300 m along the top of the limestone cliff rimming the El Hoyazo volcanics.

Stop 6: The north-western edge of El Hoyazo (738 913; Campohermoso 1 : 25 000)

Seen from here, El Hoyazo (Joyazo) volcano looks like a giant 'caldera' but in fact it was a volcanic cone (see below). Stop 6 is located at the NE rim of the 'caldera-like' relief.

In this area corals settled directly on top of the volcanic rocks. A sandstone level with fragments of volcanic material intercalated in the reef facies demonstrates that the centre of the volcanic relief was emergent during the reef development. During the Quaternary the volcanic rocks proved to be less resistant to erosion than the surrounding reef carbonates and a major bowl in the area of the former volcanic cone has resulted.

Transfer to Stop 7

Continue walking round the rim of El Hoyazo to reach Stop 7, located on a small hill *c.* 0.5 km to the east of Stop 6.

Stop 7: The north-eastern edge of the El Hoyazo (743 913; Campohermoso 1 : 25 000)

Stop 7 shows oolitic bars that were deposited on top of the eroded fringing reef during the latest Messinian. Wave-rippled cross-lamination and cross-bedding are the dominant sedimentary structures. Microbial bioherms locally developed on channels between the bars (Fig. 4.3.9). Both stroma-tolite (laminated) and thrombolite (massive, unlaminated) domes are present (Fig. 4.3.10). There are also microbial and oolitic breccias formed on the sides of the channels by syn-sedimentary slumping. These latter features are seen on the western side of the hill, on the slope of a small ravine perpendicular to the El Hoyazo volcano and trending to the NE.

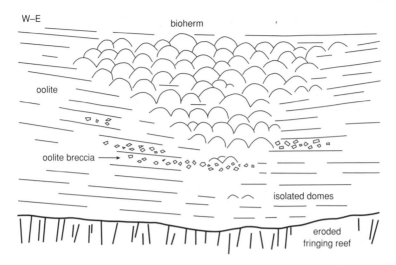

Fig. 4.3.9. Messinian stromatolite/thrombolite dome bioherm at the El Hoyazo (Stop 7). The bioherm is surrounded by oolite. Oolite breccias are locally present. (From Riding *et al.* 1991b.)

Fig. 4.3.10. Composite thrombolite–stromatolite domes, El Hoyazo (Stop 7). Hammer is 33 cm long.

The return walk crosses the eroded cone of El Hoyazo volcano (see Excursion 3.3b, Stop 2, for a description of the volcanics). The internal structure of the volcano consists of stratified layers of volcanic ashes, lavas and volcanic agglomerates. Once you leave El Hoyazo take the track and walk for 0.5 km to the SE, followed by 2 km to the west along the unpaved service road of the motorway to get back to the vehicle(s).

4.4 Excursion: Tropical carbonates of Sorbas (one day)

JOSÉ M. MARTÍN & JUAN C. BRAGA

Introduction

In the Sorbas Basin there are excellent exposures of the Messinian reef units. The bioherm unit is especially well represented along the southern margin of the basin, at Hueli, whereas the best outcrops of the fringing reef unit are at the northern margin, near Cariatiz (Fig. 4.4.1).

Directions to Stop 1

The Hueli hamlet, now abandoned, is 4 km to the SSE of Sorbas town. To drive to Hueli, there is a paved road leading to the gypsum quarries situated *c.* 2 km to the NNW of El Cerrón de Hueli, branching from the Sorbas to Venta del Pobre–Carboneras (Al-104) road at KM 3.6. From the gypsum quarries there is a dirt road to Hueli, which is *c.* 2 km to the SW. Once in Hueli, a track suitable for a four-wheel drive vehicle to the 'Alto de La Cantona' area is followed to the SW for another 1 km before reaching the outcrops. If travelling by coach, park before entering the gypsum quarries. For detailed description of access see Excursion 2.3.

First observations are to be made from a vantage point (788 018) by the track to Cantona, from which two small hills can be seen to the east. These hills, and the ones behind them, are the remnants of the bioherms to be visited on this trip. Near the top of the most northerly hill is a cave partially closed by a stone wall. Climb up to this to observe the internal structure of the bioherms.

Stop 1: **Halimeda** *bioherms, Hueli area (788 018 for the panoramic view, and 792 019 for the 'La Cueva' bioherm; Polopos 1 : 25 000)*

Many lenticular bioherms occur in the Hueli area (Fig. 4.4.2), surrounded by bioclastic sands, gravels and silty marls (Braga & Martín, 1996b). Three distinct types of bioherm are present, occupying east–west-trending belts. They formed in a shelf–basin slope setting during a cycle of relative

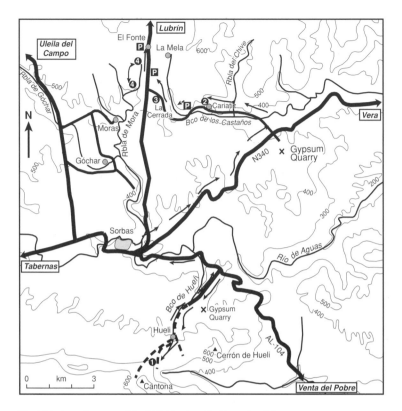

Fig. 4.4.1. Location map of the field-trip stops for Excursion 4.4.

Fig. 4.4.2. Stop 1, La Cueva bioherm seen from the west. A three-phase *Halimeda* reef with a *Porites* base, a *Halimeda*–microbial core and a bioclastic–microbial crust. Overall thickness is 24 m.

sea-level change. They are (from south to north): (i) *Porites* patch reefs; (ii) *Halimeda* reefs; and (iii) bivalve–bryozoan–serpulid reefs (Fig. 4.4.3a). During the lowstand stage, *Porites* patch reefs grew near the shelf margin; *Halimeda* reefs developed on the midslope and bivalve–bryozoan–serpulid reefs formed on the lower slope (Fig. 4.4.3b). During the transgressive stage, coral patch reefs developed near the shelf break and were overgrown by *Halimeda*. During the highstand stage, cap facies prograde basinwards as a sheet connecting many of the midslope patch reefs (Fig. 4.4.3c).

Halimeda reefs are the largest and most complex of the patch reefs (Fig. 4.4.2). They may locally have a *Porites* base (Fig. 4.4.3c). On the upper- and midslope they are up to 40 m thick and 400 m long, becoming smaller downslope. The core consists of jumbled *Halimeda* segments (Fig. 4.4.4), released by spontaneous disaggregation of the algae. The segments were stabilized close to their sites of growth and rapidly lithified by micritic and peloidal microbial crusts. Residual cavities were further veneered by isopachous marine cements (Fig. 4.4.5). Flank facies, consisting of bedded packstones to rudstones, form wedge-shaped units adjacent to the mounds. Cap facies consist of bioclastic calcarenites/calcirudites and microbial carbonates.

The key outcrops are in a ravine *c.* 1 km to the SSW of the Hueli hamlet. The western side of the ravine shows a good cross-section of the platform and its frontal slope from a vantage point situated just in front (786 015). At this point major bioherms were located at the shelf edge and on the upper slope. They are three-phase *Halimeda* bioherms with a *Porites* base, a *Halimeda* core and a calcarenite/stromatolitic cap on top. One of these *Halimeda* bioherms is superbly exposed at the eastern side of the ravine and can be easily identified as it has a small cave just in the middle ('La Cueva' bioherm; Fig. 4.4.2). Well-preserved *Halimeda* samples can be observed inside the cave.

Transfer to Stop 2

Return to the vehicle(s) and retrace your route towards Sorbas. At the junction with the Vera–Sorbas road (old N340) turn right and proceed for *c.* 8 km. Cariatiz is located 7 km to the NNE of Sorbas. To get to Cariatiz, take the turning to the left at the cross-roads located at KM 179.7. Cariatiz is *c.* 3 km to the WNW. The outcrop to be visited is on the southern side of a hill 0.5 km to the north of Cariatiz. Park at the main square in the village (by the church at 811 115) and walk from here. Proceed NW through the village streets to a fountain. Turn right here and climb the hill. On the way up note the reef-slope facies (*Halimeda* rudstones, coral blocks and breccias).

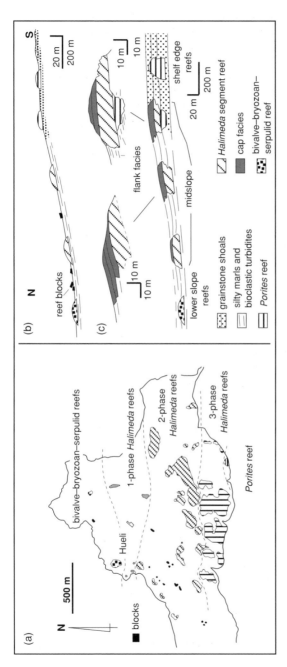

Fig. 4.4.3. (a) Spatial distribution of reef types within the bioherm unit at Hueli, Stop 1.
(b) Interpretation of the palaeoenvironmental setting of the bioherm patch reef complex at Hueli. The *Halimeda* reefs formed on the midslope at depths of 20–65 m. (c) Patch reef types and their lateral distribution. (i) Proximal shelf edge: *Porites* reefs. (ii) Midslope: *Halimeda* reefs subdivided into: (a) upper midslope: three-phase *Halimeda* reefs with a coral base and bioclastic–microbial cap; (b) mid midslope: two-phase *Halimeda* reefs with a thick bioclastic–microbial cap; (c) lower midslope: one-phase *Halimeda* reefs with local, thin, bioclastic–microbial caps. (iii) Lower slope: bivalve–bryozoan–serpulid reefs. The upper two diagrams show sketches of actual examples of upper midslope and mid midslope *Halimeda* reefs. Accretion (time) lines shown in the lowermost diagram separate lowstand, transgressive and highstand stages. (From Martin *et al.* 1997.)

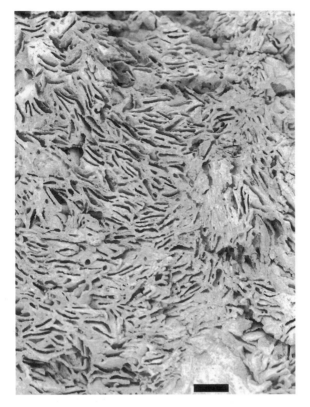

Fig. 4.4.4. *Halimeda*–microbial boundstone showing chaotic fabric of densely packed *Halimeda* plates (partly dissolved), Stop 1. Scale bar is 1 cm.

Stop 2: Cariatiz, fringing reefs: reef-core facies (807 120; Sorbas 1 : 25 000)

Stop 2 shows one of the best preserved outcrops of the Messinian fringing reefs in the area. As mentioned above, in the Messinian fringing reefs, two main zones can be distinguished: the reef core and the slope. The reef-core framework consists of corals and stromatolites (Fig. 4.3.5). *Porites*, together with very minor *Siderastrea*, are the only corals building the reefs. Coralline algae and encrusting Foraminifera line the coral skeletons and are in turn coated by stromatolitic, micritic crusts, which are responsible for the early lithification of the reefs. A bioclastic matrix fills the voids left by the crusts in the framework. The reef core can be divided into several vertical zones. In the basal zone (pinnacle zone), coral colonies concentrate in pinnacles, up to 15 m high and a few tens of metres wide, flanked by

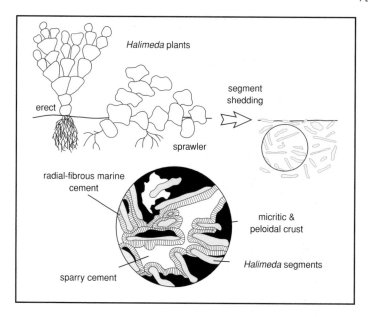

Fig. 4.4.5. Development of *Halimeda* segment reef fabric, from living *Halimeda*, through segment-shedding during life and after death, to rigid segment rock lithified by micritic and peloidal microbial crusts and marine cement. Individual segments are approximately 1 cm across. (From Martín *et al.* 1997; modified after Braga *et al.* 1996a.)

calcarenites. In the thicket zone (3–4 m thick) coral growth is laterally more continuous. Vertical sticks connected by laminar bridges are the dominant growth forms of *Porites* in these two zones. Above a transition zone, the reef crest (4–5 m thick) caps the vertical succession. In the reef crest, *Porites* colonies consist of thin, irregular, contorted laminae (Fig. 4.4.6). A complete section of the reef core can be observed with specific sites at the pinnacle, thicket, transition and reef-crest zones. The top of the reef also affords excellent views over the Cariatiz valley (for description see Stop 3).

Transfer to Stop 3

Stop 3 is a view point above the north–south orientated Barranco de los Castaños *c.* 2 km west of Cariatiz. Head west on the surfaced minor road west of Cariatiz village. This road is not passable for buses. Buses should access the next stop by returning almost to Sorbas, taking the Lubrin road northwards from the junction *c.* 500 m before Sorbas for *c.* 7 km, and parking alongside that road near 788 128, where the minor road to the east

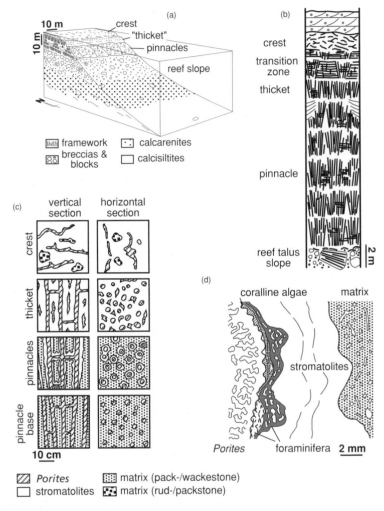

Fig. 4.4.6. (a) Reconstruction of the coral–stromatolite reef at Cariatiz showing the pinnacle, thicket and reef-crust zones and reef-derived blocks on the fore-reef slope. (b) Fringing reef sequence. The reef shows a thick pinnacle zone overlain by a thinner thicket zone, both of which were constructed principally by erect stick-like *Porites* coated with stromatolitic crusts. This erect form of *Porites* passes upwards at the reef crest into a laminar platy form. Oolites from the Sorbas Member unconformably lie on top of the reef sequence. (c) Vertical increase in crust development and change in style of coral growth from base of pinnacle zone to reef crest. (d) Sequence of reef-frame sedimentation from coral to bioclastic matrix. Stromatolitic crusts are often preceded by layers of coralline algae and foraminiferas. (From Riding *et al.* 1991a.)

would provide access on foot to Stop 3. Cars and minibuses can take the direct road west from Cariatiz. The road first crosses the barranco then twists obliquely up the valley side. The road at first heads west then, after *c.* 1 km, veers towards the north. Park at the roadside opposite where the canyon to your right (east) is at its narrowest.

Stop 3: La Cerrada, Barranco de los Castaños, at the western view point of the Cariatiz reefs (795 118; Sorbas 1 : 25 000)

The view point affords an overview of the geometry of the reef, and the relationships between the various facies (Fig. 4.4.7a,b).

1 Reef-core framework (described at Stop 2) and reef-talus blocks and breccias.

2 Framework blocks and coral breccias, together with fragments of bivalves, serpulids, coralline algae and *Halimeda* accumulated on the upper part of the reef slope.

3 Reef-slope calcarenites, rich in *Halimeda*, serpulids and coralline algae, which pass into calcisiltites and silty marls downslope.

4 Lagoonal facies. Subhorizontal beds of back-reef calcarenites with coralline algae, gastropods and bivalves, including small *Porites* patches.

5 Inverted wedges. Well-bedded, onlapping, wedge-like units that pinch out landwards and consist of bivalve–bryozoan–coralline algae packstones to rudstones.

6 Siliciclastic conglomerates and sands. Fan-delta deposits that disturb the reef carbonates.

The reef itself extends along the hill in front from 799 129 to 799 114. The total progradation of the reef system from the northern margin was at least 1.1 km. Upwards and downwards, shifts of the reef facies at the Barranco de los Castaños section indicate high-resolution, relative sea-level changes during reef progradation at two different orders of cyclicity (Fig. 4.4.7a). Within the section, the lowest order of cyclicity (C1) is represented by one cycle and the beginning of a second one interrupted in its ascending phase. Lagoon deposits overlie talus-slope breccias and proximal-slope calcarenites in the northern two-thirds of the section. The surface below the lagoonal deposits is karstified and encrusted by iron deposits. This facies relationship implies erosion/karstification of previous reef deposits during the low sea-level stage of the first cycle and subsequent deposition of lagoonal facies on top of the erosion surface during the sea-level rise of the following cycle. The sudden interruption of reef development in an ascending phase can only be explained by tectonic causes and thus supports Weijermars' (1988) hypothesis concerning the origin of the 'Messinian Salinity Crisis'. Weijermars (1988) suggested that the isolation of the Mediterranean Sea (which finally resulted in its desiccation) was initially triggered by tectonics.

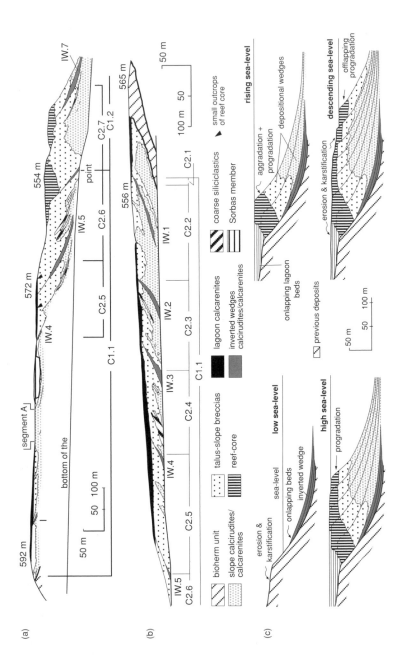

A higher order cyclicity (C2) modifies these C1 cycles. The higher order cycles are evidenced by the facies distribution between two consecutive inverted wedges. This pattern is exemplified in the southern half of the Barranco de los Castaños section (Fig. 4.3.3). Talus-slope breccias lying on top of the lowest inverted wedge shift upwards and then downwards again below the next inverted wedge, with the reef-core facies following the same trend (Fig. 4.4.7a,c).

The estimated relative sea-level change in the complete C1 cycle is about 100 m. Sea-level oscillations in C2 cycles have an amplitude of several tens of metres. Biostratigraphical and magnetostratigraphical data indicate that the Cariatiz reef developed in less than 0.36 Ma. If eustasy was the major factor controlling relative sea-level change, this temporal range, together with the observed amplitudes and relative frequencies, suggests that the C1 and C2 cycles may represent short eccentricity and precession cycles, respectively.

At this view point you can also see aspects of the general geology of the northern margin of the Sorbas Basin. Note no gypsum is present here and rocks of the Terminal Carbonate Complex (Sorbas Member) rest unconformably on reef. These are succeeded by the end Messinian–Pliocene Cariatiz Formation, which is exposed in the road sections on the southern valley side west of Cariatiz (see Excursion 5.3, Stop 4). The succession is capped by the Marchalico System of the Plio-Pleistocene Góchar Formation conglomerates (Mather, 1991, 1999; see also Section 5.1). The end Góchar Formation depositional surface forms an extensive fan radiating southwards from here. Since then the drainage has been captured by the west–east subsequent drainage of the Rambla de los Castaños; however, the sequence of capture/diversion was complicated by post-Góchar tectonic deformation, to the SE of Cariatiz (Mather, 1999; see also Section 3.5).

Transfer to Stop 4

Stop 4 affords an examination of the various reef and intercalated facies described in Stop 3, along sections in the upper part of the Rambla de

---

Fig. 4.4.7. (a) Barranco de los Castaños section, Stop 3. Vertical shifts of reef facies during reef advance are evident at two orders (C1 and C2) of cyclical relative sea-level change. Inverted wedges (IW) mark the commencement of C2 cycles and are thought to have formed at the lowest sea-level in each C2 cycle. (b) Barranco de la Mora section. (c) Model of reef-advance geometries in C2 cycles. Inverted wedges mark the beginning of cycles and are the deposits that formed at lowest sea-level. Reef aggradation combined with progradation took place during sea-level rise. Lagoon beds onlapped the eroded and karstified previous deposits. Reefs prograded during highest sea-level and offlapped during sea-level fall, at which time reef deposits from former phases started to be eroded. The relative proportions of aggrading, prograding and offlapping geometries inside C2 cycles were controlled by the interference of C1 and C2 cycles. (From Braga & Martín, 1996a.)

Fig. 4.4.8. Lowstand wedge pinching out to the upper right (northwards) between talus-slope breccias. Note the onlapping geometry of beds inside the wedge. Barranco de la Mora section.

Mora, to the SW of El Fonte/La Mela (Fig. 4.4.1). Return to the coach or, if in cars or minibuses, continue north and west along this minor road. Turn right (north) at the junction with the Sorbas–Lubrín road and continue *c.* 1 km to the twin villages of El Fonte/La Mela. Park where convenient, then walk SE from El Fonte for *c.* 500 m to the floor of the rambla.

Stop 4: Rambla de la Mora, El Fonte (783 135–782 124; Sorbas 1 : 25 000)

In the Rambla de la Mora section (Fig. 4.4.7b) the inverted wedges very clearly exhibit onlapping relationships with respect to the reef facies of the previous cycle (Fig. 4.4.8). They also locally show erosive basal surfaces, bar morphologies and cross-bedding directed landwards.

4.5 Excursion: Evaporites and stromatolites of the Sorbas Basin (one day)

JOSÉ M. MARTÍN & JUAN C. BRAGA

Directions to Stop 1

Starting from Sorbas, drive eastwards for *c.* 1 km along the old N340 road towards Vera, and turn right towards Venta del Pobre/Carboneras on the Al-104. Stop 1 is some 300 m to the north of Los Molinos del Río de Aguas in the canyon of the Río de Aguas (Fig. 4.5.1). Park at the roadside at KM 5. Follow a path from the north of the village into the Aguas canyon.

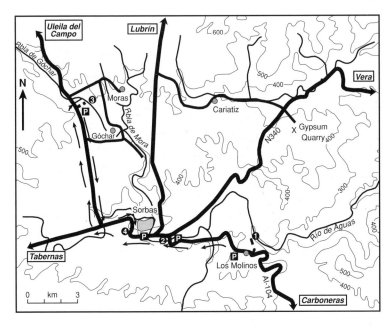

Fig. 4.5.1. Location map of the field-trip stops for Excursion 4.5.

Stop 1: The Aguas canyon, Los Molinos del Río de Aguas (823 058; Sorbas 1 : 25 000)

Pre-evaporitic, laminated diatomites containing fish remains can be seen immediately below the gypsum. Further to the north there are some very good exposures of the gypsum beds of the Yesares Member. In the lower part of the sequence, the massive gypsum beds (up to 20 m thick) consist of selenitic gypsum exhibiting the typical swallow-tail twins (Dronkert, 1977). They are separated by thin silt–marl interbeds. Higher in the sequence the interbeds become more prominent, exhibiting similar thicknesses to those of the gypsum banks (up to several metres). The vertical transition to the overlying Sorbas Member is gradational, with the gypsum beds gradually diminishing in thickness and the silt–marl interbeds increasing progressively. This vertical change can be seen upstream to the north of the canyon, in the area where it suddenly opens.

Transfer to Stop 2

From Stop 1 drive back towards Sorbas. Park at the river bridge (at KM 0.5). Walk for *c.* 0.5 km to the SW, along the rambla bed, to reach the outcrop.

(a)

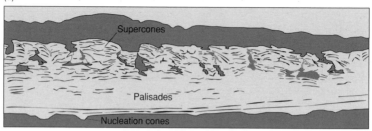

(b)

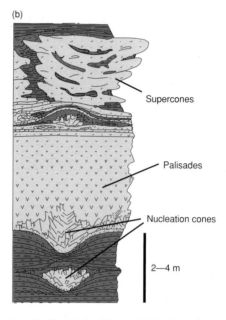

Fig. 4.5.2. (a) Gypsum cycles at the Río de Aguas, Stop 2. (b) Detailed structure of one of the gypsum beds showing the nucleation cones at the base, the selenite palisades in the middle and the supercones on top, with 'pockets' of silts inside the supercones and lateral to them. (From Dronkert, 1977.)

Stop 2: Rambla de Río de Aguas selenite gypsum (790 056; Sorbas 1 : 25 000)

At Stop 2 the uppermost beds of the Yesares Member crop out in vertical cliffs on the southern side of the rambla. Three gypsum cycles, each 5–10 m high, can be recognized (Fig. 4.5.2a). They show spectacular selenite gypsum growth, known as supercones, associated with other gypsum structures such as nucleation cones and selenite palisades (Dronkert, 1977; Fig. 4.5.2b).

Bunches of curved selenite crystals (up to 1.5 m in size), in the shape of 'scimitar', form the branches of the cauliflower-like, supercone growths. Pockets of finely laminated, fine-grained terrigenous sediments, which occur between the branches of these 'tree-like' growths, record internal sedimentation. However, close inspection reveals that the gypsum crystals and the terrigenous sediments accumulated simultaneously, as evidenced by the fact that some of the lutite laminae can be traced laterally inside gypsum crystals where they gradually thin and disappear.

Transfer to Stop 3

Take the main road west through Sorbas then 1 km beyond Sorbas turn right towards Uleila del Campo. After 4.75 km turn right onto a track which takes you *c.* 500 m towards the Rambla de Góchar. Park at the side of the track before it descends into the valley. Follow the track on foot into the rambla and turn left. Cross the rambla using the irrigation canal and climb up the opposite side along the bedding plane behind the pump house, to the top surface.

Stop 3: *Rambla de Góchar, siliciclastic microbial carbonates (stromatolites and thrombolites) of the Late Messinian, post-evaporite marine unit (Sorbas Member) (762 107; Sorbas 1 : 25 000)*

At Stop 3 siliciclastic microbial carbonates, intercalated with oolitic limestones and small *Porites* patch-reefs, occur within siliciclastic fan-delta conglomerates that pass laterally into sands (Dabrio *et al.* 1985; Martín *et al.* 1993, 1998; Braga *et al.* 1995; Braga & Martín, 1996b). All these sediments formed a parasequence and were deposited on an erosional surface on top of Messinian fringing reefs (Fig. 4.5.3a). Three distinct units can be distinguished within the parasequence.

1 The lower one (lowstand-stage deposits) consists of conglomerates that change downslope to sands and onlap the reef-slope facies (unit 1).

2 The intermediate one (transgressive-stage deposits) is made up of conglomerates deposited directly on top of the eroded fringing reef passing laterally into sands (unit 2).

3 In the upper one (highstand-stage deposits) there is clear evidence of basinwards progradation of conglomerates and extensive development of microbial carbonates (unit 3).

This stratigraphic relationship can be observed on both sides of the rambla, but it is clearer along the western side. A good overview can be obtained by crossing the rambla bed using the irrigation wall and ascending up the bedding plane. Looking back across the rambla there is an excellent view of the Sorbas Member sediments.

Microbial carbonates extend laterally from shallow platform areas, to

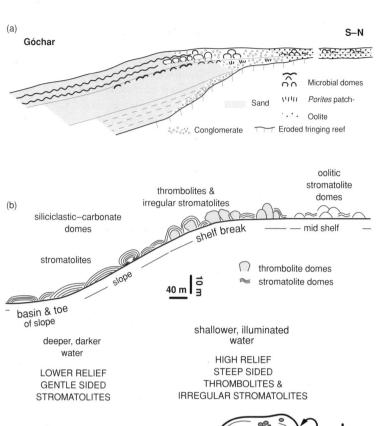

(a)

Góchar S–N

Microbial domes

Sand *Porites* patch-

Oolite

Conglomerate Eroded fringing reef

(b)

siliciclastic–carbonate domes

thrombolites & irregular stromatolites

oolitic stromatolite domes

shelf break mid shelf

stromatolites

slope

thrombolite domes

stromatolite domes

40 m 10 m

basin & toe of slope

deeper, darker water

LOWER RELIEF GENTLE SIDED STROMATOLITES

shallower, illuminated water

HIGH RELIEF STEEP SIDED THROMBOLITES & IRREGULAR STROMATOLITES

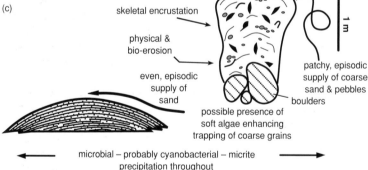

(c)

skeletal encrustation

physical & bio-erosion

even, episodic supply of sand

1 m

patchy, episodic supply of coarse sand & pebbles boulders

possible presence of soft algae enhancing trapping of coarse grains

microbial – probably cyanobacterial – micrite precipitation throughout

Fig. 4.5.3. (a) Góchar section showing a thin fan-delta, oolitic shoal and coral patch-reef shelf sequence deposited on top of an eroded fringing reef. Microbial carbonate (stromatolite and thrombolite) beds extend from the shelf area, down the steep palaeoslope, into the basin for distances of 0.5 km or more (Stop 3). (Modified after Martín *et al.* 1993.) (b) Composite cross-section of shelf–basin dome distribution at Góchar showing variations in dome type, shape, grain size and grain composition. (From Braga *et al.* 1995). (c) Deep, slope–basin stromatolite domes and shallow-water, shelf break stromatolite–thrombolite domes contrasted. (From Braga *et al.* 1995.)

the adjacent basin, down to depths of *c*. 40 m. They consist of both stromatolites and thrombolites (Fig. 4.5.3b). The stromatolites present a very
distinct internal lamination, whereas the thrombolites are massive and
contain abundant irregular fenestrae. Both the stromatolites and the
thrombolites incorporate abundant terrigenous sand and lesser amounts
of terrigenous pebbles. They also exhibit typical carbonate microbial structures (i.e. dense, clotted and bushy micrite) produced by cyanobacterial
calcification. Their marine origin is indicated by an associated biota of
coralline algae, vermetids and encrusting Foraminifera. Borings of
bivalves are present within the sequence.

Microbial bioherms, up to 9 m wide and 3–4 m high, occur as mounds
separated by channel-fill conglomerates at the shelf break. Biostromes occur
throughout the slope deposits. At the shelf break, thrombolite domes, up
to 2 m high, are predominant. Within the upper slope, microbial domes
occur on top of debris flows. In this area, steep-sided thrombolite domes
(Fig. 4.5.4) are up to 1.5 m high and 2 m wide, whereas the stromatolites
(Fig. 4.5.5) have a more gentle relief (1–2 m high and up to 5 m wide).
Lower-slope stromatolites (up to 11 m across and 3 m high) are ellipsoidal
in plan view and have a delicate, finely laminated internal fabric (Fig. 4.5.3c).
The outcrops at the eastern side of the rambla, and those in the next small
ravine immediately to the east, are the best for viewing the different styles
of microbial-dome growth.

The dome shape and internal structure of the microbial carbonates are
closely related to the depth at which they grew and appear to have been
controlled by a series of factors, such as water clarity, currents and sediment supply. Dome formation largely reflects micrite precipitation by
cyanobacteria, but was significantly augmented by the trapping of siliciclastic grains. Strong currents at the shelf break delivered coarse sediment
to the crests of the domes. This, together with erosion of the sides of the
domes, resulted in tall, vertically elongated forms. Gentler downslope currents resulted in lower relief, less steep-sided domes (Fig. 4.5.3c). At the
shelf break, photic conditions and the episodic nature of sediment supply
promoted the colonization of dome surfaces by encrusting, grazing and
boring organisms, which resulted in patchy thrombolitic and irregular
stromatolitic macrofabrics in these domes. More regular cyanobacterial
mat growth, and the supply of better sorted sediment into downslope
domes produced regular, continuous stromatolitic layering. The younger
rocks at this locality are examined in Excursion 5.3, Stop 1.

Transfer to Stop 4

Return to Sorbas and park in front of Bar Fátima on the left-hand side of
the main road, south of the town.

Fig. 4.5.4. Steep-sided thrombolite dome with stromatolite cap placed on top of a large metamorphic boulder, upper slope, Góchar (Stop 3). Hammer is 33 cm long.

Stop 4: Sorbas town, stromatolites associated with beach deposits of the Sorbas Member (776 062; Sorbas 1 : 25 000)

Another example of siliciclastic stromatolites occurs in the Messinian, post-evaporitic Sorbas Member near Sorbas town (Roep *et al.* 1979, 1998; Dabrio *et al.* 1985, 1998; Braga & Martín, 2000). The stromatolites occur in association with beach deposits, lying on top of sand and conglomerate shoreface bars and covered by wave-rippled sands. They grade laterally into micaceous silts. The stromatolite domes grew at the transition from the lowermost shoreface to the shelf (Fig. 4.5.6). The domes are made up of dense, peloidal, clotted and bushy micrite interpreted as the result of microbial precipitation, and siliciclastic particles which may constitute up to 40% of the rock volume. Within a single stromatolite bed there is a variation in the composition and morphology of the domes. Proximal

Fig. 4.5.5. Stromatolite dome, upper slope, Góchar (Stop 3). Scale in centimetres.

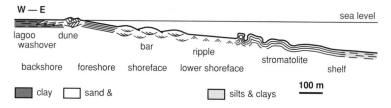

Fig. 4.5.6. Palaeoenvironmental setting and schematic lateral changes in shape of the stromatolites from the Sorbas Member at Sorbas (Stop 4). (From Braga & Martín, 2000.)

stromatolites contain sand-grade siliciclastics, and exhibit distinct lamination, steep sides and high synoptic relief. Downslope they grade into large, flattened, silty, gentle-sided stromatolites. Microbial mats were able to develop at the bottom of the lowermost part of the beach but were inhibited by stronger wave energy in shallower settings. Deeper waters on the shelf were probably too dark for mat growth and subsequent dome formation.

Chapter 5
Marine to Continental Transition

ANNE E. MATHER & MARTIN STOKES

5.1 Introduction

With continued uplift of the Betic area many of the Almería basins became elevated above sea-level, despite a global rise in sea-level during the Pliocene (Haq *et al.* 1988). As a function of the continuing deformation associated with the regional tectonics (outlined in Chapter 3), the basins became increasingly isolated from each other. The record of the marine to continental transition is held dominantly within the upper Messinian, Pliocene and Pleistocene record of the sedimentary basins. However, the timing of the transition varied between the basins as a function of different relative uplift rates. In addition, this uplift and associated deformation also stimulated erosion (discussed in Chapter 6) which affected preservation of the continental sequences. Within the Sorbas Basin the marine to continental transition was initiated in the Late Messinian, with the development of marginal alluvial fans and coastal plain deposits in the basin centre (examined in Excursion 5.3). The later fluvial systems developed as a function of the early emergence of the basin-fed, low-relief, Plio-Pleistocene fan-deltas in the Almería Basin to the south (Mather, 1993b). In the south of the Tabernas Basin, in the Vera and Almería Basins, the earlier Pliocene is marked by the development of fan-deltas, mainly Gilbert-type (Postma, 1984; Postma & Roep, 1985; Boorsma, 1992; Stokes & Sendra, 1996; Stokes, 1997), suggesting low-energy, restricted basins which were being supplied with a sediment from the adjacent elevated topography of the sierras (aspects of the fan-deltas are examined in Excursion 5.2). These basins did not display true continental deposition (examined in Excursion 5.4) until the ?Early Pleistocene, when alluvial-fan deposition dominated the basin margins.

Links between adjacent basins during the transition

Messinian

With the relative sea-level fall, which occurred from the Messinian onwards, regression was experienced by the main basins. This occurred

186

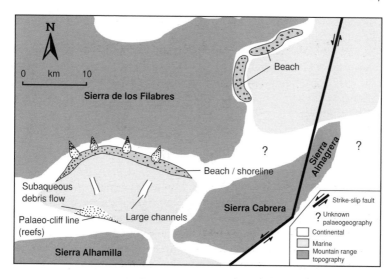

Fig. 5.1.1. Lower Pliocene reconstruction of the continental to marine transition for the Sorbas and Vera Basins. (Modified after Mather, 1991, 1999, in press a; Stokes, 1997.)

first in the Sorbas Basin after the regression of the Sorbas Member seas and was associated with the development of alluvial-fan sediments accumulating at the basin margins (Moras Member, Cariatiz Formation; Mather, 1991), and coastal plain sediments (Zorreras Member, Cariatiz Formation) in the basin centre. During the latest Messinian–Pliocene the Sorbas Basin continental sediments were affected by two basin-wide brackish water lacustrine incursions. Similar deposits have been found in the Vera Basin (Fortuin *et al.* 1995). Within the Almería Basin a more extensive, brackish water carbonate was in evidence (Van de Poel, 1991). Sediments within the Sorbas Basin show signs of more marine influence only towards the southern margin of the basin, implying a link with the Almería Basin across the Feos Gap (Fig. 5.1.1). Evidence for the links to the west (Tabernas Basin) and east (Vera Basin) is absent. In fact there is evidence to suggest that topographic highs were already established in the east at that time, restricting continuity through this area (Mather, 1991). Continued uplift (?Late Messinian–Pliocene) of the Sierra de los Filabres created an angular unconformity at the northern basin margin.

Debate rages about the timing of reflooding of the Mediterranean following the 'Messinian Salinity Crisis'. Some argue that the reflooding is of Pliocene age. Others argue for a much older reflooding event in the Messinian (see discussion in Chapter 4). The Zorreras Member of the

Chapter 5

Cariatiz Formation has become the recent focus of attention in this debate as a candidate for the 'Lago Mare' (Civis *et al.* 1977). The term means 'lake sea' and has been used to characterize a Messinian oligohaline Mediterranean sea which resulted from the 'Mediterranean Salinity Crisis' (Hsü *et al.* 1977), post-dating the deposition of evaporites. This sea is considered by some researchers to pre-date the Pliocene marine reflooding (De Dekker & Chivas, 1988).

Plio-Pleistocene

The top of the Cariatiz Formation in the Sorbas Basin is marked by a yellow, basin-wide marine unit. The fauna imply a link to more open marine conditions in the south. This incursion, of Lower Pliocene age (Roep & Beets, 1977; Ott d'Estevou, 1980) most probably reflects flooding in response to global sea-level rise recorded at this time (Haq *et al.* 1988). In the Sorbas Basin, continental sedimentation subsequently became dominated by alluvial-fan and braided-river sedimentation (Fig. 5.1.1; Góchar Formation; Ruegg, 1964; Mather, 1991). The sediment from the Sorbas Basin exited to the south across the axis of the Sierras Alhamilla and Cabrera, forming low-relief fan-deltas in the north of the Almería Basin (Fig. 5.1.1; Mather, 1993b). The Sorbas Basin had no connection with the adjacent basins apart from this one exit point and possibly through the topographic low marked by the modern Gafares river to the east.

During the deposition in the Sorbas Basin of the upper units of the Cariatiz Formation in the Pliocene, the adjacent Vera and Almería Basins were dominated by shallow marine deposition (the Cuevas Formation). The base of these deposits in the Vera Basin marks the Mediterranean-type locality for the Mio-Pliocene boundary (Fortuin *et al.* 1995). As sea-level reached its maximum, Gilbert-type fan-deltas developed where suitable conditions and accommodation space were available (i.e. adjacent to pronounced topography, in sheltered areas of the basins; Fig. 5.1.2). In the Vera Basin this is represented by the Espíritu Santo Formation (Völk & Rondeel, 1964; Postma & Roep, 1985; Stokes, 1997) and in the Rioja corridor to the south of the Tabernas Basin, the Abrioja Formation (Postma, 1984). With continued uplift and regression the fan-deltas gave way to continental sedimentation, such as the Salmerón Formation (Vera Basin; Völk, 1979; Stokes, 1997, Stokes & Mather, 2000).

With continued uplift during the Pleistocene, incision began to occur, with resulting significant modification of the existing river systems with major sediment rerouting as a function of river capture (Harvey & Wells, 1987; Mather, 1993a, 2000a). This aspect is followed up in Chapter 6.

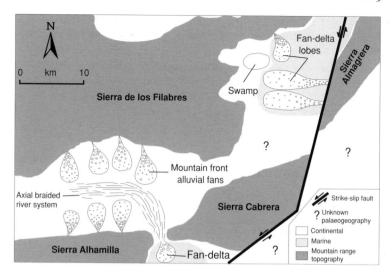

Fig. 5.1.2. Upper Pliocene reconstruction of the Sorbas and Vera Basins. (Modified after Mather, 1991, 1999, in press a; Stokes, 1997.)

5.2 Excursion: Late Pliocene Gilbert-type fan-deltas of the Vera Basin (one day)

MARTIN STOKES

Introduction

Pliocene sedimentary successions within the Vera, Almería and Tabernas Basins show significant accumulations of fan-delta deposits. The deposits are most noted for their impressive large-scale fore-sets which, in places, reach up to 40 m in height and are characteristic of Gilbert-type fan-deltas (*sensu* Gilbert, 1885). With uplift of the Betic area, fan-delta sedimentation records a transitional phase from marine to continental sedimentation within many sedimentary basins in the region.

Within the Vera and Almería Basins, fan-delta sedimentation is typically conglomeratic. Deposition appears to relate to continuing regional uplift and, more importantly, major tectonically induced reorganizations of basin palaeogeography as a result of sinistral strike-slip movement along the Palomares and Carboneras Fault Zones (Völk, 1966; Postma & Roep, 1985; Boorsma, 1992; Stokes, 1997). In the Tabernas Basin, fan-delta deposits show considerable proximal to distal variance in grain size (Postma, 1979, 1984). Although the positioning of the Tabernas fan-delta bodies appears to be tectonically controlled within an extensional graben structure,

Chapter 5

the Abrioja Corridor (Postma, 1984), the overall evolution of the Tabernas fan-delta is attributed to fluctuations in sea-level and sediment supply (Postma, 1995).

The purpose of this one-day field excursion is to examine some of the best and most easily accessible Late Pliocene fan-delta and associated deposits within the Vera Basin. The sediments belong to the Espíritu Santo Formation and are exposed within central–northern areas of the basin (Figs 5.2.1 and 5.2.2). Excellent exposure and preservation allows for the examination of the stratigraphical architecture, depositional environments and sedimentary processes, and for an evolving Late Pliocene basin palaeo-geography to be established.

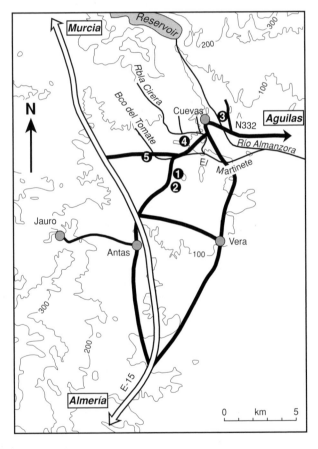

Fig. 5.2.1. Location map of the field-trip stops for Excursion 5.2.

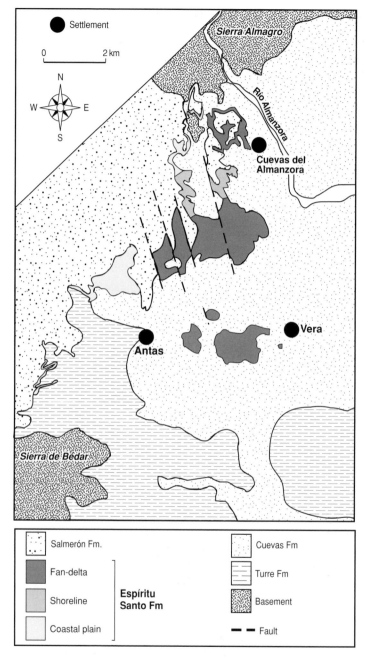

Fig. 5.2.2. Geological map of the north–central region of the Vera Basin showing the spatial occurrence of the Vera and Almanzora Members (Espíritu Santo Formation) in relation to the rest of the basin-fill and basement. (Modified after Völk, 1967.)

Directions to Stop 1

Drive northwards along the motorway from Almería, passing through
the southern and central regions of the Vera Basin. In the north of the
Vera Basin take the first exit, signposted to Cuevas del Almanzora. At the
slip-road junction turn right and head eastwards towards Cuevas del
Almanzora town. Between KM 3 and KM 4, turn right at the cement
works and follow the road for *c.* 2 km as far as the municipal rubbish
dump. On reaching the rubbish dump, drive past the main entrance and
immediately park your vehicle next to a large section of brightly coloured
sandstones. This is Stop 1.

**Stop 1: *Shallow marine fan-delta interactions, entrance to municipal
rubbish dump (977 264; Vera 1 : 50 000)***

This section provides an opportunity to examine shallow marine shelf
sediments of the Cuevas Formation, which lie directly under the fan-delta
deposits of the Espíritu Santo Formation. Lower and mid-parts of the
section comprise brightly coloured yellow, red and orange fine sands.
Sedimentary structures are abundant within the sands, comprising exceptionally well-preserved wave-rippled and horizontal laminations, together
with *Skolithos* and *Ophiomorpha* burrows, providing evidence for a shallow marine setting. The bright coloration of the sands has been attributed
to hydrothermal mineralization by Fe and Mn (Barragan, 1994), and is
typical of the Cuevas Formation sandstones located near the northern
margin of the Vera Basin.

Upper parts of the section show thick, massive sandstone units interbedded with thin, cemented sandstone units, picked out by differential
weathering. The sandstones are cut by vertical veins of gypsum whose
orientations pick out the regional N–NNW to S–SSE compressional stress
field affecting the Vera Basin during the Plio-Pleistocene. Topmost parts
of the section demonstrate the transition from the Cuevas Formation
shallow marine sandstones to the conglomerate units of the Espíritu
Santo Formation. These conglomerate units represent the first fan-delta
sediments, corresponding to the lower deltaic beds of the Espíritu Santo
Formation.

Transfer to Stop 2

Carry on up the road for 200 m and take the first track on the left, which
leads to a large quarry. The quarry is worked periodically and is best
accessed on Sundays when not operating. The track may be chained and
vehicles must be left outside. Follow the track for *c.* 100 m around a corner

into the main quarry. Warning!—the sections are unstable and extra caution will be required when examining the outcrops close up.

Stop 2: Basin centre, fan-delta sedimentology, Espíritu Santo Quarry, Cabezo Colorado (977 259; Vera 1 : 50 000)

From the previous locality, you have moved approximately 30 m up the stratigraphy from the marine sandstones of the Cuevas Formation into the main body of the coarser grained, conglomeratic Espíritu Santo Formation. This transition records a switch in depositional environment from open marine shelf to fan-delta deposition within a protected, enclosed area.

The quarry displays a series of large-scale, cross-bedded conglomerate units which form the main fore-set beds of the fan-delta. Individual beds can be traced laterally over several tens of metres and often display an asymptotic relationship with underlying lower angle units. Detailed examination of the conglomerates reveals distinct textural characteristics, comprising sub- to well-rounded gravel- to pebble-sized clasts that are generally clast-supported. According to Postma & Roep (1985), similar deposits in the nearby El Hacho region suggest deposition by mass flow processes typical of grain flows avalanching down the front of an unstable delta lobe. Many of the conglomerate units possess a distinct yellow, orange or black coloration, probably relating to hydrothermal mineralization (Barragan, 1994), similar to that observed in the marine sands of the Cuevas Formation at Stop 1.

The quarried faces allow a three-dimensional reconstruction of the internal architecture of the fan-delta. Bed orientations show minor dips to the north and south, with an overall dip of up to 28° towards the west, suggesting deposition by a unidirectional lobe of sediment. Individual fore-sets within the quarry reach up to 40 m in height and approximate to the depth of the water in central and western areas of the Vera Basin during fan-delta deposition in the Late Pliocene (Völk, 1966; Postma & Roep, 1985; Stokes, 1997).

The Espíritu Santo Formation crops out over large areas of the central and northern parts of the Vera Basin. Palaeogeographical reconstructions by Stokes (1997) (Fig. 5.2.3) suggest the occurrence of two distinct fan-delta lobes in central basin areas, of which the quarried exposures at Cabezo Largo correspond to the more northerly lobe. Both fan-delta bodies prograded from east to west, sourced from a land mass in the vicinity of the Palomares Fault Zone along the eastern basin margin. No such land mass exists now. Left-lateral strike-slip movement along the Palomares Fault Zone and displacement of the modern Sierra Almagrera could account for the occurrence of an eastern margin land mass during the Pliocene. Stokes (1997) estimated the minimum volumes of sediment

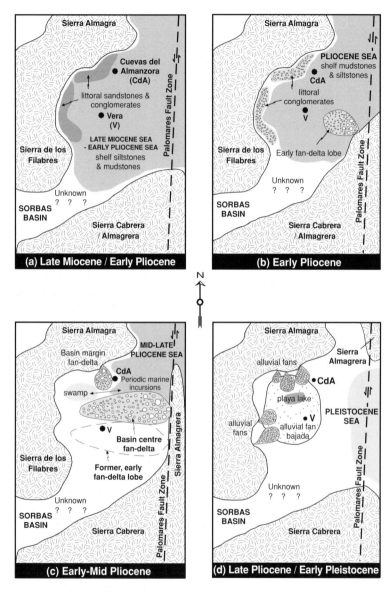

Fig. 5.2.3. Schematic representation of the changing Pliocene palaeogeography of the Vera Basin from the late Miocene (a) to Pleistocene (d). (Modified after Stokes, 1997.)

deposited by the fan-deltas within the central parts of the Vera Basin, with 60 km³ of sediment per fan-delta. To generate and transport such volumes of texturally mature sediment to the areas of fan-delta deposition would require a catchment area of at least 10 km² with a significant topographical relief (Stokes, 1997). Such conditions could be created by lateral displacement of the Sierra Almagrera mountain range. Alternatively, sediment could have been routed from source areas in the north, down through the Pulpi Corridor to the Vera Basin.

Transfer to Stop 3

Retrace your route to the cement works, turning right onto the road towards Cuevas del Almanzora. Continue towards Cuevas del Almanzora, turning left at the main road junction and proceed through the outskirts of Cuevas del Almanzora town, curving right through a major set of traffic lights, and head towards Aguilas. Cross over the Río Almanzora bridge and take the road immediately on the left. Follow this road northwards for 0.5 km, during which you will climb up onto a pronounced flat surface, a Pleistocene terrace of the Río Almanzora. Park where convenient off the road on the left and proceed to the edge of the terrace where you will get a view over the modern Río Almanzora.

Stop 3: Northern margin fan-delta stratigraphical overview, east of Río Almanzora (004 288; Vera 1 : 50 000)

Looking westwards towards the town of Cuevas del Almanzora, observe the cliff section adjacent to the modern Río Almanzora where there is a large-scale overview of a Pliocene fan-delta body sourced from the northern basin margin. Unlike the previously examined fan-delta bodies in the central parts of the Vera Basin, where only bottom-set and fore-set beds were preserved, this locality demonstrates a complete fan-delta succession of bottom-set, fore-set and top-set beds. These bottom-set/fore-set fan-delta deposits are laterally equivalent to those observed in the basin centre (Stop 2) and therefore suggest contemporaneous sedimentation from the northern basin margin.

Fan-delta bottom-set beds are exposed in lower and mid-parts of the cliff section, characterized by gently dipping, yellow-coloured marine sandstones of the Cuevas Formation. Many cave houses are constructed into these sands, from which the town of Cuevas del Almanzora has derived its name. Fore-set beds are exposed in mid-parts of the cliff section and comprise large-scale, cross-bedded boulder conglomerates of the Espíritu Santo Formation. Cross-beds reach up to 30 m in height and dip at up to 28°SW. Uppermost parts of the section form the fan-delta top-set beds, comprising gently dipping conglomerates and red-coloured, pedogenically altered sandstones which correspond to alluvial-fan deposits of the Salmerón Formation. The sequence from the Cuevas Formation through to Espíritu Santo and Salmerón Formations at this locality records a complete marine to continental sedimentary transition.

Transfer to Stop 4

Retrace your route to the main road, turning right and then passing back over the Río Almanzora road bridge and returning through the traffic lights on the outskirts of Cuevas del Almanzora. On the main road heading out of town towards Vera, turn right at the major road junction as if returning back towards the cement works and the motorway. If travelling by coach carry on directly to Stop 5. If in a car or minibus take the first road on the right, signposted to El Martinete. After 150 m turn left, following the twisty road which climbs past the town cemetery, and proceed until the road forks. Take the left fork and proceed until the road passes over a concrete reinforced rambla. Immediately after the rambla turn left onto a gravel track. At this point, a wall and a large gate to a farmhouse should be on your left. You are advised to park your vehicle at this point near the track entrance, although it is possible to drive up the track into the river bed at the risk of becoming stuck in soft sediment following recent floods. Walk along the track and up into the river bed, examining small riverside exposures as you progress, until reaching the entrance to the Rambla Cirera canyon, marked by a series of large fallen boulders.

Stop 4: Northern margin fan-delta sedimentology, Rambla Cirera (983 282; Vera 1 : 50 000)

A short walk up the Rambla Cirera canyon allows the internal architecture of the northern margin fan-delta to be examined in detail. This had previously been observed at a distance from the overview locality at Stop 3.

Canyon entrance

Before entering the canyon, small exposures of grey-coloured Cuevas Formation marine sandstones and siltstones can be observed adjacent to the Rambla Cirera river bed. These sediments are stratigraphically beneath the main fan-delta body and correspond to open marine shelf conditions. Detailed examination yields a sparse marine invertebrate fauna including the Pliocene bivalve *Amusium cristatum*. From the canyon entrance, *c.* 100 m towards the north, an impressive cliff section of large-scale, 40 m high

Fig. 5.2.4. Three-dimensional exposure of the northern margin fan-delta lobe adjacent to the Rambla Cirera canyon entrance (Stop 4). Palaeocurrent coming out of the photo towards the reader. Exposed cliff is *c.* 25 m high.

cross-beds provides a three-dimensional overview of the prograding fan-delta body about to be viewed in detail (Fig. 5.2.4).

Lower canyon

On entering the canyon, follow the river bed upstream. Both sides of the canyon wall display laterally persistent, cross-bedded conglomerates and sandstones of the Espíritu Santo Formation and correspond to some of the main fore-set beds of this northern margin fan-delta body. The more numerous conglomeratic beds comprise poorly sorted angular to sub-angular boulder- to gravel-sized clasts of locally derived metamorphic basement. Some beds and fallen blocks contain numerous disarticulated oyster shells of *Crassostrea gryphoides*, up to 30 cm in length. Sedimentation probably occurred by mass flow processes, with the size and textural immaturity of sediments suggesting small transportation distances from a significant topographic relief along the basin margins.

Mid-canyon

Approximately 100 m upstream from the canyon entrance, the rambla opens out into a less confined area. At this locality, a small and complex succession of Cuevas, Espíritu Santo and Salmerón Formation sediments exists with stratigraphical relationships complicated by deformation features. Upstream of the tributary junction, fore-set beds are not present. Instead, sections on both sides of the Rambla Cirera comprise top-set beds of the Salmerón Formation. Sediments are characterized by inter-bedded red conglomerates and sandstones deposited within an alluvial-fan environment by debris-flow and sheet-flood processes. Many exposures of alluvial-fan sediments along Rambla Cirera and adjacent areas display extensive deformational features, typically characterized by NNW–SSE to NNE–SSW orientated extensional faults with displacements of up to 60 m. The deformational features correspond to an Early Pleistocene basin-wide phase of deformation in the Vera Basin (Stokes, 1997).

Upper canyon

By continuing upstream for several hundred metres, you will pass from the continental Salmerón Formation back into Cuevas Formation marine sandstones. These sediments do not appear to form bottom-set beds of the fan-delta but simply correspond to a shallow marine shoreline area, typical of much of the basin margin during Cuevas Formation times. The sand-stones are quite coarse, often calcarenitic in nature and contain a rich and diverse marine fauna of invertebrates (bivalves, brachiopods, echinoids, coralline algae) and, less commonly, marine vertebrates (dolphins, whales and dugongs). On reaching greyschists of the metamorphic basement you are advised to return to the vehicle(s) and continue on to Stop 5 to com-plete the excursion.

Transfer to Stop 5

Retrace your route to the main road junction by turning right out of the gravel track, taking the right-hand road fork, passing the cemetery and turning right back towards the main road junction. At the junction turn right as if heading back towards the motorway, passing the KM 4 and KM 3 signs. Immediately after KM 3, look for an abandoned road section and disused bridge on the right. Park your vehicle(s) in the grass area next to the main road and walk along the abandoned road section until you get a good overview into Barranco del Tomate below.

Stop 5: *Marginal marine–brackish sedimentology, Barranco del Tomate (971 268; Vera 1 : 50 000)*

This final locality permits an examination of marginal marine Espíritu Santo Formation sediments which have been deposited in areas, distal and adjacent to the main fan-delta bodies. From the abandoned road alongside Barranco del Tomate, a cliff section to the NE displays an 80 m thick sedimentary succession which comprises a series of distinct, laterally persistent beds which are tectonically tilted up to 20°NE. The succession records a transition from an open marine to a continental depositional environment.

Time permitting, a short examination of the sedimentary succession can be undertaken from the exposures along the abandoned road section. However, some of the most important sedimentary units from the upper part of the succession are exposed only within the river bed and therefore any detailed examination of this locality should also involve a traverse up the barranco.

Cuevas Formation

Basal parts of the succession are best exposed in the barranco adjacent to the road bridge. Sediments comprise the typical grey siltstones containing *Amusium cristatum* bivalves which correspond to deposition within a mid–outer shelf environment (Stokes, 1997).

Espíritu Santo Formation

Unlike the previously examined fan-delta deposits (Stops 2 and 4), large-scale, cross-bedded fore-set conglomerates are absent from this locality. Instead, the Espíritu Santo Formation begins with 7 m of laminated gypsiferous siltstones, which are best exposed in the abandoned road-cut. These sediments contain a rare fossil fauna of fish and plant material, with preservation characteristics typical of a Konservat Fossiles Lagerstatten (Sendra *et al.* 1996). Above the laminated sediments, the remaining 25 m of the Espíritu Santo Formation consists of interbedded pebbly sandstones, fossiliferous muds and sandstones, which are best exposed in the abandoned road section. The sedimentary and faunal characteristics of these deposits suggest deposition within a marginal marine, coastal plain environment fluctuating between open marine and brackish water conditions (Stokes & Sendra, 1996; Stokes, 1997).

By walking 200 m northwards, in an upstream direction from the road bridge, the uppermost part of the Espíritu Santo Formation can be observed in a tight meander bend of the barranco. Here, a distinct green-coloured clay horizon, rich in *Melanopsis* gastropods, marks the onset of freshwater conditions and thus a continental environment.

Salmerón Formation

Stratigraphically above the Espíritu Santo Formation *Melanopsis* bed, in an upstream direction, the sedimentary succession is characterized by un-fossiliferous beds of pink siltstones and fine sandstones. These sediments represent distal alluvial-fan/outer fan apron sediments of the Salmerón Formation and can be traced laterally into much coarser grained conglomeratic alluvial-fan sediments closer to the basin margins (Stokes, 1997). These deposits thus reflect the transition to terrestrial sedimentation and are dealt with in Excursion 5.4.

5.3 Excursion: Mio-Pliocene marine to continental transition of the Sorbas Basin (one day)

ANNE E. MATHER

Stratigraphy

Within the Sorbas Basin the last fully marine sequence was that of the Messinian Sorbas Member (see Excursion 4.5). Continued uplift progressively weakened the marine connection to the south (Mather, 1991). This is reflected in the deposition of the overlying Cariatiz Formation (Messinian and Lower Pliocene), which records the transition from these marine conditions to the fully terrestrial environment of the overlying Góchar Formation (Plio-Pleistocene). Within the Cariatiz Formation the sedimentary succession records three distinctive stratigraphical markers which can be used for basin-wide correlation. These are two white carbonate beds representing lacustrine sedimentation, and a yellow marine unit which caps the sequence. The carbonate marker beds contain ostracods of Messinian age (Civis *et al.* 1977; De Dekker & Chivas, 1988). The marine unit has been identified as Lower to Middle Pliocene on the basis of marine fauna (Ott d'Estevou, 1980). These markers allow two laterally equivalent, but lithologically very different members to be identified. These are the Moras Member, which crops out along the northern basin margins and is dominated by alluvial-fan conglomerates (Mather, 1991; Mather & Harvey, 1995), and the Zorreras Member, which crops out in the central and southern areas of the basin. This member is dominated by sandstone and siltstone and reflects deposition in a coastal plain environment.

Structure

Structural controls were significant in generating topographic highs within the Sorbas Basin during the Messinian (Mather, 1991). In the case of the Cariatiz Formation this is recorded by thickness variation within, and between, the two carbonate marker beds (Fig. 5.3.1).

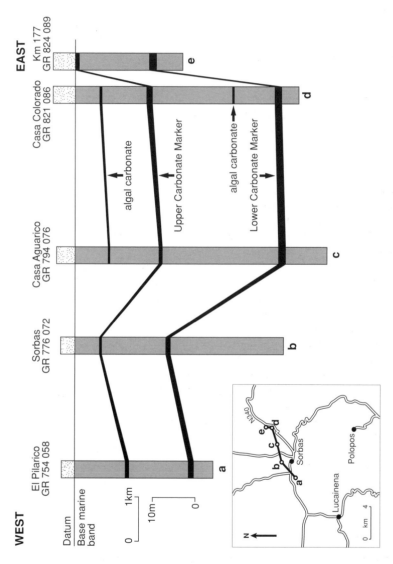

Fig. 5.3.1. Thickness variations in the Cariatiz Formation across the Sorbas Basin. Inset location map indicates line of correlation. (Modified after Mather, 1991.)

At the end of the Messinian and during the Early Pliocene, uplift of the sierras tilted the Cariatiz Formation sediments within the Sorbas Basin towards the basin centre, developing an angular unconformity of some 35° at the basin margins. This occurred after deposition of the upper carbonate marker bed and some way below the Lower Pliocene marine unit which caps the sequence.

The aim of this excursion is to examine the range of palaeoenvironments associated with the marine to continental transition of the Cariatiz Formation. Most previous work has primarily been concerned with the fine-grained sediments of the basin centre, the Zorreras Member, and many visitors are unaware of the palaeoenvironmental variability of the sediments of this unit. The excursion will take a transect from north to south across the basin (Fig. 5.3.2) from the more proximal continental units of the Moras Member to the dominantly marine coastal plain to basinal sequence of the Zorreras Member in the south. To allow stratigraphical correlation, the main marker beds (the two dominant white lacustrine carbonates and the capping marine unit) can be used at each site.

Directions to Stop 1

West of Sorbas take the road to Uleila del Campo. Proceed along this road for 4.75 km. Before a blind corner, turn right (if you come to the bridge over the Rambla de Góchar you have gone too far) and follow the track *c.* 200 m to a wide area where vehicles can turn; park here. Descend into the rambla bed on foot, following the goat track. This is the Rambla de Góchar (see Excursion 4.5, Stop 3).

Stop 1: *Rambla de Góchar (762 106; Sorbas 1 : 25 000)*

Walk to the hill south of the track for an overview (760 104; la on Fig. 5.3.2b). Looking east from here to the opposite river cliff an angular unconformity of *c.* 30° can be observed (Fig. 5.3.3). This is within the Moras Member, but occurs several metres below the marine unit which caps the Cariatiz Formation. This latter unit forms a prominent cemented horizon just above the abandoned building. The Moras Member is some 160 m thick here (Mather, 1991). Several metres above the unconformity a slightly more prominent bed can be identified. This is the marine unit and it can be traced, gently dipping, onto the prominent reef outcrops further upstream (to the left), but not beyond them. It is clear that the dead reefs provided a topographic high around the basin beyond which this final marine incursion did not reach (Fig. 5.3.4). The Moras Member sediments are infilling the accommodation space below the topographic high of the dead Messinian reefs, which provided a topographic niche for fan development.

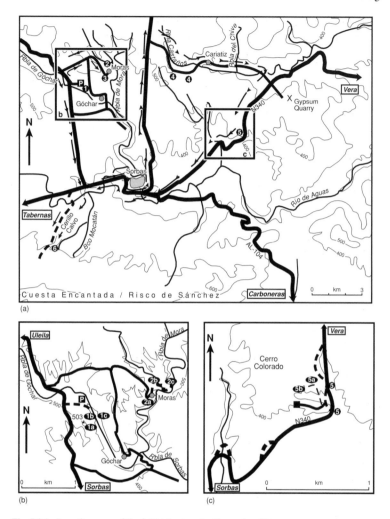

Fig. 5.3.2. Location map of the field-trip stops for Excursion 5.3.

From where the vehicles are parked descend along the goat track into the rambla. As you descend you are walking just above the contact between the Sorbas Member carbonates (described in Excursion 4.5, Stop 3) and the basal Moras Member. On your right note the faulting, combined with sediment liquefaction. If you traverse round to the right to 761 107, just before entering the river bed, bioturbation (*Beaconites* and *Thalassinoides*) can be observed a few metres below the lower carbonate unit, reflecting

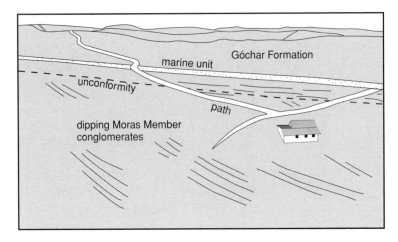

Fig. 5.3.3. Sketch illustrating the view from Stop 1 of the angular unconformity at the top of the Moras Member, Rambla de Góchar (764 104). (Modified after Mather, 1991.)

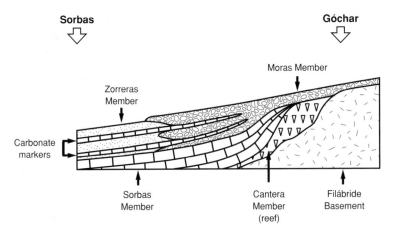

Fig. 5.3.4. Schematic stratigraphical relationship between the Messinian carbonates and the Cariatiz Formation on the northern margin of the Sorbas Basin. (Modified after Mather, 1991.)

burrowing similar to that described for crayfish (Ekdale *et al.* 1984; Mather, 1991; 1b on Fig. 5.3.2). In the topmost sequence pedogenically modified sands are evident. Across the rambla at 762 106 (1c on Fig. 5.3.2) an excellent clastic dyke is evident, within the alluvial sediments of the Moras Member. Returning to the base of the track where it intersects the stream channel, head upstream (west). Just below the main goat track, by which you originally descended, on the path by the channel, note the contact between the Sorbas and Moras Members. The top Sorbas Member is exposed in the path as a series of small (< 1 m diameter) stromatolite domes. The tops of these appear to have been brecciated prior to the deposition of the conglomerates of the Moras Member.

The total thickness of the Moras Member here (some 160 m) is much greater than total thicknesses observed at other localities (90 m at Stop 4 and 60 m at Stop 5). This, together with the evidence for early (prelithification) deformation, suggests deposition in an actively subsiding area. This is further supported when considering that the Moras Member alluvial fans, which would have originally formed positive topographic features, were twice inundated by lacustrine conditions.

Transfer to Stop 2

If in a coach proceed directly to Stop 5 by retracing your route back towards Sorbas, continuing in the direction of Vera (NE, Fig. 5.3.2) on the N340 to KM 505, parking on the shoulder here and following the directions to Stop 5. If in a car or minibus rejoin the Uleila del Campo road and turn right (north). Just over the bridge of the Rambla de Góchar turn right towards Moras (beware of the hairpin bend some 200 m along the road). After *c.* 1.5 km you will reach a junction. Take the road to the left. Drive the vehicle(s) to the bus shelter in the village of Moras. Park here and retrace your steps to the northern end of the village.

Stop 2: Moras village, type locality for the Moras Member (774 114; Sorbas 1 : 25 000)

Leave the village by the goat path to the WNW that contours around the rambla. At Stop 2a (Fig. 5.3.2b) in the cliffs to the north, note the same angular unconformity in the conglomerates that was observed at Stop 1, below an indistinct yellow/white marine unit. This unit can be traced to the lip of the dry waterfall in the Rambla de Mora, which comprises the well-cemented marine unit that marks the top Moras Member. This site is also visited to examine the incisional sequence of fluvial terraces as described in Excursion 6.2, Stop 2.

Follow the path round to the NE. Note the bedding plane below you to your right, dipping towards the SE. On your left, note the stromatolites of the top of the Sorbas Member, intercalated with conglomerates. Leave the path and head north to Stop 2b (774 114; Fig. 5.3.2b). This is the type locality for the Moras Member. Descend into the barranco (a tributary of the Rambla de Mora). Heading downstream, you will walk through shallow water carbonates of the Sorbas Member, which abut Messinian reefs at the point of entry to the barranco. Carrying on downstream, note the progression from stromatolites (exposed as a dry waterfall in the barranco bed) to conglomerates. This is the transition to the Moras Member. As you move downstream, note the outcrop of a (?lower) lacustrine marker bed. Locally it contains small intercalated debris-flow lobes. This carbonate unit contains *Melanopsis* (freshwater gastropods) indicating the proximity of significant freshwater inputs at the basin margin. Note that the sections left and right of the river are affected by a fault which strikes approximately parallel to this section of the stream bed.

If you follow the path to the left and scramble to the upper rim of the main dry waterfall in the Rambla de Mora you will now be standing on the top of the Moras Member, on the marine unit. The marine unit here comprises coarse, angular, beach zone material with *balanus* still attached to individual clasts. If you walk upstream to Stop 2c (Fig. 5.3.2b) you will encounter a collapse feature in the marine unit to your left at 777 113. The marine unit appears to have been undercut by erosion, failed, and then been subsequently buried by sands and conglomerates. The sediments in the upper parts of the section are clearly fluvial in origin and belong to the overlying Góchar Formation.

Transfer to Stop 3

Retrace your steps back to the vehicle(s). Return along the road to KM 4, just past the hairpin bend into the village. Note the well-preserved white channel feature. This is Stop 3.

Stop 3: Moras Member channel (773 110; Sorbas 1 : 25 000)

This channel, whose infill forms part of the top marine marker unit of the Cariatiz Formation, is cut into alluvial-fan sediments of the Moras Member containing calcreted pedogenic profiles. The channel-fill is a shallow, brackish, lagoonal carbonate containing charophytes and ostracods (*Cyprideis, Loxochoncha, Cytherura, Aurilla* and *Neocyprideis*) which indicate warm waters (up to 26°C; Van Morkhovan, 1962). The channel reflects the infilling of the topography on the Early Pliocene fan surface at the time of the final marine incursion into the basin.

Note the presence of large (10–20 cm) angular, subvertical and vertical clasts of amphibole mica schist in the channel-fill. In the centre of the channel they can locally be observed to deform the underlying sediments. Most of the clasts are concentrated around the margins of the channel. These clasts represent dropstones (*sensu* Bennett *et al.* 1994, 1996), their origins most probably related to rafted kelp. Kelp commonly uses large clasts as 'holdfasts'. During storms both kelp and holdfasts can be swept into lagoonal settings. The holdfasts would become grounded on the channel margins but could occasionally break loose, falling into the lagoon sediments in mid-channel, thus creating the distribution of 'dropstones' observed in this section (Bennett *et al.* 1994).

Transfer to Stop 4

Return to the Uleila del Campo road and turn south towards Sorbas. At the N340 junction turn left, and continue past Sorbas town (the type locality of the Sorbas Member). Carry on past the town and over Río de Aguas bridge. Take the first turning on the left towards Lubrín. As you pass, note the quarries on the left and right (quarried for rough-grade clays for local pottery and roofing tiles) and that the hill on the right (Alto de Zorreras) is the type locality of the Zorreras Member (791 074). Note the presence of only one white marker here. After *c.* 7 km on this road you will pass a junction (788 127) on the right, signposted for Cariatiz. Take this road. After *c.* 1 km you will join the gorge of the Rambla de los Castaños. As you drive (carefully as the ravine is deep!) note the steeply dipping talus beds of the reef complex to your left (see Excursion 4.4, Stop 3). Note also the irregular terrain on the south side of the rambla. This reflects the abundance of landslips in the area related to the river incision. Most of the Cariatiz Formation sediments here are not *in situ*, but preserved as part of relict rotational slips. Once the road dog-legs from a SE to WNW bearing, park the vehicle. Stop 4 begins where the pale grey/white sediments of the Sorbas Member have passed into grey/brown silts and sands of the Cariatiz Formation, exposed on a hairpin bend in a cutting.

Stop 4: Type locality for the Cariatiz Formation (802 112–797 113; Sorbas 1 : 25 000)

This section shows part of the type locality for the Cariatiz Formation. The succession at this locality is some 90 m thick (Mather, 1991) and is composed of sands and conglomerates which reflect the transition between the marginal Moras Member (Stops 1 and 2) and the basinal Zorreras Member (Stops 5 and 6).

It is possible to traverse through the succession using the road-cuttings by following the road upwards from this point for some 700 m, although the section is disturbed by landslips.

The sands at the start of the succession comprise channelized and sheet sands and gravels. Some of the channels are filled by the deposits of viscous flows containing 'floating' clasts. The clasts are typical of the Nevado–Filábride basement, dominated by amphibole and chlorite mica schists. Moving up through the succession you will observe a prominent, poorly sorted, coarse sand and conglomerate bed. Some 2 m below this at the road base you will observe the poorly exposed upper white carbonate unit. The unit is laminated on a millimetre scale.

Further up the track the succession demonstrates more evidence of channelization of flows and better organized, imbricated conglomerates. Red sands then dominate the sequence to the next clear road-cutting where the top of the unit is exposed. The base of the cutting comprises sheet conglomerates, which here include clasts of tourmaline gneiss. The top of the cut (797 113) comprises a 3-m section through the top marine unit, which here comprises white and buff-coloured sands. The sands are thoroughly bioturbated by simple burrows with spreite, and contain shell fragments (bivalves).

Note that overlying the marine unit to the west and south of here are the Góchar Formation conglomerates. At this locality their relict alluvial-fan morphology can be appreciated from aerial photographs, and the remnant radial surface drainage on the 1 : 25 000 map (Mather, 1999). This east–west section of the Rambla de los Castaños is an aggressive subsequent drainage which has captured former north–south drainage, such as the upper section of the Castaños to La Mela and the Rambla del Chive to the east of here (Mather, 1999).

Transfer to Stop 5

Continue east along the road which descends into the village of Cariatiz, and follow the 'main' road out to the N340. At the junction with the N340, turn right. Continue along the main road for *c*. 2.5 km. Park 50 m after KM 505 marker on the abandoned road section on the right.

Stop 5: Cerro Colorado (824 088; Sorbas 1 : 25 000)

In the road-cutting the gradational contact between the Sorbas Member (grey/white) and the Zorreras Member (red) can be observed. The Zorreras Member is dominated by laminated silts and fine sands with rare ripple lamination. These sediments pass into nodular red sands which appear to have been pedoturbated. Leaf imprints have been found in these sediments (Mather, 1991). Up the sequence, and exposed in the main road-cut south of the abandoned road, is the lower white carbonate unit. This contains small shell moulds (bivalves) and plant fragments.

If time permits, it is possible to see the complete Zorreras Member sequence by descending into the valley by a track and then crossing the field by the track in the valley bottom. Walk up the stabilized gully system to the agricultural terrace on the northern side of the valley at *c.* 822 092. From here head for the pronounced boulders on the hillside (Stop 5a, Fig. 5.3.2c). The boulders are composed of well-cemented beach gravels. Note the sphericity of the clasts and good shape and size sorting. Locally large-bored oysters can be found. This is the top marine unit of the Zorreras Member. The sequence at 821 086 is 60 m thick (Mather, 1991) and includes the two white carbonate units, with the addition of two thinner white units composed of algal mat carbonates.

Head west, contouring round the slope on the tracks, and descend into the valley (Stop 5b, Fig. 5.3.2c). In this valley some 2–3 m of well-exposed compound calcretes which are developed on the marine unit can be observed. These form part of the base of the Góchar Formation and the final switch to terrestrial deposition in this part of the basin. Return to the vehicle(s).

Transfer to Stop 6

Stop 6 can be reached only by those travelling by car or minibus. Head towards Sorbas on the old N340 and drive past the town. At 300 m past the turning for Uleila del Campo (764 064, on the right) take the dirt track to the left. Follow the track down to the river bed, turn left along the rambla bed, then after 400 m bear right to climb out of the southern side of the rambla. This track is not passable after heavy rain. Follow the track for 2.7 km, until you see a yellow/white hill to your left. Park and walk to the top of this hill (Cerrillo Calvo).

Stop 6: Cerrillo Calvo (755 041; Polopos 1 : 25 000)

This hill exposes the upper sequence through the most southerly outcrops of the Zorreras Member. The complete sequence can be examined in the gully system at 758 039 and is some 30 m thick. It is dominated by silts and fine sands with rare, coarser (gravel) intercalations. These sediments are dominantly laminated on a millimetre-scale but locally reveal ripples and low-angle cross-bedding. Weak pedogenic features can be observed in the silts at 758 039 (mottles indicating pseudogley entisols and carbonate nodules). Both white marker beds are present. The lower white marker bed is 1 m thick, but the upper one only 0.2 m thick and is associated with more clay than the equivalent units found further north. The top 6 m of the sequence (exposed in the hillside) is dominated by yellow silts and sands that contain shell fragments including *Ostrea*. At adjacent sections, barnacles and echinoderm fragments (Ott d'Estevou, 1980) have also been found. The marls in the upper part of the sequence in this area of the basin contain Foraminifera, indicative of an Early Pliocene age and indicating water depths of 20–40 m (Ott d'Estevou, 1980). The section indicates more prolonged periods of marine influence than the sections seen further north, although the local mottling and carbonate fragments indicate localized emergence and exposure to pedogenesis which suggests that some significant topography existed (Fig. 5.3.1).

Synthesis: a depositional model for the Cariatiz Formation

Using some of the observations made in this excursion, combined with additional data from Mather (1991), it is possible to construct a depositional model for the Cariatiz Formation. During Cariatiz Formation times the Pliocene sea occupied the Almería Basin to the south, and part of the southern portion of the Sorbas Basin (Fig. 5.3.5). In the basin centre the Cariatiz Formation is represented by the Zorreras Member (coastal plain fines with rare conglomerate intercalations). Across this wide, low-gradient coastal plain rare conglomerate and sand-filled channels with little evidence for lateral migration, preferentially concentrated in the topographic lows east of Sorbas (Fig. 5.3.5). On well-drained, topographically higher areas (such as near Sorbas at 769 067) weak palaeosols developed.

Along the northern margins of the basin a sequence of alluvial fans (the Moras Member) developed, dominated by unconfined sheet-floods. The basal parts of the fan sequence are represented by wide, shallow channels. These delivered coarse clastic deposits onto the coastal plain of the Zorreras Member east of Sorbas. The fans were typically of limited radial extent, reaching little more than 2 km basinwards (south) from the palaeo-reef cliff-line over which they developed.

Sedimentation was punctuated by the development of two lacustrine units. Mg/Ca and Sr/Ca values indicate that the host water was fresh to brackish (De Dekker & Chivas, 1988). The lakes were probably fairly shallow and unstratified (Mather, 1991) but may have had connections with the sea towards the south. The upper lake was less extensively developed. This may in part reflect continuing tectonic control of the basin topography (Mather, 1991).

Deposition of the Cariatiz Formation was terminated by the final marine incursion into the Sorbas Basin. This created a shallow, well-oxygenated, low-energy inland sea. The sea maintained continuity with the Mediterranean through a gap in the sierras at the southern margin of the basin. After the regression of this sea, fully continental conditions emerged and are recorded in the deposits of the Góchar Formation (see Excursion 5.4).

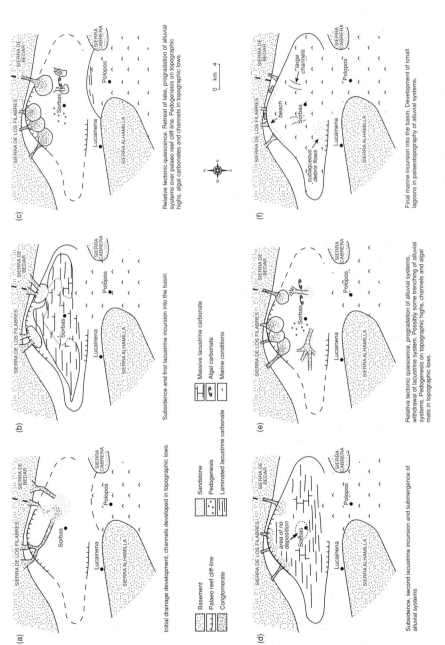

Fig. 5.3.5. Synthesis of the evolution of the Sorbas Basin during deposition of the Cariatiz Formation. (Modified after Mather, 1991.)

5.4 Excursion: Plio-Pleistocene alluvial environments of the Sorbas and Vera Basins (one day)

ANNE MATHER & MARTIN STOKES

Introduction

As the final marine phase withdrew from the Neogene basins (first in the Early–Middle Pliocene from the Sorbas Basin, then in the Plio-Pleistocene from the Vera Basin) alluvial sediments were deposited (Fig. 5.1.2). Palaeoenvironmental reconstructions using evidence from clast assemblages, sedimentology and palaeocurrent information facilitate reconstruction of the alluvial systems. In the Sorbas Basin such an analysis indicates the presence of a basinally convergent drainage network which exited from the south of the basin, across the sierras into the Almería Basin. The deposits are represented by the Góchar Formation, which can be broken down into four fluvial subsystems, three of which drained the margins of the basin and one which axially drained the basin centre from west to east (Mather, 1993a; Mather & Harvey, 1995), exiting the basin to the south (Fig. 5.1.2). This drainage system represents the initiation of the modern drainage.

Within the Vera Basin, the continental sequence was initiated by alluvial fans at the basin margins, which later became dissected by major river systems. These converged towards the basin centre. Upper Pliocene alluvial sediments of the Salmerón Formation represent the onset of continental conditions and the establishment of a convergent drainage network which flowed eastwards towards a Pliocene Mediterranean coastline. This initial drainage network comprised a series of distinct coalescent mountain front alluvial fans which prograded over the former Pliocene shelf and shoreline areas, as recorded by the underlying Cuevas and Espíritu Santo Formations. Modification of the original drainage network occurred towards the end of Salmerón Formation times during the Early Pleistocene, when extensional tectonic activity resulted in significant changes in sediment load and routing (Mather *et al.* 2000; Stokes & Mather, 2000). Later modifications took place by localized river capture in the Vera area (Excursion 6.4, transfer to Stop 6).

This excursion aims to examine the range of alluvial environments established in the basins (Fig 5.4.1), contrasting the Góchar Formation of the Sorbas Basin with the Salmerón Formation of the Vera Basin. It is interesting to note the contrast in depositional process, with debris-flows important in the Vera Basin (Stop 5), and dominantly fluvial, hyperconcentrated flows in the Sorbas Basin (Stops 1–3). This difference is in part a function of the preservation of the systems, with the Vera Basin sediments still attached to the mountain front, thus preserving the more proximal

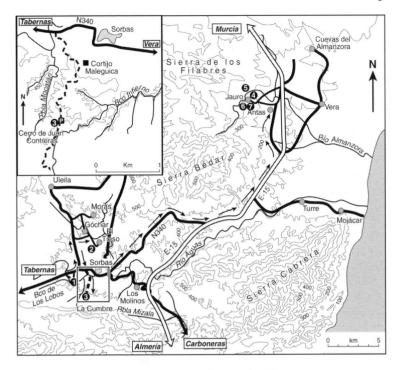

Fig. 5.4.1. Location map of the field-trip stops for Excursion 5.4.

portions of the sequence, whereas the preserved sequence in the Sorbas Basin occupies more central parts of the basin. Catchment lithology may also have been important with the large, schist-rich catchments of the Sorbas Basin (Mather, 1999) providing flows with higher water/ sediment ratios than those derived from the Vera Basin mixed basement (Stokes & Mather, 2000). Styles of deformation also differ between the basins. The Sorbas Basin deformation is concentrated in the south and east of the basin and was most active along the NE–SW orientated left-lateral strike-slip lineament, the Infierno–Marchalico lineament (Mather & Westhead, 1993; Excursion 3.5b, Stop 2). Deformation continued throughout the Plio-Pleistocene. In the Vera Basin the deformation was more spatially and temporally restricted to extensional faulting in the area of the Sierra Bédar (Stokes, 1997; Mather & Stokes, 2000).

Transfer to Stop 1

From the N340, 2.5 km west of Sorbas, take the abandoned road at

753 062 to the south (Fig. 5.4.1). Park at the junction if in a coach. If in a car or minibus drive *c.* 700 m along the abandoned road to the bridge. Park here. This site is also visited to examine the deformation in Excursion 3.5b, Stop 3. Stop 1 is the sequence of conglomerates exposed in the road-cutting.

Stop 1: Type locality of the Los Lobos System, Góchar Formation (748 059; Sorbas 1 : 25 000)

Walk back to the main cutting exposed on the north side of the abandoned road. A tilted section of conglomerates and sands is exposed. The conglomerates display cross-bedding and weak channelization. Careful examination of the conglomerates reveals that gently (< 10°) upstream-dipping back-sets are preserved. The bars are dominantly longitudinal (gravel sheets) with rare cross-bedding associated with lateral bars, indicating low-sinuosity channels (Mather, 1991). Cross-bedding indicates that flow depths were probably less than 1 m.

The interbedded sands display pedogenic carbonate accumulation in the finer sediments and some mottling, indicative of poorly drained conditions (pseudogleys). The succession is 120 m thick. The clast composition is a mixture of Messinian carbonates, amphibole mica schist from the Filabres and rare, purple metacarbonates from the Sierra Alhamilla. The presence of abundant sand in the section probably reflects a Tortonian sandstone provenance from the west and south of the basin. Palaeocurrents derived from imbrication are predominantly to the east. These sediments represent part of the axial drainage system for the basin, the Los Lobos System (Mather, 1991; Mather & Harvey, 1995).

At this section there is abundant evidence of syn-sedimentary deformation including soft sediment deformation (which is examined in Excursion 3.5b, Stop 3; Fig. 3.5.7), tilting and syn-sedimentary unconformities. These produce cumulative thickness variations of up to 130 m within the succession.

Transfer to Stop 2

Return to the N340 and turn right towards Sorbas. After 1 km take the turning on the left towards Uleila del Campo. The deep red sediments and conglomerates exposed in the river-cutting and bulldozed agricultural terraces are all part of the Góchar Formation. After 4 km take the paved turning on the right towards Góchar/El Tieso. Follow the road 1.75 km to a cross-roads. Carry straight on for a further 1.5 km. Note that the flat red surface over which you are now driving is a Pleistocene river terrace (Terrace B of Harvey *et al.* 1995) which is inset into the Góchar Formation conglomerates. Park on the road shoulder and descend into the rambla along a concreted path which lies to the east of the road. Once in the rambla, head upstream to the large cliff section.

Stop 2: El Tieso cliff section (Góchar System, Góchar Formation) (783 085; Sorbas 1 : 25 000)

You should be looking at the steep cliffs exposed in the meander bend of the Rambla de Sorbas. To the south the cliff is lower and capped by a deep red soil on a Pleistocene terrace (Terrace B, Excursion 6.2). The main cliff section, however, is composed of organized, cross-bedded conglomerates of the Góchar Formation. Transverse and lateral bars are evident (Fig. 5.4.2), associated with abundant cross-bedding. Some of the cross-bedding has tangential bases, indicating flow separation on the bar tops. The cross-bedding indicates water depths of up to 2 m. Palaeoflow is variable but dominantly towards the SE. Clast composition is dominated by amphibole mica schists from the Sierra de los Filabres located to the north, and is typical of the Góchar System of the Góchar Formation (Mather, 1991; see also Excursion 2.2, Stop 6). Clearly, this is a large, braided fluvial system with a high water/sediment ratio facilitating the dominance of fluvial activity (Mather, 1991). The variable palaeocurrents indicate a complex network of small channels occupied at different flow stages.

Transfer to Stop 3

Retrace the road back to the main N340. Turn left at the junction. At

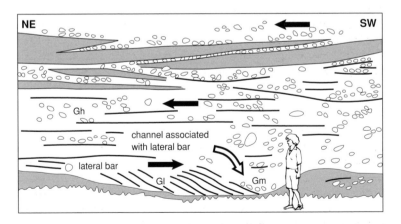

Fig. 5.4.2. Sketch from photograph of the basal part of the El Tieso section (Stop 2). Gh, horizontally bedded conglomerate; Gl, low-angle bedded conglomerate; Gm, massive conglomerate. Arrows indicate palaeocurrent (from imbrication). (Modified after Mather, 1991.)

c. 100 m past this junction, on the right, are a group of white buildings (at Larache, 767 063). If in a car, turn right here, taking the track to the left of the house, between the white buildings (the track is signed to La Cumbre). This road is unsuitable for coaches. If in a coach proceed on along the N340 and follow the directions to Stop 4.

Proceed along the track, bearing right onto a plateau on the Sorbas Member. This was probably a previous (Stage A of Harvey *et al.* 1995) river course of the Río Aguas. At Cortijo Maleguica bear right again (at 773 055) up the hillside. Take care; this section of the track can be in bad repair after rain. Pause at the top of the slope for the view to the north, across the old river course towards Sorbas and across the basin. Continue south along the track. You will pass over a series of abandoned drainage cols—former SW–NE drainage lines which have been captured by the aggressive streams of the south–north Barranco del Mocatán (Mather, 2000b).

Continue along the track. Note the badlands to the right developed in the basal sandstones of the Mocatán System (Góchar Formation). Park by the olive tree terrace to the left (Stop 3, 768 042). Note that the deformation affecting the Góchar Formation here is examined in Excursion 3.5b, Stop 2, and the badland geomorphology in Excursion 6.4, Stop 3.

Stop 3: The Mocatán System of the Góchar Formation, drainage evolution (769 042; Polopos 1 : 25 000)

Walk westwards from the parking place to where the track swings south. From this corner continue west to the head of the vegetated ridge to gain a view over the Zorreras Member and lower Góchar Formation. The top of the Zorreras Member can be identified by a white and yellow band (the top marine unit) in the lower valley to the SW. In this area the Mocatán System of the Góchar Formation can be divided according to clast type into Triassic metacarbonate-rich unit (TRU) and an upper Messinian carbonate-rich unit (MRU; Mather, 1993a, 2000a). The TRU dominates the succession west of here, and the MRU dominates the sequence to the east, overlying the TRU at this point (Fig. 3.5.8).

Up to the track the deposits mainly belong to the TRU. Typically these have sheet geometries and are dominated by sandstone with rare, thicker conglomerate units. To the north at 767 044 large channel bodies belonging to the TRU are evident, indicating the development of significant fluvial trenching (5–10 m). The sedimentary environment of the TRU is interpreted as a fluvial distributary system (Mather, 2000a), dominated by unconfined flows, but with some channelization. Palaeocurrents, together with the presence of the Triassic metacarbonate clasts, indicate that part of the catchment area was located in the basement of the Sierra Alhamilla.

The abundance of reworked benthic Foraminifera found in the sands suggests that the remainder of the source area was located on Tortonian deepwater sediments.

Walk 100 m up the track, note as you go the low-angle shear plane exposed in the road-cut sediments. This is the slip plane of an exhumed Pleistocene landslip, which can be truly appreciated looking north from the view point on Juan Contreras at 757 040. The rotation of the MRU beds is clearly evident, dipping into the hillside to the left of the track. After 100 m along the track to the north, a section through the distal portions of the MRU is exposed. This reveals of lateral accretion conglomerates (Fig. 5.4.3), commonly terminated by channels plugged with the poorly sorted deposits of hyperconcentrated mud flows (contrast this with the deposits of the TRU observed below the track). The final channel exposed in this section is a small, meandering channel (additional excellent sections through the MRU can be observed at 768 049, containing superb plugged channels). The sediments of the MRU are dominated by sheet geometries in the eastern, more proximal parts of the sequence. The succession is also interpreted as a fluvial distributary system, but smaller in scale than the TRU (Mather, 2000a). Palaeocurrents and the dominance of Messinian carbonate clasts suggests a catchment to the SE, north of Cantona, rich in Messinian carbonates (these are examined in Excursion 4.4, Stop 1).

The dominance of the MRU in the latter stages of alluvial system development is interpreted as the result of the beheading of the source areas to the TRU by the east–west orientated aggressive subsequent river, the Rambla de Lucainena, which lies to the south of the escarpment of the Risco de Sánchez/Cuesta Encantada (Mather, 1993a, 2000a; Fig. 5.4.4). The small, fluvial distributary system of the MRU was unaffected by the capture and continued to develop at the expense of the TRU (Fig. 5.4.4).

The hills to the south in the distance are the Risco de Sánchez, which also contain a series of cols of former south–north drainages. While the Barranco de Mocatán maintains a more or less south–north orientation, the Barranco de los Contreras turns a dog-leg at 768 038. This is an elbow of capture where the SW–NE orientated, aggressive, lower part of the Contreras captured former consequent south–north drainages (Mather, 2000b). This upper section forms a canyon transverse to the structure of the Sorbas Member carbonates.

Note the extensive badland development in the sandstone below you. The processes are dominated by piping of the valley-fill. Major gullies developed along strike (east–west), capturing former south–north drainage (such as that in the col to the north, 767 045). More recent capture by aspect-controlled gullying and piping is evident in the valley directly to the west and SW. This is covered in more detail in Excursion 6.4, Stop 3.

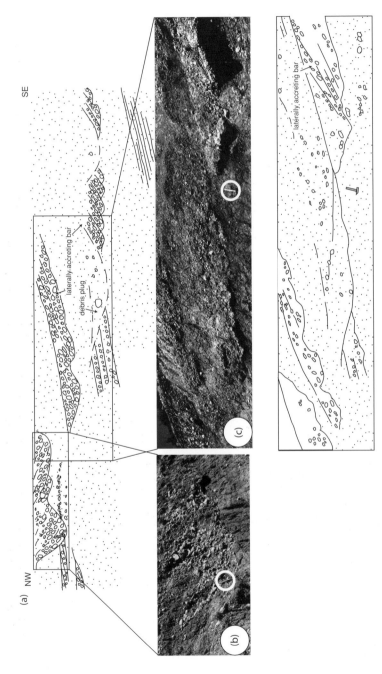

Fig. 5.4.3. Channels within the Messinian carbonate-rich unit (MRU), Mocatán System, Góchar Formation (Stop 3). (Modified after Mather, 1991.)

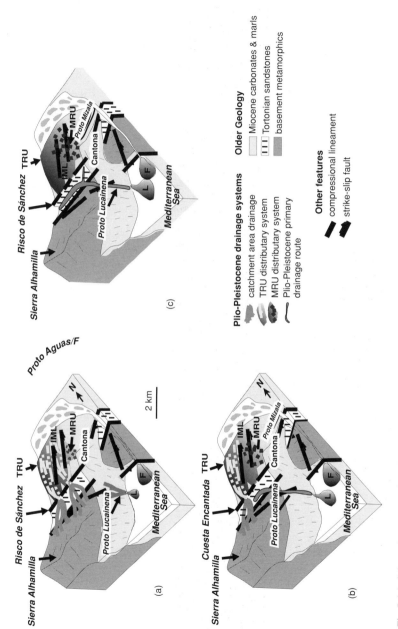

Fig. 5.4.4. Schematic representation of the palaeogeographical evolution of the Lucainena drainage, the Triassic metacarbonate-rich unit (TRU) and Messinian carbonate-rich unit (MRU) at the southern margin of the Sorbas Basin. (Modified after Mather, 2000b.)

Transfer to Stop 4

Return to the N340 and head east to join the motorway near Alfaix, heading northwards into the Vera Basin. Take the exit signposted to Antas. Turn left at the slip-road junction and drive 5.5 km into the town of Antas. Within Antas town take a left turning signposted to Jauro. Proceed along this road for *c.* 3.5 km, taking the left fork at the Roman aquaduct remains and the left fork at the bus stop, continuing onward towards Jauro village. After the bus stop, the road will begin to climb steeply up a hill of red sediments and will pass through a pronounced road-cut through grey conglomerates which cap the hill. Park your vehicle immediately on the right after the road-cut. Leave the vehicle and scramble onto the hilltop to the north, walking along the ridge until you get an unobstructed view northwards towards the Río Antas below and Sierra Lisbona beyond.

Stop 4: *Salmerón Formation stratigraphical and geomorphological overview, road to Jauro village (928 234; Vera 1 : 50 000)*

Some general palaeogeographical and stratigraphical considerations of the Salmerón Formation can be made from this view point as an introduction to this part of the excursion. Along the western margins of the Vera Basin, Salmerón Formation alluvial sediments occur between the Sierra Lisbona to the north, the Sierra de los Filabres to the west and the Sierra de Bédar to the south. Early stage palaeogeographical reconstructions demonstrate the occurrence of two coalescent alluvial-fan bodies with distinct provenance and palaeocurrent signatures sourced from the Sierra Lisbona and Sierra de Bédar (Fig. 5.4.5). Late-stage reconstructions illustrate alluvial-fan abandonment and the establishment of a braided river system sourced from the Sierra de Bédar (Fig. 5.4.5). In the south, braided river sediments are incised by up to 100 m into the southern alluvial fan, whereas in the north, incision is negligible and fan sediments are conformably overlain by braided river sediments (Stokes & Mather, 2000).

From the view point, a cliff section exposed north of the modern Río Antas records an 80 m succession of Salmerón Formation sediments. Basal and mid-parts of the section comprise red, interbedded conglomerates and sandstones. These sediments correspond to deposition by the northern alluvial-fan body sourced from the Sierra Lisbona range (observed in the distance to the north). Mid- to upper parts of the succession are characterized by grey-coloured conglomerates which correspond to deposition by a braided river. Provenance and palaeocurrent signatures suggest a Sierra de Bédar source area some distance to the south of this view point. The topmost part of the succession is capped by a well-developed pedogenic calcrete (Stage IV+, after Machette, 1985), locally up to 1.5 m thick. The calcrete covers an extensive geomorphological surface which grades basin-

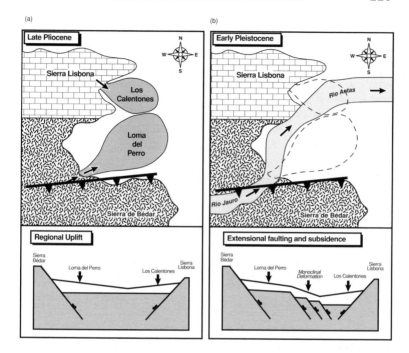

Fig. 5.4.5. A proposed depositional model for the final stages of the Jauro system drainage.
(a) Late Pliocene progradation and coalescence of alluvial fans from the basement.
(b) Early Pleistocene incision following deformation of the fan surfaces, coupled with catchment area expansion of the Río Jauro leads to the abandonment of the Pliocene fan system and development of a braided river system. (Modified after Stokes, 1997; Stokes & Mather, 2000; Mather *et al.* 2000)

wards from the Sierra Lisbona mountain front for several kilometres. The surface corresponds to the highest and oldest fluvial terrace in this part of the Vera Basin, marking the final stages of sedimentary fill prior to basin dissection by the drainage network later in the Pleistocene.

On returning to the vehicle(s), examine the grey, braided river conglomerates exposed in the road-cut section. Note the clast imbrication and evidence for low-angle cross-bedding, which are indicators of fluvial deposition. Also look for the presence of tourmaline gneiss clasts, which are clear indicators of a Sierra de Bédar source area as this is the only area in which this basement material of the Nevado–Filábride nappe can be found. Red-coloured, distal, alluvial-fan sediments can also be observed in the road sections near the vehicle park. Note the grey mottling of these fan sediments, which suggests pedogenic modification through reduction from ponding of water following deposition.

Transfer to Stop 5

Continue along the road into Jauro village. If in a coach go directly to Stop 7 (Fig. 5.4.1). If in a car or minibus enter Jauro village and follow the road round to the left and into the river bed. Turn right along the river bed and continue for 200 m until you reach an impressive 100 m high section in a bend along the river. If you miss this section you will be able to continue for only a few hundred metres before reaching an impassable, dry waterfall section of the river.

Stop 5: Río Jauro alluvial-fan sedimentology (923 238; Vera 1 : 50 000)

This locality provides an opportunity to examine proximal alluvial-fan sediments sourced from the Sierra Lisbona region. Basal parts of the section are characterized by metacarbonate basement clasts belonging to the higher Nevado–Filábride unit of the Nevado–Filábride complex (outlined in Section 3.2). Overlying the basement, and forming the majority of the section, are a series of red-coloured, laterally persistent, sheet-like conglomerates. The lowermost part of these conglomerates comprises a distinct 3 m thick bed of poorly sorted, matrix-supported conglomerate deposited by debris-flow processes. The conglomerate is dominated by gravel- to cobble-sized clasts of locally derived metacarbonate and, most notably, contains several oversized boulders of reworked, older fan material up to 2.5 m in diameter (Fig. 5.4.6). The presence of such locally reworked fan material could be accounted for by the occurrence of a fan-head trench, from which collapsed sediment could be incorporated into subsequent debris-flows and onto the fan surface. Mid- to upper parts of the section are dominated by conglomerates corresponding to deposition by sheet-flood and debris-flow processes typical of alluvial-fan environments (Stokes & Mather, 2000).

Transfer to Stop 6

Retrace your route to the junction of the river bed and the road from Jauro village. From this junction continue upstream along the river bed for *c.* 300 m. You will pass an impressive cliff section of red conglomerates on your left and in front of you. As the river bends upstream to the right, park your vehicle(s) off the track and view the cliff section (at 922 230) on the south side of the river in front of you.

Stop 6: Río Jauro deformation overview (922 233; Vera 1 : 50 000)

This locality provides an overview of the large-scale deformation which affects alluvial-fan sediments along the western basin margins. The 100 m

Fig. 5.4.6. Debris-flow incorporating reworked fan material in the Río Jauro alluvial fan, Stop 5.

high cliff section comprises a series of tectonically tilted alluvial-fan sediments which dip 22°NE (Fig. 5.4.7) and can be mapped over an area covering approximately 0.5 km^2 (Stokes & Mather, 2000). Clast provenance analysis of these fan sediments demonstrates a mixing of metacarbonate and tourmaline gneiss lithologies. This suggests that this locality was in an area of fan coalescence, receiving sediment from the north from the Sierra Lisbona and from the south from the Sierra de Bédar.

The zone of deformation delimits a well-defined, uplifted alluvial-fan body to the south from a less clear, subsided alluvial fan to the north. This spatial distribution is clearly demonstrated from geological mapping and palaeogeographical reconstructions of the area (Fig. 5.4.5).

Transfer to Stop 7

Return to the river/road junction and turn right, driving through Jauro village and following the road to the right out of the village. On leaving the village, look for the first olive grove on the right-hand side of the road. Leave your vehicle here and find the track adjacent to the olive grove. Follow the track southwards near the cliff edge with the Río Jauro valley on your right until reaching a fork in the track. The right fork follows a steep, rocky path down to the Río Jauro. Climb the small hilltop above this rocky track and look for a series of deformed rocks in the cliff section in front of you.

Fig. 5.4.7. Detailed view of the faulted fan sediments, Río Jauro (Stop 7). (Photo M. Stokes.)

Stop 7: Río Jauro deformation detail (924 233; Vera 1 : 50 000)

Stop 7 is located at the down-dip end of the previously considered tilted deformation zone. This locality allows a detailed examination of a fault zone affecting the alluvial-fan sediments in this area. The sections show that well-defined, sheet-like conglomerates have been dragged into the fault zone (Fig. 5.4.7). An approximate NW–SE orientation suggests that this fault forms part of the deformational phase which affected the Vera Basin during the early Pleistocene (Stokes, 1997). Although an exact determination of fault displacement from this locality is difficult, a 100-m height difference exists between the approximate upper surfaces of the southern and northern alluvial fans. Such a large-scale displacement of the alluvial-fan bodies would result in incision and entrenchment of the southern alluvial fan (Stokes & Mather, 2000) which eventually caused fan abandonment and a switch to braided river deposition. Rapid expansion of the former southern fan drainage network in the Sierra de Bédar region via river capture (Mather *et al.* 2000) would have increased flows and sediment loads. This could account for the switch in depositional style to a braided river. It would also account for the change of sediment provenance observed at Stops 4 and 5.

Chapter 6
Uplift, Dissection and Landform Evolution: the Quaternary

ADRIAN M. HARVEY

6.1 Introduction

The processes that were initiated from the Late Messinian to the Pliocene, whereby the basins changed from marine to terrestrial environments, continued into the Pleistocene. The basins were uplifted and the local conditions switched from net deposition to net erosion. The river systems, initiated in the Pliocene, cut through the Neogene basin-fill sediments to produce the dominantly erosional landscape seen today. Quaternary depositional zones became restricted to the modern coast, the main river valleys and alluvial fans in mountain-front zones. The pattern of incision reflects patterns of Quaternary uplift (outlined in Section 3.5).

Drainage evolution

During the Pliocene the main continental drainage systems developed (Fig. 5.1.2). Drainage lines transverse to structure were superimposed and became antecedent following uplift (Harvey, 1987a; Harvey & Wells, 1987; Mather & Harvey, 1995; Stokes, 1997). In the Sorbas Basin the Aguas/Feos river became established as the master stream, to which the smaller, also transverse, Lucainena system was tributary. This combined system must have drained to the Alias and eastwards through the Carboneras Fault Zone.

Uplift of the Tabernas Basin during the Pliocene caused dissection of the upper part of the basin by river systems feeding into the Rioja corridor (between the Sierras de Gádor and Alhamilla). These rivers deposited large volumes of fan-delta sediments in the Rioja corridor (Postma, 1984). Sustained uplift of the Tabernas Basin during the Pleistocene caused deep incision in the basin centre (Harvey, 1987a) and dissection of the Pliocene rocks in the Rioja corridor, as the Andarax system prograded towards the modern coast at Almería.

The Vera Basin, with less uplift, was a receiving basin for drainages originating outside the basin. The lower Río Aguas system drained the zone between the Sorbas and Vera Basins. Its tributary, the Río Jauto,

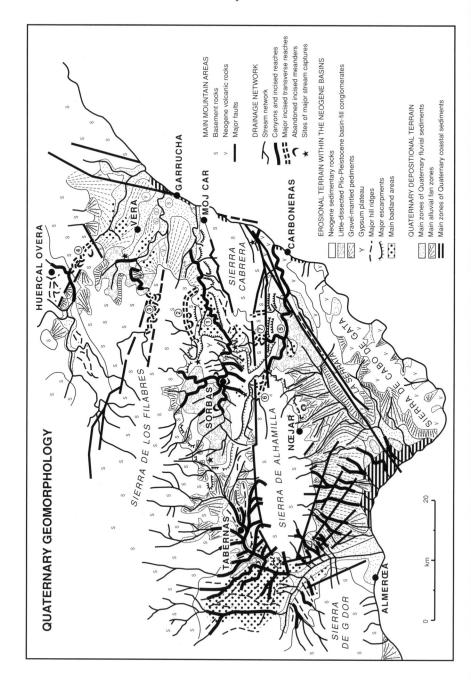

QUATERNARY GEOMORPHOLOGY

HUERCAL OVERA

GARRUCHA

VERA

MOJ CAR

CARBONERAS

SIERRA DE LOS FILABRES

SIERRA CABRERA

SORBAS

NIJAR

SIERRA DE ALHAMILLA

TABERNAS

SIERRA DE GATA

CABO DE GATA

SIERRA DE G DOR

ALMERIA

0   km   20

MAIN MOUNTAIN AREAS
s Basement rocks
v Neogene volcanic rocks
— Major faults

DRAINAGE NETWORK
Stream network
Canyons and incised reaches
Major incised transverse reaches
Abandoned incised meanders
★ Sites of major stream captures

EROSIONAL TERRAIN WITHIN THE NEOGENE BASINS
Neogene sedimentary rocks
Little-dissected Plio-Pleistocene basin-fill conglomerates
Gravel-mantled pediments
Gypsum plateau
Major hill ridges
Major escarpments
Main badland areas

QUATERNARY DEPOSITIONAL TERRAIN
Main zones of Quaternary fluvial sediments
Main alluvial fan zones
Main zones of Quaternary coastal sediments

is transverse across the structures of the Sierra de Bédar, presumably through a combination of superimposition and antecedence. The lower Río Aguas itself is transverse across the Sierra Cabrera and the Palomares Fault Zone near the present coast. The Río Antas drains the eastern part of the Sierra de los Filabres, north of the Sierra de Bédar and is also transverse across basement structures. The Almanzora, the main drainage of the area north and west of the Vera Basin, is another major transverse drainage, crossing the Almagro uplift between the Huercal Overa and Vera Basins (Fig. 6.1.1). Both rivers had deposited large alluvial fans at the basin margins during the Plio-Pleistocene which became dissected during the Pleistocene (see Excursion 5.2; Stokes, 1997).

Two main drainages developed within the Almería basins. To the east the Alias system formed the distal part of the Aguas/Feos drainage from the Sorbas Basin, and followed the Carboneras Fault Zone, although little is known of its evolution. In the west, the basin is drained axially towards the coast east of Almería, with tributaries from the Sierra Alhamilla. South of the Carboneras Fault Zone, a small west-flowing axial system drains the minor basin between La Serrata and the Sierra de Cabo de Gata.

During Quaternary incision, drainage modification took place by capture, especially at the basin margins. Tabernas drainage captured the western headwaters of the Sorbas drainage. During and since the Early Pleistocene (Mather & Harvey, 1995), the Lucainena system has progressively extended headwards as a strike stream, capturing basinal Sorbas drainage (Mather, 1993a; Mather, in press a). In the north-east of the Sorbas Basin, the river development involved interaction between Sorbas drainage and the Jauto/Castaños system tributary to the Lower Aguas/Vera drainage (Mather, 1991, 1999).

In terms of drainage rearrangement the most important river capture was the Late Pleistocene (Harvey & Wells, 1987; Harvey *et al.* 1995) capture of the Aguas/Feos master drainage near Los Molinos. This took place by the lower Aguas working its way headwards from the Vera Basin, along the outcrop of the weak Abad Member marls. Incision following this capture led to the deep dissection of the central parts of the Sorbas Basin (Harvey *et al.* 1995; Mather, 2000b), complicated by the Late Pleistocene deformation at Urra (related to the lineament discussed at Excursion 3.5b, Stop 2; Mather *et al.* 1991), and also led to other local minor captures in the southern part of the basin (Excursion 5.4, Stop 3; Mather & Harvey, 1995).

Fig. 6.1.1. Drainage patterns, landforms and Quaternary sediments. 1, Río Aguas; 2, Río Jauto; 3, Río Antas; 4, Río Almanzora; 5, Río Alias; 6, Rambla Lucainena; 7, Rambla Feos. (Modified after Harvey, 1987a.)

Various other modifications of the drainage pattern have taken place. Within the Vera Basin, the lower Antas has captured the Rambla de Ballabona/Cajete, leaving an abandoned valley through Vera town. Within the mountain areas, there are suggestions of local drainage modifications by capture. The most impressive of these is at Sopalmo in the Sierra Cabrera, where the steep Granatilla appears to have captured the more gently graded, fault-aligned Río Macenas (Harvey, 1987a; see also Excursion 3.5a, Stop 3).

Quaternary erosional development

Landform evolution within the basins took place through erosion, controlled by base levels which were related to the incising river systems. The resultant erosional landform patterns therefore partly reflect the regional tectonic patterns, and partly the resistance of the bedrock lithologies. The much more restricted depositional landscape is also controlled by these same patterns. Zones dominated by erosion include three main types of landscape (Harvey, 1987a): (i) canyon landscapes; (ii) deeply dissected soft rock areas; and (iii) scarplands.

Canyons

Deep canyons and incised valleys developed where rapid vertical incision coincided with the outcrop of more resistant rocks (Fig. 6.1.1). They occur within uplifted mountains both on rivers originating within the mountains and on superimposed/antecedent transverse rivers (e.g. on the Lucainena and Feos rivers crossing the Sierras Alhamilla and Cabrera; on the Jauto and upper Antas rivers crossing the Sierra de Bédar; on the Almanzora crossing the Sierra Almagro). They also occur where vertical incision coincides with the outcrop of more resistant rocks within the basin-fill sequences (e.g. the canyons in the Tabernas Basin, cut in Serravallian and Tortonian conglomerates and sandstones; the Aguas and Sorbas canyons cut in Azagador Member limestones, in Yesares Member gypsum and in Sorbas Member sandstones).

An interesting characteristic associated with many of these areas of rapid incision into resistant rocks is the presence of numerous abandoned incised meander loops. There are several on the Río Almanzora (now partially inundated by the waters of the reservoir) and on the Río Antas and Río Jauto. In the upper Aguas system there are cut-offs in the headwaters at Moras, and on the Guapos headwater two at Sorbas, including that which isolates the 'island' mesa of Sorbas town, and others along the Barranco de Hueli near Urra. There is one on the Feos within the trans-mountain section, and another on the Alias where it crosses the Carboneras

Fault Zone. These features do not appear to be associated with any particular phase of dissection, and their relationships to regional incision rates or local deformation are not clear.

Dissected erosional landscapes and badlands

Where rapid incision coincided with the outcrop of weaker rocks, selective erosion of soft rock areas has produced deeply dissected erosional landscapes, characterized by gullying and, in extreme cases, by badlands (Fig. 6.1.1; see also Calvo-Cases *et al.* 1991a).

The Tabernas badlands, perhaps the most spectacular in Europe, dissect a great thickness of Tortonian marls in a zone of maximum tectonically-induced dissectional relief.

Various badlands in the Sorbas Basin dissect Abad Member marls and Zorreras Member and Góchar Formation sands and silts, and in the García Alta area, weak, highly altered Triassic rocks. Each major badland area is associated with deep dissection. Badlands occur along the Aguas valley between La Huelga and Los Molinos, in the south of the Sorbas Basin where the capture-induced dissection is cutting into Góchar Formation sands and silts, and at García Alta, in response to uplift along the Cabrera Northern Boundary Fault.

The badlands at Antas and Vera are developed in Messinian marls (Harvey, 1987a). At Antas they developed in response to river incision and at Vera in relation to dissectional zones between successive pediments.

The styles of modern processes operating within the badland areas reflect interactions between geological, topographic and climatic factors (Harvey 1982; Calvo-Cases *et al.* 1991a,b; Alexander *et al.* 1994). The badlands themselves range from apparently simple zones of relatively recent dissection, some of it undoubtedly human-induced, to zones of complex multiple sequences of badland development and stabilization. In the Tabernas area these probably date back to the Pleistocene (Alexander *et al.* 1994).

Scarplands

Where erosion has coincided with alternating strong and weak rocks, the resistant bands form ridges or escarpments (Fig. 6.1.1). The main resistant rocks and scarp formers are:

1 Serravallian sandstones and conglomerates, forming the folded terrains west and east of Tabernas town and in the Gafarillos area;
2 Tortonian sandstones, forming escarpments in the Tabernas Basin and the southern part of the Sorbas Basin;
3 Azagador Member calcarenites and Cantera Member reef limestones, forming escarpments in the Níjar–Polopos areas, along the south rim of

the Sorbas Basin east to Mojácar, and along the north rim of the Sorbas Basin east to Bédar; and
4 Yesares Member gypsum, forming escarpments in the Sorbas Basin and south of the Sierra Cabrera (Harvey, 1987a).

Zones of less intense erosion (relict surfaces and pediments)

Zones of less intense erosion include two main types of landscape: relict depositional surfaces and uplifted pediments. In basin centres, away from tectonically induced dissection, are relict landscapes dominated by the remnants of the last stages of basin filling. Little-dissected Plio-Pleistocene depositional surfaces occur on conglomerates in the west of the Sorbas Basin, in the eastern part of the Almería Basin and in the northern part of the Vera Basin (Harvey, 1987a).

In some soft rock areas where vertical incision is limited, erosional pediment landscapes have developed. These occur marginal to uplifted mountain areas in the Almería Basin, and in the eastern part of the Vera Basin. In many places the pediments are mantled by a thin depositional veneer of early Pleistocene gravels.

Zones of Quaternary deposition

Because of uplift and the regional switch from deposition to erosion, there are only restricted zones of Quaternary sedimentation. There are three important groups of Quaternary sediments and depositional landforms: (i) coastal sediments; (ii) river terrace sediments; and (iii) alluvial-fan sediments (Fig. 6.1.1).

Coastal sediments

Post-Pliocene uplift has restricted Quaternary coastal deposits largely to modern coastal locations. Only in the Almería area is there a sequence of Quaternary coastal deposits, spanning Early to Late Quaternary time (Zazo *et al.* 1981; Goy & Zazo, 1986; Goy *et al.* 1986). Offlap and interdigitation between terrestrial and coastal sediments in this zone reflects the interaction of the Alhamilla uplift and the Almería/Rioja downfaulting with Pleistocene sea-level change (Ovejero & Zazo, 1971; Zazo *et al.* 1981). Extensive Quaternary coastal sediments occur between Almería and Cabo de Gata, and on the Cabo de Gata coast itself there are isolated patches of beach and dune sediments. In the eastern part of the Almería Basin, marine regression after the Pliocene marine and Plio-Pleistocene fan-delta sequence (Mather, 1993b) has restricted Quaternary marine sediments to the modern shoreline zone near Carboneras. Along the Sierra Cabrera coastline there are excellently preserved Quaternary coastal sediments at Macenas (Harvey, 1987a). The scenario in the Vera Basin appears to be similar to that in the Carboneras Basin, with Quaternary marine sediments restricted to the Mojácar/Garrucha areas (Stokes, 1997).

With the exception of the Almería area referred to above, most Quaternary marine sediments relate only to Tyrrhenian time (late Pleistocene). Tyrrhenian I sediments have been identified in the Vera Basin (Völk, 1979), but most of the dated sediments along this coast relate to Tyrrhenian II (Isotope Stage 5) times (Thurber & Stearns, 1965; Goy & Zazo, 1986). The sediments are exposed at a number of coastal locations, with perhaps the best site at Macenas (Harvey, 1987a). Some locations show deformation by the Carboneras faults (e.g. east of Almería; Excursion 3.5a, Stop 6), or by the Palomares faults (e.g. at Garrucha and Mojácar; Angelier *et al.* 1976; Goy & Zazo, 1986).

The modern shoreline ranges from an erosional cliffline coast throughout most of the Cabo de Gata and Sierra Cabrera zones, to shingle and sand beaches.

River terrace sediments

The main river valleys record a Quaternary sequence of progressive but intermittent incision as an erosional response to regional uplift, punctuated by major phases of aggradation. This has resulted in well-developed river terrace sequences along the major rivers. The major aggradation phases, resulting in sustained fluvial deposition, appear to relate broadly to cold, dry but stormy climates which equate temporally with the northern European glacials. The major dissection phases, producing the terraces, appear here, as elsewhere in the western Mediterranean region, to relate broadly to the milder interglacials (Amor & Florschutz, 1964; Butzer, 1964; Sabelberg, 1977; Rhodenburg & Sabelberg, 1980; Harvey, 1987a; Harvey *et al.* 1995). However, there are local variations, some of which may be related to tectonic activity (e.g. the Tabernas lake sequence, see below). Other variations can be related to topographical and base-level changes following capture, and include the complex and deformed sequence at Urra (Mather *et al.* 1991).

On the Aguas/Feos river system (Fig. 6.1.1), draining much of the Sorbas Basin, two groups of river terraces can be recognized (Harvey, 1987a; Harvey & Wells, 1987; Harvey *et al.* 1995). The older group (terraces A, B, C) pre-date the Aguas/Feos river capture (see above), and the younger group (terraces D, E) post-date the capture.

The terraces have been correlated on the basis of soil development (Harvey *et al.* 1995; Table 6.1.1). Terraces A, B, C (dating from ?Early to Middle Pleistocene) have carbonate-accumulating red soils, with

Table 6.1.1. Summary of soil properties on terrace surfaces of the Río Aguas, Sorbas Basin (modified after Harvey *et al.* 1995; $CaCO_3$, stages after Gile *et al.* 1966; Machete, 1985)

Relative age	Holocene	Late Pleistocene	
Terrace	E	D3	D1
Characteristic soil property			
Approx. depth (cm)	< 50	< 80	> 150
$CaCO_3$ Stage*	0	I	I–II
B horizon			
Hue†	10YR	10YR	7.5YR
Redness index‡			
mean	< 1.0	< 1.0	4.2
approx. range			1–7
Percentage clay, max.	na	13	13
Fe_d percentage§ range	0.1–0.5	0.3–0.6	0.6–1.0
Activity ratio§	0.34–0.66	0.31–0.57	0.29–0.35
Mineral magnetic characteristics, B horizon¶			
X (range)	23	76	135
X_{fd} (range)	4.5	9.8	1.2
ARM (range)	34	121	145
SIRM (range)	2620	7601	15 052

* After Gile *et al.* (1966) and Machette (1985).
† Munsell colour hues.
‡ After Hurst (1977); modified by Harvey *et al.* (1995).
§ See text.
¶ See Thompson & Oldfield (1986) for definitions.

well-developed Bt horizons (colours generally reaching 5YR Munsell colour hues or stronger), and well-developed Bk or K horizons with at least Stage III carbonate accumulation (terminology after Gile *et al.* 1966; Machette, 1985). The soils on the younger (Late Pleistocene and Holocene, D and E) terraces show much less maturity. Only on terrace D1 does the Bt horizon reach 7.5YR coloration, and carbonate accumulation reach Stage II. On D2, D3 and E terraces, colour (10YR hues) is little different from that of the parent materials, and on the younger D soils, carbonate accumulation reaches only Stage I, and is not evident on the soils on terrace E (Harvey *et al.* 1995). Terrace E has been radiocarbon dated to the Holocene (Harvey & Wells, 1987). The terraces of the Aguas/Feos system have been mapped (Harvey *et al.* 1995), and in the Urra area have been used to clarify the Late Pleistocene sequence of deformation (Mather *et al.* 1991).

In the northern part of the Almería Basin, along the Río Alias, there is a sequence of several terraces, the gravels of the oldest of which have been deformed in the Carboneras Fault Zone near El Argamasón (see

Late–Mid Pleistocene		
C	B	A
II–III	150–200 III–IV	IV
5YR	2.5YR	2.5YR
9.1	12.7	14.0
7–14	9–16	12–16
35	46	na
1.0–2.2	1.4–3.0	1.3–2.6
0.17–0.34	0.13–0.27	0.13–0.27
135–220	83–173	105–110
2.2–7.8	4.9–7.7	5.2–9.2
128–368	84–145	74–150
13 259–	7270–	9179–
16 378	8132	10 192

Units: X (magnetic susceptibility), $10^{-8} \, m^3 \, kg^{-1}$; X_{fd} (frequency dependent susceptibility percentage), $X_{LF} - X_{HF} \times 100$; ARM (anhysteretic remnant magnetization), $10^{-6} \, Am^2 \, kg^{-1}$; SIRM (saturated isothermal remnant magnetization), $10^{-6} \, Am^2 \, kg^{-1}$ at 1 Tesla. For discussion of magnetic properties of Sorbas Basin terrace soils see Harvey *et al.* (1995). For further discussion of magnetic properties of Cabo de Gata soils see Harvey *et al.* (1999b).

Excursion 3.5a, Stop 4). However, the relationships between this sequence and that of the Aguas/Feos upstream is unknown.

Terrace remnants can be traced throughout the Tabernas Basin. In the upper part of the basin, alluvial fans (see below) feed into terrace sediments (Harvey & Mather, 1996). In the central and lower parts of the basin two or more major terrace groups are present together with small younger terraces. The older terraces can be differentiated on the basis of calcrete development (Nash & Smith, 1998). Groundwater calcretes characterize the contacts between the terrace gravels and the underlying Tortonian mudrocks, often forming erosional terrace-like mesa surfaces. Locally, especially to the west of Bar Alfaro, this effect is enhanced by travertine precipitation (Mather *et al.* 1997; Mather & Stokes, 1999). In the central part of the basin, the lower of the two main terraces comprises palustrine and lake sediments rather than fluvial gravels (Harvey & Mather, 1996). These sediments now form two dissected 'lake' floors, one extending from Bar Alfaro upstream to Tabernas, and the other along the Rambla de

Los Molinos upstream of Tabernas. The lake sediments have been dated, using U/Th isochron dating of rodent teeth, to 150 ka (± 50 ka) (Delgado-Castilla *et al.* 1993). During deposition, drainage through the lower lake zone was impeded by tectonic uplift occurring downstream of Bar Alfaro, where the 'lake' sediments thin out and are deformed. The upper 'lake' zone appears to have formed in response to damming of the Tabernas canyon by tributary junction fans, especially that of the Rambla de la Sierra (Harvey & Mather, 1996; Harvey *et al.* 1999a).

In the Vera Basin, there are major terrace sequences along the Antas and Almanzora rivers, which record the sequence of fluvial dissection following Early Pleistocene alluvial-fan sedimentation, where the two main rivers issue into the basin (Stokes, 1997; Stokes & Griffiths, 1999; Stokes & Mather, 2000).

Along most of the river valleys the modern floodplain sediments range between silts on the smaller streams, to sands and gravels, especially on rivers fed from mountain catchments. The sediments record relationships between lateral and vertical accretion, in a number of places (e.g. on the Río Aguas at Urra). The modern river channels tend to be wide, shallow, ephemeral, braided rivers although locally meandering does occur.

Alluvial-fan sediments

Quaternary alluvial fans have accumulated at several mountain-front locations, where streams from mountain catchments emerge into the sedimentary basins (Fig. 6.1.1). They occur in zones where tectonically induced incision is limited, and are best developed within the basin margin areas in the divide zones between major drainages (Harvey, 1987a). They are particularly well developed in the Tabernas Basin, where coalescent fans occupy the whole of the upper part of the basin. Fans are also well developed in the Níjar area and, to a lesser extent, at the northern margin of the Sorbas Basin near Uleila, and also at the northern margins of the Vera Basin. Small fans also occur around the Cabo de Gata ranges (Harvey *et al.* 1999b).

The alluvial-fan sedimentary sequences show complex sedimentary successions, with the balance between aggradation and dissection changing through time. The earlier phases (probably during the Middle Pleistocene) were dominated by episodic but progressive aggradation, and the later phases (probably during the Late Pleistocene and Holocene) were dominated by episodic dissection (Harvey, 1984a, 1990). Throughout the duration of fan evolution, shorter periods of aggradation were related to periods of excess sediment supply, whereas periods of dissection resulted from a net deficit of sediment from the source areas. The controls over aggradation and dissection appear to be related primarily to climatic factors, modified by tectonic context and base-level change. Fan aggradation caused by sediment excess occurred primarily during 'glacial' equivalents, and dissection caused by sediment deficit during 'interglacials' (Harvey, 1990; Harvey *et al.* 1999b).

Sedimentary facies exposed in sections within the fans show the deposits of debris-flow and fluvial processes, including channel and sheetflood deposits (Harvey, 1984b). The sequences also include buried soil horizons and are often capped by pedogenic calcretes. Overall there is evidence of a progressive trend from debris-flow dominance through fluviatile deposition to dissection, within which the major controls of facies type are drainage basin area, relief and geology (Harvey, 1992a).

The catchment characteristics and depositional facies in turn influence fan morphology. Morphometric relationships can be established between catchment area (drainage area and, to a lesser extent, basin slope or relief) and fan variables, particularly fan area and fan gradient (Fig. 6.1.2). The relationships differ between fan groups, reflecting geology, tectonics, base-level characteristics and dominant sediment transport processes (Harvey, 1987b, 1992b; Harvey *et al.* 1999b).

Fan dynamics are also influenced by these factors, most significantly through their impact on fan gradient (Harvey, 1987b, 1992b). At present most of the fans in this region are proximally trenched, distally aggrading fans, and act as buffers within the modern fluvial system, trapping most of the sediment derived from the mountain catchments (Harvey, 1997). Some fans, however, are trenched throughout, as a result of either intrinsic characteristics or tectonically induced basal dissection. In these cases, the mountain catchments are coupled with the axial drainages, and coarse sediment moves through the modern fluvial system. The Cabo de Gata fans show two tendencies, reflecting the interaction between proximal catchment controls and distal controls (Harvey *et al.* 1999b). The inland fans and those buffered from base-level influences by the Cabo de Gata lagoon are weakly dissected prograding fans; those facing the east coast have been influenced by base-level change following Quaternary sea-level change and are dissected. Interestingly, periods of dissection followed coastal erosion during high sea-levels, rather than, as might be expected, dissection during low sea-levels (Harvey *et al.* 1999b).

The landforms and the Quaternary geology of the Almería region exhibit relationships between tectonics and patterns of erosion and deposition with a clarity rare elsewhere in Europe. This is also essentially a dryland landscape, resulting from both Quaternary and modern semiarid climates. There are few places in Europe where the relationships between fan sedimentation, morphology and dynamics are so well shown and with the variety of erosional landscapes, especially in the badland areas. The long-term relationships between tectonics and drainage evolution are well

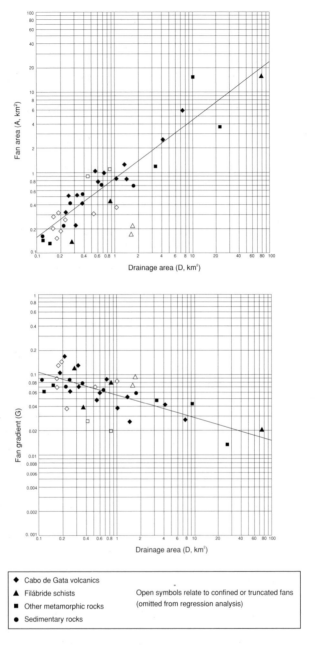

Fig. 6.1.2. Alluvial-fan morphometry for alluvial fans in the Almería region, showing relationships of fan area and gradient to drainage area. Open symbols relate to confined or truncated fans: data not used in the regressions. Regression analyses show:
$A = 0.822D^{0.722}$; n, 28; s, 0.217; r, 0.924; $P < 0.0001$; $G = 0.058D^{-0.271}$; n, 28; s, 0.154; r, 0.786; $P < 0.0001$ (A is fan area, D is drainage area, both in km²; G is fan gradient).

exhibited here, and the coastal zone preserves an excellent record of the Quaternary coastal sequence.

In this guide, the excursions that follow deal thematically with the various aspects of Quaternary geology and geomorphology described above. Excursion 6.2 deals with the drainage evolution of the Sorbas Basin. Excursion 6.3 deals with the alluvial fans within the context of the regional geomorphology of the Tabernas Basin. Excursion 6.4 deals with badland environments in the Tabernas, Sorbas and Vera Basins. Excursion 6.5 deals with Quaternary coastal sequences and the geomorphology of the coast between Almería and Garrucha.

6.2 Excursion: Drainage evolution and river terraces of the Sorbas Basin (one day)

ADRIAN M. HARVEY & ANNE E. MATHER

Introduction

The purpose of this excursion is an examination of the river terraces of the Río de Aguas, which record the Quaternary dissectional sequence of the Sorbas Basin. The excursion covers three zones from the upper reaches of the system above Sorbas, through the zone of deformation and the capture site, to the beheaded master stream in the southern part of the basin (Fig. 6.2.1).

At the end of Góchar Formation times, uplift caused a switch from deposition to incision in the Sorbas Basin. The original drainage was southwards by the ancestral Aguas/Feos river system. The terraces of the Aguas record a sequence of episodic incision and aggradation, reflecting tectonic and climatic controls during the Pleistocene (Harvey, 1987a; Harvey & Wells, 1987; Harvey *et al.* 1995). The terraces have been mapped (Fig. 6.2.2a) and the first three terraces (A, B, C) can be traced south from the upper part of the system through the Feos valley.

A major capture took place after terrace C, of the ancestral Aguas/Feos by the lower Aguas, at Los Molinos (Fig. 6.2.2a), taking the main drainage east into the Vera Basin. Terraces D and E post-date the capture. During terrace D times, deformation took place in the Urra area, downstream of Sorbas. This may have been caused by erosional off-loading, following capture-induced incision, and causing diapiric disturbance of the underlying Messinian gypsum (Mather *et al.* 1991). Alternatively, it may have been tectonically induced (see Excursion 3.5b, Stop 2), or related to subsidence following karstic collapse of the gypsum.

Correlation of the terraces (Figs. 6.2.2a,b, 6.2.3 and 6.2.4) has been achieved using a combination of relative elevation and soil evidence. Soil

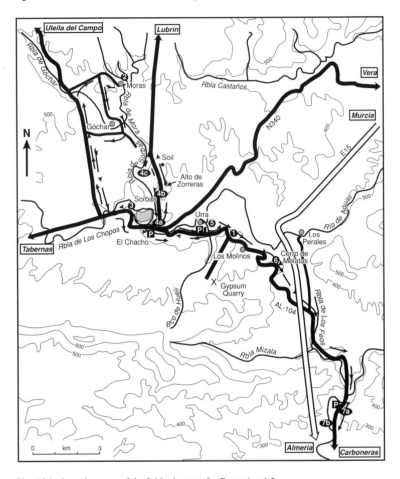

Fig. 6.2.1. Location map of the field-trip stops for Excursion 6.2.

horizon thicknesses, carbonate characteristics, colour, clay content, iron oxide properties and magnetic mineral properties all show age-related trends (Table 6.1).

The capture of the main drainage has resulted in three zones within the basin, with different histories of dissection expressed by long profiles (Fig. 6.2.4) and terrace relationships (Fig. 6.2.5; Harvey *et al.* 1995).

The excursion is designed as a full day, covering the Aguas/Feos system from the centre of the Sorbas Basin, through the deformation and capture zone, to the beheaded Feos valley at the southern margin of the basin. The

(a)

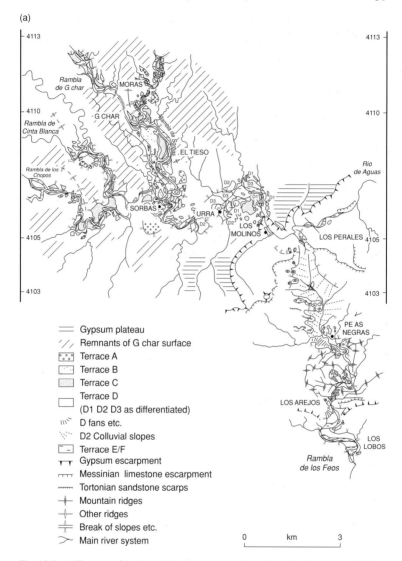

Gypsum plateau
Remnants of G char surface
Terrace A
Terrace B
Terrace C
Terrace D
(D1 D2 D3 as differentiated)
D fans etc.
D2 Colluvial slopes
Terrace E/F
Gypsum escarpment
Messinian limestone escarpment
Tortonian sandstone scarps
Mountain ridges
Other ridges
Break of slopes etc.
Main river system

Fig. 6.2.2. (a) Terraces of the Aguas/Feos river systems (modified after Harvey *et al.* 1995).

(b)

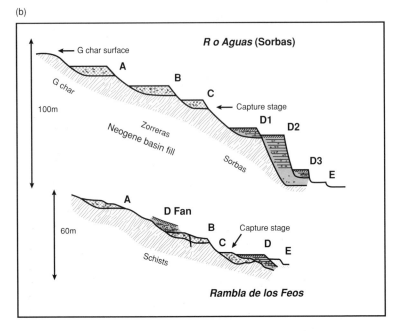

Fig. 6.2.2. (b) Schematic section from the Rambla de los Feos (modified after Harvey, 1987).

excursion starts at the same site as Stop 3 on the introductory excursion because this site forms a useful overview of the terrace sequence in relation to the river capture. However, if you wish to omit this stop, go directly to Stop 2 (if using cars or minibuses, or Stop 3 if travelling by coach). For a shorter excursion we recommend Stops 4b, 5 and 6. For those who prefer a hike, an alternative to Stops 2–4 is suggested, involving a 5-km walk, beginning and ending in Sorbas, and involving Stops 4a–c.

Most stops are fairly short (unless the hiking alternative is chosen at Stop 4), but allow at least one and a half hours for Stop 5. The excursion starts near Los Molinos, SE of Sorbas (Fig. 6.2.1).

Directions to Stop 1

From Sorbas proceed east on the old N340 for *c.* 800 m, then turn right onto the Al-104 (signed for Los Molinos). After 3.5 km, opposite the junction at 816 056 for the gypsum quarry, turn left onto a gravel area where there is plenty of space to park. Walk up the hill to the south, to the crest of the gypsum escarpment to Stop 1 (818 055).

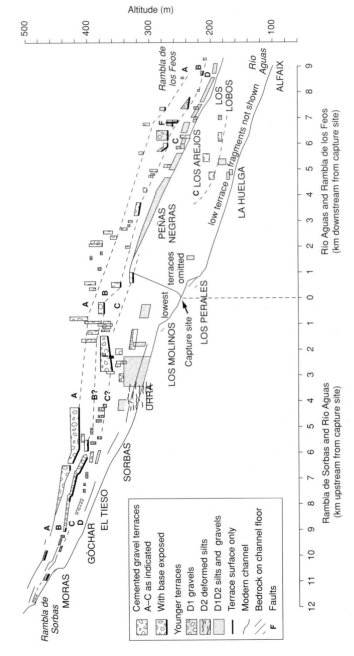

Fig. 6.2.3. Long profiles of the terraces of the Aguas/Feos river systems. (Modified after Harvey *et al.* 1995.)

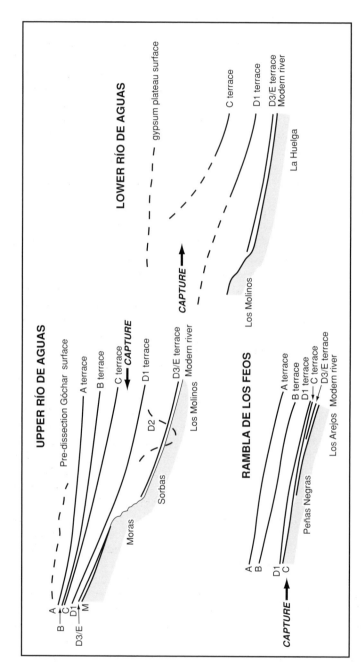

Fig. 6.2.4. Profile relationships in the three zones of the Aguas/Feos river systems. (Modified after Harvey *et al.* 1995.)

Stop 1: Los Molinos viewpoint (819 057; Sorbas 1 : 25 000)

This is the same locality as Stop 3 on the introductory excursion (Excursion 2.2), but also forms a useful overview of the Quaternary dissection of the basin and a useful starting point for this excursion. The site provides a spectacular panoramic view north into the centre of the basin and south across the Los Molinos area and the site of the Aguas/Feos river capture. For a full description of the geology see the notes for the introductory excursion (Excursion 2.2, Stop 3). These notes relate only to the Quaternary dissection sequence.

At the top of the hill, resting unconformably on the gypsum, are two conglomerate units. The older disturbed unit is interpreted to represent the preserved base of the Góchar Formation (see Excursion 2.2, Stop 3). The second conglomerate unit is *in situ* within a channel, cut into the first set and the underlying gypsum. We interpret these to be an eroded remnant of terrace A gravel, representing the first dissectional phase of the basin. Note the abundance of Filábride clasts in the deposit, and the general north–south palaeocurrents suggested by the imbrication.

Below us, to the NW, and forming a bench on the ridge immediately below us to the east, are degraded remnants of terrace B. Below that and forming an extensive surface to the north, marked by a small olive grove, are remnants of terrace C. The same terrace forms benches on the opposite side of the river to the NE, and a lower bench on the ridge below us to the east. The valley is incised into terrace C (beyond the olive grove), with terraces D and E within the incision. To the east the river cuts a deep canyon through the gypsum escarpment.

The view south from Stop 1 takes in the deeply dissected southern part of the basin, where the incising river system has cut through the older part of the basin-fill. From where you stand the gypsum escarpment can be traced towards the SW and across the valley to the east. Below the gypsum, in a belt stretching to the SW, is the lower ground of the Abad Member marl outcrop. This terrain is deeply dissected by gullies and badlands. Beyond is the dip slope of the escarpment formed by the Azagador Member calcarenite, to the east of the motorway and to the south of Los Molinos at Cerro Molatas (834 045). The Tortonian marls, below the basal Azagador Member unconformity, are exposed in the hillside below Cerro Molatas. In the distance to the SE is the Sierra Cabrera.

Immediately in front of us is Los Molinos (Fig. 6.2.5) and the site of the major river capture of the proto Aguas/Feos by the modern Aguas (Fig. 6.2.6). The course of the proto Aguas/Feos can be traced through this area. Remnants of terrace A occur at this site, on the flanks of Cerro Molatas and capping isolated hills between there and Peñas Negras *c.* 8 km to the south. A remnant of terrace B can be seen across the valley from Los

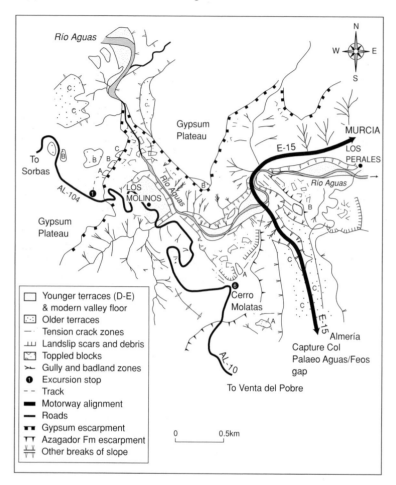

Fig. 6.2.5. Los Molinos area: Stops 1 and 6. Geomorphic map of the Río Aguas canyon and capture site, Los Molinos.

Molinos. Another caps the hill beyond the motorway at 836 054. Other remnants occur towards Peñas Negras.

Terrace C can be traced flooring the gap east of Cerro Molatas, now followed by the motorway (see Stop 6 for a description of terrace C). Terrace C is the youngest fluvial terrace that can be traced southwards through this gap into the Feos drainage. D terraces are below the floor of the gap and follow the modern course of the Aguas to the east. Capture took place between terraces C and D, which we suspect relates to the

Fig. 6.2.6. View south from Los Molinos (Stop 1) towards the abandoned col, previously utilized by the proto Aguas/Feos river prior to capture by the modern lower Aguas. Heavy arrow indicates path of the former south-flowing Aguas/Feos. The capture col is floored by terrace C gravels. Small arrow indicates post-capture course of the Río Aguas, flowing towards the east. l, landslipped terrain on the dip slope of the Azagador Member limestones, undercut by the incised Río Aguas; g, gullied terrain within the Abad Member marl. (Photo A.M. Harvey.)

Late Quaternary, occurring perhaps *c.* 100 000 BP (Harvey & Wells, 1987; Harvey *et al.* 1995), a date also supported by U/Th chronology of the calcretes developed on the terrace (Kelly *et al.* 2000). The total depth of dissection here is *c.* 160 m, with *c.* 60 m prior to the capture (Stage C) and *c.* 100 m after.

The deep incision following the capture has led to accelerated slope instability. On the outcrop of the Abad Member marl, base-level related gullying and badland development have occurred, here particularly evident in the tributary drainage that joins the Río Aguas from the west at Los Molinos. Where erosion of the Abad Member marl has cut back into the gypsum escarpment, there are huge topple failures in the gypsum, evident across the canyon to the east. Where the incision has cut through the unconformity at the base of the Azagador Member into the underlying Tortonian marls the Azagador Member limestone is subject to down-dip planar slides and associated tensional failures (Hart, 1999). These are evident on the sides of Cerro Molatas to the south of Los Molinos (Figs 6.2.5 and 6.2.6).

Transfer to Stop 2

Return to the vehicles and back to the road. Turn right and retrace the route to Sorbas. Coaches should omit Stop 2 and proceed directly to Stop 3 at the west end of Sorbas. Cars and minibuses should proceed beyond Sorbas on the old N340 for *c.* 1 km, and take the right-hand turn towards Uleila del Campo. Continue along this road to KM 5, then turn right onto a minor road towards Moras. Follow this road, bearing left when confronted with a road junction, to Moras village. Entering the village via a hairpin bend, park by the bus shelter (Stop 2, 774 108).

The route from Los Molinos takes you up the geological succession from the gypsum at Stop 1 and along the road to Urra, then through the Sorbas Member from Urra west of Sorbas, through the Zorreras Member and finally onto the conglomerates of the Góchar Formation on the Uleila road. Topographically, it follows the Aguas valley upstream to Sorbas, then through an abandoned incised meander loop past Sorbas (equivalent to terrace D). Beyond Sorbas the route at first follows then crosses the Rambla de los Chopos (Guapos on the 1 : 50 000 maps), the main western headstream of the Aguas, then climbs up onto eroded remnants of the end-Góchar Formation depositional surface. This surface represents the final phase of basin-filling. To cross the Rambla de Góchar, and again at Moras, the route descends into valleys of the northern headstreams of the Río Aguas, cut below the basin-fill surface.

Stop 2 of this excursion also forms Stop 2 in Excursion 5.3. A more detailed description of the Neogene geology is given there.

Stop 2: Moras, Quaternary dissection through the upper Messinian–Pliocene sequence of the northern part of the basin (774 110; Sorbas 1 : 25 000)

Walk to the edge of the gorge for an overview of the incised meanders of the Rambla de Moras (Fig. 6.2.7), cut into the Neogene basin-fill sedimentary sequence. The Neogene geology comprises a succession from the Cantera Member reefs to the Góchar Formation conglomerates (described more fully in Excursion 5.3, Stop 2).

Several stages of Quaternary incision are preserved here in the river terrace sequence. The summits represent dissected remnants of the end-Góchar depositional surface, the final stage of basin-filling. To the SE, across the valley behind Moras village, several metres of terrace gravels rest unconformably on the Góchar Formation in a sinuous former course of the Rambla de Góchar, cut into the upper surface. These gravels culminate in a mature red soil with hues of 2.5–5YR coloration, and represent an early stage (terrace A) in the Quaternary dissection of the end-Góchar

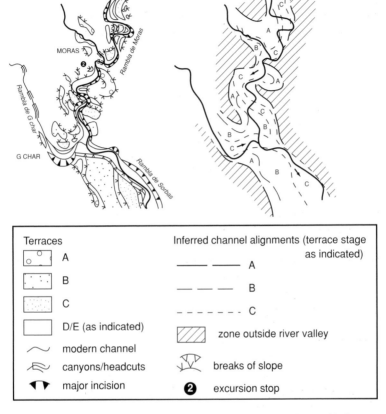

Fig. 6.2.7. The Moras area: Stop 2. (a) Map of the terrace sequence on the Rambla de Moras. (b) Interpretation of the sequence of valley development during incison on the Rambla de Moras. (Modified after Harvey *et al.* 1995.)

Formation depositional surface. To the NE, the much younger terrace C, with a less mature soil profile reaching only 7.5YR hues, sits within the valley of the Rambla de Moras. In this area terrace B appears only as erosional remnants, but further downvalley both terraces B and C are visible as depositional terraces.

The total depth of dissection upstream of the dry waterfall is < 50 m below the top-Góchar Formation surface, increasing to > 60 m downstream from Moras village. Downstream of Moras there has been a complex sequence of meandering during dissection (Fig. 6.2.7). Upstream of Moras the incised meanders have become progressively larger since dissection of terrace C.

If you wish to examine terrace C more closely and to see something of the underlying geology, a short walk to the north is worthwhile. Take the goat path towards the north around the apex of the incised meander bend. As the ground flattens out, cut across the flat, north for 100 m until you meet a small NW–SE tributary of the Rambla de Moras.

Unconformably over a fault (at 775 111), exposed on the far side of the tributary rambla but with a junction poorly exposed, are Quaternary terrace C gravels, which culminate in a red soil. The soil, though somewhat degraded at the surface, preserves a red B horizon with 7.5YR hue, above a Stage II pedogenic carbonate horizon. Both properties indicate terrace C age.

Transfer to Stop 3

Return to the vehicles and retrace your route out of Moras. At the first intersection, turn left, rather than continuing towards the Uleila road. Follow this minor road south through Góchar village, across the rambla and onto a ramp out of the valley.

At this point (772 095) it is worth stopping to view the geomorphology. Note the steeply dipping, folded Góchar Formation conglomerates exposed in the valley sides. Across the valley, to the north, these are unconformably overlain by horizontal Quaternary (Stage B) gravels.

Climbing out of the valley to the cross-roads, note the extensive terrace surface (terrace B) forming a valley-in-valley form, stretching away to the SE towards El Tieso and Sorbas. At the cross-roads, turn right, then follow that road for *c.* 1.6 km to rejoin the Sorbas–Uleila road. There, turn left and return towards Sorbas.

About 200 m after the left turn onto the old N340, note the section exposed behind the 'Taller Mecánico' station on your left. This site forms Stop 5 on Excursion 2.2 (see there for more detail). The Sorbas Member at this site is capped unconformably by terrace C gravels, which include large blocks of Góchar Formation conglomerates. Folded Góchar conglomerates are also exposed to the west of this section, resting unconformably on the Sorbas Member. Terrace C forms an extensive surface in this lower part of the Chopos valley, extending upstream for *c.* 1 km to the west.

At the west end of Sorbas, just beyond the petrol station on the right, take a left turn as though you are going into Sorbas, but then turn left and park behind the crash barriers. Walk 100 m north to the overlook on the edge of the canyon, beside the Red Cross station. This is Stop 3.

Stop 3: West of Sorbas, incision and meander cut-off (776 065; Sorbas 1 : 25 000)

From this view point the confluence of the Rambla de los Chopos with the

Rambla de Sorbas can be seen, and the sequence that led to the Sorbas meander cut-off can be identified (Fig. 6.2.8). Immediately across the canyon is a remnant of terrace B, capped by a degraded red soil. Sorbas is also on the same terrace.

Small remnants of terrace C occur in both the main valley east of the town and the former valley of the Chopos, through the now abandoned incised meander, SW of Sorbas. Cut-off took place during terrace D times (probably during the Urra D2 phase, see below), by two meanders, one on the Chopos and another on the Rambla de Sorbas, cutting back into each other. Since cut-off, the meander to the SW of Sorbas has been abandoned. The former neck can be identified in the cliffs to the north of Sorbas.

With further incision and headwards erosion, the Chopos, with an over-steepened lower course, cut the inner canyon in front of you to your left.

Transfer to Stop 4

There are two alternative locations for Stop 4 (Fig. 6.2.1): Stop 4a for those using a coach; and Stop 4b for those in cars or minibuses.

Return to the vehicles and rejoin the old N340, heading east.

For Stop 4a park outside Bar El Chacho, on the right of the old N340, at the east end of Sorbas (783 060), immediately over the Río Aguas bridge. If taking the hiking alternative, also park here. Walk east for about 400 m along the main road to the junction with the Lubrín road (786 060). Turn left up the hill and, where convenient, leave the road and climb to the top of the hill, onto terrace A at Stop 4a (787 063).

For Stop 4b continue beyond Bar El Chacho 400 m to the Lubrín turn. Turn left up the road towards Lubrín. Continue for *c.* 1.1 km to a tight right-hand hairpin bend (Stop 4b, 787 072). It is best to proceed beyond the bend, turn where visibility permits, and park on the west side of the road a little to the north of the hairpin bend, where there is room to park a couple of cars, with care!

Stop 4a: Terrace A, above Sorbas (787 072; Sorbas 1 : 25 000)

Here, gravels of terrace A rest unconformably on the Zorreras Member. There is a rather degraded soil visible on the terrace surface in places. From this view point the whole of the terrace sequence of the Sorbas area can be seen (Fig. 6.2.8). You are standing on a remnant of terrace A, another can be seen *c.* 2 km to the NW, marked by almond trees. Sorbas is built on terrace B, and other remnants can be seen to the NW across the lower Chopos canyon. A degraded remnant of terrace C caps the hill below you to the north; another occurs to the south near the industrial buildings. You can make out several surfaces which relate to the various Stage D terraces,

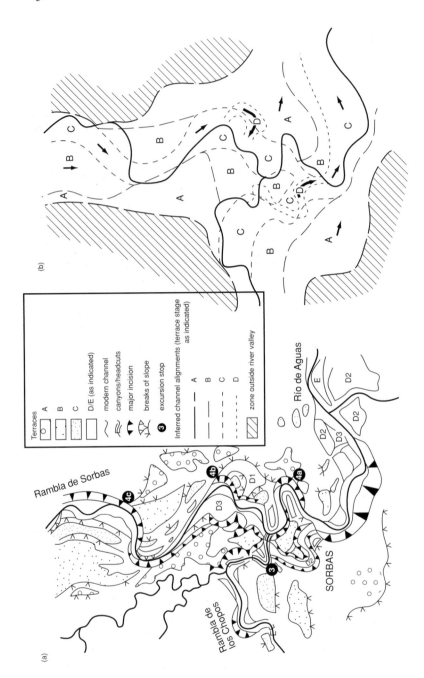

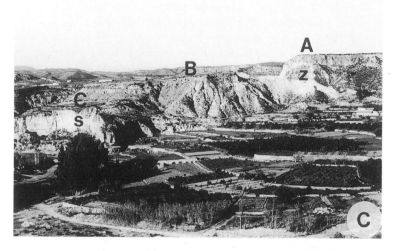

Fig. 6.2.8. The Sorbas area: Stops 3, 4a–c (for detailed locations see Fig 6.2.1 inset).
(a) Map of the terrace sequence on the Río Aguas feeder streams at Sorbas. (Modified after
Harvey *et al.* 1995.) (b) Interpretation of the sequence of valley development during incision
of the Río Aguas at Sorbas. Note cut-off valley meander loops. (Modified after Harvey
et al. 1995.) (c) View west across the Rambla de Sorbas from Zorreras Hill (Stop 4b).
Bedrock dips gently north. The sequence includes upper Sorbas Member sandstones (s),
overlain by the red silts and sands of the Zorreras Member (z), then the conglomerates of
the Góchar Formation (to the right, but not shown on photo). Terrace sequence (A, B, C)
indicated. Terraces A and B here comprise near-horizontal gravels (capped by red soils)
unconformably overlying the Neogene rocks. The fragment of terrace C indicated here is an
erosional remnant only. (Photo A.M. Harvey.)

including an abandoned cut-off channel of the Rambla de Sorbas to the
north, the abandoned course of the Chopos to the south of Sorbas, and
the spur tops within the meander bend at the east end of Sorbas and to the
north.

If you wish to continue on foot to Stop 4b, walk NE and climb down to
the Lubrín road again. Walk north *c.* 1 km along the road to a tight right-
hand hairpin bend. This is Stop 4b.

Stop 4b: Alto de Zorreras, overlooking Sorbas (787 072; Sorbas 1 : 25 000)

At this site there is a panaromic view (Fig. 6.2.8c) over the Sorbas valley,
exposing the Upper Messinian–Pliocene rocks from the Sorbas Member,
through the Zorreras Member to the Góchar Formation. The complete
history of Quaternary dissection can be followed through the river terrace

sequence (Fig. 6.2.8; Harvey *et al.* 1995). Here, the total depth of Quaternary dissection is *c.* 80 m.

The grey–white rocks outcropping on the far side of the valley below Sorbas are the Sorbas Member. These are overlain by the Zorreras Member, the reddish silts and sands outcropping on this side of the valley and visible on the opposite side, which include a white marker bed and the yellow sands of a marine unit which cap the Zorreras Member sequence. This is exposed just below you to the right (north) and can be traced across the valley to Zoca, where it can be seen capping a pillar. Above this are the Góchar Formation conglomerates. The whole sequence dips gently to the north.

Little remains here of the top-Góchar Formation surface into which the Quaternary valley is cut, but it forms the ridge above the valley to the north of Zoca. The oldest terrace gravels of terrace A rest unconformably on the Zorreras Member and Góchar Formation sediments, capping Zoca Hill, the hill behind us and the hill to the south (the location of Stop 4a). Terrace B caps the spurs between Zoca and Sorbas and underlies Sorbas itself. Both terraces A and B here are characterized by strong red soils. A degraded remnant of terrace C gravels caps the hill below you to the left. There is an abandoned meander of the rambla of an early D-age terrace (D1) immediately below you to the left, but with no good exposure of the sediments, and infilled by younger colluvial material. The other abandoned meander loop behind Sorbas, that formerly of the Chopos, is of a younger D-age. It is a little more difficult to distinguish other younger D and E terrace remnants because of agricultural modification, but they commonly occupy bend sites, such as that below Zoca.

Transfer to Stop 5, via Stops 4c and 3: alternative route involving a 5-km hike back to Sorbas

If you are proceeding on foot and following the 5-km hike back to Sorbas, continue north along the Lubrín road for just over 1 km until the road begins to rise above the terrace surface (terrace B) at *c.* 786 081. *En route* you will pass an excellent roadside section exposing the terrace B soil, with an argillic B horizon of 2.5YR hues, over a Stage III pedogenic calcrete. The whole section shows evidence of secondary carbonate accumulation. From point 786 081 head west across the terrace surface to a point at the edge of the incised valley of the Rambla de Sorbas (Stop 4c, 782 081). If driving here from Stop 4b, stay on the Lubrín road until near the top of a rise at 785 084, where there is rather a hazardous turn onto a cliff-top track.

Optional Stop 4c: Terrace B above El Tieso (782 081; Sorbas 1 : 25 000)

From terrace B above El Tieso, there is an extensive view of the terraces of the Rambla de Sorbas. Terrace B is visible as an extensive surface on the far side of the valley, set below the Góchar Formation summits. Within the valley are remnants of terraces C, D and E.

At this site, several metres of terrace B gravels rest unconformably on flexured and faulted Góchar Formation conglomerates (examined in the cliff below you in Excursion 5.4, Stop 2) though it is difficult to locate the unconformity itself. At the terrace surface is a well-developed soil, including a red Bt horizon (colour 2.5YR 3/6) over a Stage III–IV carbonate horizon, characteristic of terrace B soils.

If travelling by car or minibus, return to the vehicle. If taking the hike, descend to the valley bottom by the narrow goat path down the near-vertical cliffs! Follow the rambla floor downstream, for *c.* 3 km back to Sorbas. *En route* note the bedrock exposures, going down the sequence from the Góchar Formation, through the Zorreras Member into the Sorbas Member. Note the occasional exposure of river terrace gravels, unconformably on the Neogene bedrock. Also note the landslipped terrain in the area upstream of Zoca.

Back near Sorbas, note the high cliffs breached by the cut-off that isolated Sorbas during terrace D times (780 065). From here, climb the track on the right of the channel, that rises obliquely up the slope to the east end of Sorbas. This brings you out in a small square with a church in the pottery quarter of the town (the main pottery 'shops' are down a street on the left, by the church). Fresh drinking water is available at the tap below the road which climbs steeply up to town on the right. Carry straight on across the square (due south) until you come to a junction where you turn left to parallel and rejoin the old road over the Río Aguas. Cross the main N340 to rejoin your vehicle.

Transfer to Stop 5

From Stop 4a return to your vehicle at Bar El Chacho and head east on the old N340 for *c.* 800 m to the junction with the Al-104 (signed for Los Molinos). From Stop 4b/c return south along the Lubrín road to the junction with the old N340. Turn left for *c.* 500 m to the junction with the Al-104. At that junction turn right and continue for another *c.* 800 m to the Urra Field Centre track on the left. Drive down to the field centre and park (804 062). Seek permission from the owner (Mrs Lindy Walsh) for access to her land.

Stop 5: Urra, Quaternary deformation (804 062; Sorbas 1 : 25 000)

This section of the excursion examines the terrace sequence developed around Urra (Fig. 6.2.9). The area reveals an unusual number of terrace

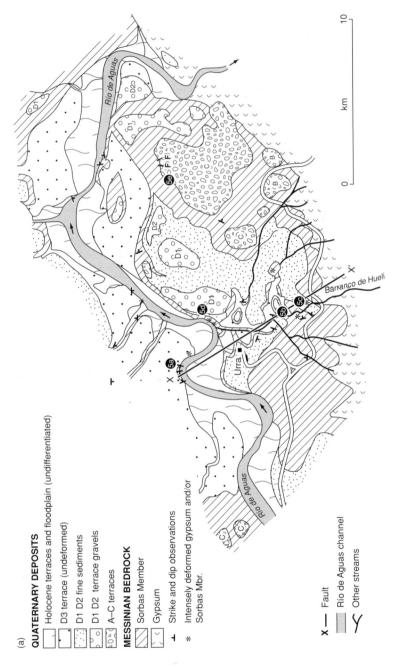

Fig. 6.2.9. The Urra area: Stop 5. (Modified after Mather *et al.* 1991.) (a) Map of the terraces of the Urra area. X–X' indicates line of Fig. 6.2.9d.

(a)

QUATERNARY DEPOSITS

Holocene terraces and floodplain (undifferentiated)

D3 terrace (undeformed)

D1 D2 fine sediments

D1 D2 terrace gravels

A–C terraces

MESSINIAN BEDROCK

Sorbas Member

Gypsum

Strike and dip observations

Intensely deformed gypsum and/or Sorbas Mbr.

x — Fault

Rio de Aguas channel

Other streams

Rio de Aguas

Barranco de Hueli

Urra

km

0   10

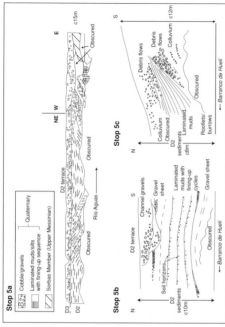

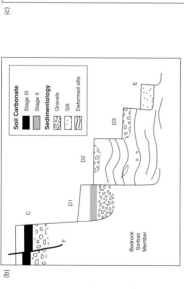

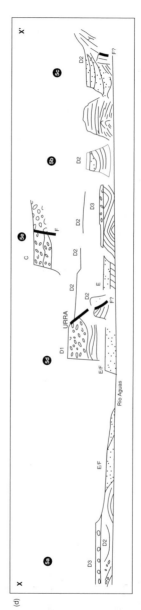

Fig. 6.2.9. (b) Summary sequence of terraces at Urra. (c) Detail from Stops 5a–c. (d) Relationship between Stops 5a–e at Urra. Position indicated by X–X′ in Fig. 6.2.9a.

levels relating to the terrace D phase, when compared with other locations. The sequence at Urra is complicated by deformation. This deformation may have resulted from: (i) diapiric activity of the underlying gypsum, following erosional off-loading which was caused by the rapid incision in early terrace D times (Mather *et al.* 1991); (ii) tectonically induced deformation along the Marchalico–Infierno lineament of Excursion 3.5b, Stop 2; or (iii) karstic collapse of the underlying gypsum. Both the main river (Nevado–Filábride-rich gravels) and the Hueli tributary (Messinian carbonate-rich gravels) were ponded at the time, resulting in impeded drainage, then deeply dissected to near modern river level.

Walk first to the promontory east of the field centre. Look directly east across the Rambla de Hueli, to the surface above the Río Aguas. Several metres of terrace D1 gravels are evident overlying Sorbas Member marls (Fig. 6.2.9) capped by a moderately developed red soil. This terrace is cut into terrace C (exposed at Stop 5e, Fig. 6.2.9) but pre-dates the most recent deformation in this area. The gravels dominantly comprise Nevado–Filábride clasts derived from the main Río Aguas drainage.

Looking south there is an extensive area of red/brown coloured silts, which comprise the D2 (deformed) sediments. Their base is near or below modern stream level; their upper surface is a little below the level of the D1 terrace (Fig. 6.2.9). Five metres above the Barranco de Hueli is an inset surface, formed by several metres of undeformed gravels, and which correlates with the extensive terrace surface seen in the Río Aguas to the north. This is the (undeformed) terrace D3. Immediately below the former river cliff in front of you is a younger cut of the former Hueli drainage, one of a series of abandoned meander loops on the Barranco de Hueli. The present straight course of the Barranco de Hueli through this area may have been artificially created by irrigation modification (?Moorish) and subsequent abandonment (the modern barranco follows a ?Moorish irrigation canal at a similar height to the terraces).

Proceed down the spur of the hill and follow the path which drops down into the barranco. Walk downstream to the main Río de Aguas. Head upstream along the river bed for *c.* 400 m to the meander bend. The cliff section in front of you exposes 4 m of deformed terrace D2 deposits (Fig. 6.2.9, Stop 5a). From SW to NE they comprise silts which thicken into a syn-sedimentary low. At the margins of the section, conglomerates are tilted and locally liquefied. To the east the deposits overlay deformed Sorbas Member. The D2 sediments are truncated by a gravel surface relating to the youngest, post-deformation terrace D3 of the main river.

Head back to the Barranco de Hueli. Note the sections in the modern floodplain showing a stacked sequence of vertical accretionary deposits.

Entering the Barranco de Hueli from the junction with the Río Aguas, note the deformed D2 sediments overlying deformed Sorbas Member marls in the stream bed. Further upstream on the left-hand side of the rambla, liquefied D2 conglomerates can be seen overlying deformed Sorbas Member marls. The conglomerates relate to main river (Nevado–Filábride) gravels. Another 100 m further upstream a section is cut through the D2 sediments which here comprise a lower sequence of debris-flows composed of gypsum clasts and fining upwards flood cycles in silts. The deposits form part of the northern limb of a small syncline.

A large cliff section is encountered 200 m further upstream on the left, which shows a thick section through the D2 sediments (Fig. 6.2.9, Stop 5b). They comprise silts and fine sands, thickening into the centre of the section. Some darker pedogenic horizons are evident. The section is capped by D2 gravel composed of marl clasts which characterize drainage from the Barranco de Hueli.

Further upstream, the barranco narrows and the southern limb of the syncline can be observed, comprising deformed Sorbas Member marls containing small overthrusts and reverse faults on the right of the barranco channel.

As the road bridge comes into view, a cliff section on the left exposes D2 sediments (Fig. 6.2.9, Stop 5c). The upper fine sediments show syn-sedimentary thickening towards the barranco, into another synform. They overlay and, at the southern end of the section, are interbedded with debris-flow deposits of marl. Underneath the road bridge, gypsum crops out which locally shows crystal reorientation along its margins (?faulted).

Return down the barranco to the vehicles parked at the field centre.

Transfer to Stop 6

Return up the drive to the Al-104. Turn left and continue for *c.* 8 km, past the turn-off for Stop 1, down the hairpin bends through Los Molinos, and up the hairpin bends on the other side. About 400 m beyond the last hairpin, stop at Cerro Molatas (832 047) on the left-hand verge 200 m before a tight right-hand bend.

From the turn-off for Stop 1 the route runs down the face of the gypsum escarpment onto the Abad Member marl, and beyond Los Molinos climbs up the dip slope of the Azagador Member escarpment.

Stop 6: Cerro Molatas, above the capture site (832 047; Polopos 1 : 50 000)

This is the same locality as Excursion 2.2, Stop 2. It affords an excellent view over the Aguas canyon (Fig. 4.1.3) and the site of the Aguas/Feos capture. Back to the north, looking up the Aguas valley, the view takes in the gypsum escarpment, above the deeply dissected Abad Member marl terrain. Beyond Los Molinos you can see the escarpment capped by gravels of terraces A, B and C (Stop 1 was on the top of the ridge, adjacent to the terrace A gravels, see Fig. 6.2.5). Note the topple failures that characterize the gypsum escarpment, including the huge block of gypsum in the canyon (for a description of the geology see Excursion 2.2, Stop 2).

Note the slope instability on the Azagador Member at this locality, whereby down-dip planar slides characterize the whole of the southern valley side. Just to the east are slipped masses of Azagador Member limestone. Just below you are large tension cracks. Others can be seen in the wall of rock above the river to the east (Hart, 1999).

Terrace fragments allow the dissection sequence to be traced through this area (Fig. 6.2.5). Little remains of terrace A; the next outcrop is *c.* 1 km to the south in the Campico area of the upper Feos valley. There is a patch of terrace B gravel, capped by a red soil, resting unconformably on the gypsum, across the Aguas valley from where you stand. Another, also marked by a strong red soil, forms the ridge top on the other side of the abandoned valley, below you to the east. Terrace C floors this abandoned valley. Prior to the construction of the motorway there were excellent sections revealing *c.* 10 m of coarse imbricated fluvial gravels, set into the Tortonian marls. Palaeocurrent observations confirm north–south flow (Harvey & Wells, 1987). Fragments of terrace D are below the level of the floor of the abandoned valley and follow the modern Aguas valley downstream to the east.

From this viewpoint the course of the ancestral Aguas/Feos master stream can be traced southwards through the now abandoned valley. During terrace C times, that stream was captured by the lower Aguas, working headwards from the east, exploiting the outcrop of the weak Abad Member marl.

Along the lower Aguas, downstream from here, the approximate extent of incision post-terrace C times can be assessed from the lower limit of red soils on the hillslopes of the south side of the valley. The only terrace fragments in this section of the valley are of D-age or younger. There are no traces of terraces equivalent to C-age or older for a considerable distance downvalley. Near La Huelga there are terraces capped by soils which suggest C-age. The clast content of these terrace gravels is local, from the Sierra Cabrera, with no evidence of transport of Nevado–Filábride clasts from the upper parts of the Sorbas Basin (Harvey *et al.* 1995).

Transfer to Stop 7a

Return to the vehicles and, with care, rejoin the Al-104 road heading south, past the motorway intersection. Beyond the motorway intersection (842 029) it is possible to drive north on the road signposted for Los Perales for 2.5 km to where the terrace C gravels in the floor of the capture col can be seen, though exposure has been altered by motorway construction. The main excursion continues along the Al-104 towards Peñas Negras.

En route from Cerro Molatas to Peñas Negras the road descends the Azagador Member escarpment onto the Tortonian rocks, with marls forming the lower ground, and sandstones and conglomerates the isolated hills.

This area forms the Campico valley, the beheaded valley of the ancestral Aguas/Feos system. Some of the isolated hills in this area are capped by Aguas/Feos terrace A and B gravels, carrying Nevado–Filábride clasts southwards. The lower slopes are mantled by thick (D-age) colluvium, presumably related to Late Pleistocene hillslope erosion, feeding sediment into the main valley, from which there was no means of removal. In places there are gravels (?terrace C) buried beneath the colluvium.

At Peñas Negras there is an excellent view south and west of the southern margin of the Sorbas Basin (for a full description of the geology see Excursion 2.2, Stop 1).

At Peñas Negras the diminutive Campico stream which you have been following is joined by the Rambla de Mizala, flowing from the west. This stream, which now forms the main headwater of the Feos, developed as a subsequent stream, exploiting the outcrop of the Tortonian rocks, and capturing earlier north-flowing basinal drainage. This process is still going on at Cantona at the head of the valley (788 016), where steep, east-draining gullies are about to capture a less steep, north-flowing drainage. D-age and younger terraces are well developed in this area (see Excursion 2.3).

From Peñas Negras, continue south on the Al-104 *c.* 2 km to Stop 7a in the Sierra Alhamilla, midway between KM 14 and KM 15 (856 095). There is ample parking space on the right-hand shoulder. *En route* note the multiple mountain-front faults, south of Peñas Negras. The faults have not disturbed the D-age terrace sediments. South of the faults, you are now on upper Nevado–Filábride black, low-grade schists.

Stop 7a: Transverse course of the Palaeo Feos across the Sierra de Alhamilla (856 994; Polopos 1 : 25 000)

At this site you can trace the transverse course of the Palaeo Feos river through the Sierra de Alhamilla, through the remnants of the precapture terraces (A, B and C) (Fig. 6.2.10). These terrace sediments include clasts derived from the Sierra de los Filabres (coarse, grey/spotted green hornblende schists, of higher metamorphic grade than the local, low-grade blackschists). There has been no post-Stage C incision in this valley, in contrast with the Aguas valley (Fig. 6.2.4). Prior to the construction of the

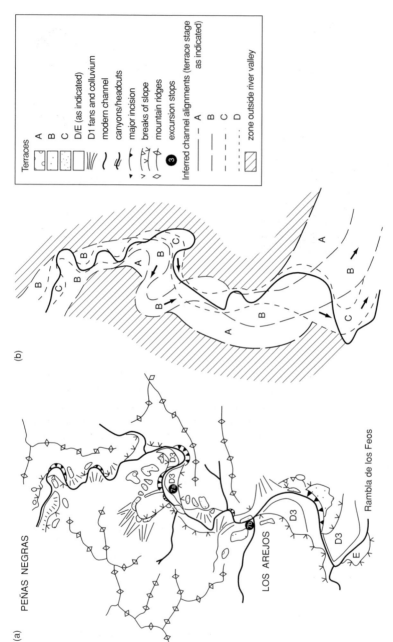

Fig. 6.2.10. Transmontane Feos valley, Stops 7a,b. (a) Map of the terrace sequence on the Rambla de los Feos. (Modified after Harvey *et al.* 1995.) (b) Interpretation of the sequence of valley development during incision on the Rambla de los Feos. Note cut-off meander of terrace B. (Modified after Harvey *et al.* 1995.)

(c)

Fig. 6.2.10. (c) The Feos valley, abandoned meander loop (view north from Stop 7a). Cemented gravels of terrace B (B) floor the abandoned loop, overlain by younger fan deposits (f). Arrow indicates palaeoflow direction.

motorway, Stage C gravels were visible locally on the floor of the valley. Stage D deposits, which rest on Stage C gravels, include only local clasts, and are restricted to low alluvial terraces in the valley bottom, colluvial hillslope deposits, and alluvial-fan deposits at tributary junctions. The terrace D deposits are indicative of the reduced power of the post-capture Rambla de los Feos.

During terrace B stage, the river made a loop to the north of this site, which was later abandoned by cut-off. If time permits, cross the motorway through a culvert, hike up to 853 997 to where terrace B gravels can be traced through the abandoned meander loop, and in the loop are overlain by small, younger (D-age, Late Quaternary) alluvial fans (Fig. 6.2.10). It is also worth making one more stop 1.2 km south of this site at Cortijada los Arojos (Stop 7b), where the Feos crosses the Alhamilla Southern Boundary Fault, and flows into the Almería Basin.

Transfer to optional Stop 7b, if time permits

Return to the vehicles and drive south another 1.2 km to Stop 7b at Cortijada los Arejos (853 984). Park on the shoulder just before KM 16.

Optional Stop 7b: Southern Boundary Fault of the Sierra de Alhamilla at Cortijada los Arejos (853 984; Polopos 1 : 25 000)

The Southern Boundary Fault of the Sierra Alhamilla crops out in a road-side section at 852 987. On the north side of the fault, Nevado–Filábride schists are overlain by Azagador Member basal conglomerate, followed by subaqueous debris-flows. This sequence is capped by cemented terrace B conglomerates. On the south side of the fault neither the schists nor the base of the Azagador Member are seen. There is a thicker sequence of the debris-flows, including a richly fossiliferous *Ostrea* bed. This sequence is capped by cemented conglomerates of terrace B, which include Nevado–Filábride clasts. The fault has a total throw in excess of 12 m, but deforms the terrace B gravels by *c.* 3.5 m (Harvey & Wells, 1987).

To the south there is a view of the sequence at the margin of the Almería Basin. The Azagador Member can be traced across the valley, resting on and faulted against the schists. The overlying Abad Member marl forms a narrow belt of low dissected terrain, followed to the south by the out-crop of the gypsum. The Quaternary terrace gravels can be traced south, terrace A capping a hill and terrace B forming a valley-fill at Los Lomillas (856 978).

6.3 Excursion: Geomorphology of the Quaternary alluvial fans and related features of the Tabernas Basin (one day)

ADRIAN M. HARVEY

Introduction

The Tabernas area (Fig. 6.3.1) includes some of the most spectacular dry-region landscapes in southern Europe, with a dramatic contrast between a Quaternary aggradational landscape of coalescent alluvial fans to the east and a deeply dissected badland landscape to the west (Harvey, 1987a). The two areas are linked by a system of Quaternary river terrace and lake/palustrine sediments. Consideration of all three zones is important for the understanding of the Quaternary geodynamics of the basin.

The Quaternary landscape of the eastern part of the basin is dominated by coalescent alluvial fans. The fan sediments rest unconformably on the Serravallian and Tortonian basin-fill (see Excursion 3.4). Messinian and Pliocene sediments were stripped out by major erosional phases which followed basin uplift, probably during the Plio-Pleistocene. Late during the Quaternary, sedimentation resumed, and the erosional landscape became buried by alluvial fans. The fans have three main source areas

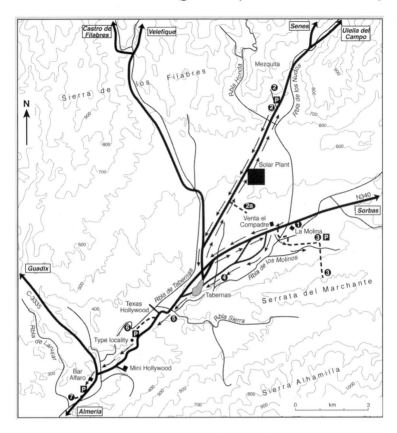

Fig. 6.3.1. Location map of the field-trip stops for Excursion 6.3.

(Fig. 6.3.2): the graphite schists of the Sierra de los Filabres to the north; the Serravallian mica schist-rich conglomerates of La Serrata del Marchante ridge in the centre; and the various basement rocks of the Sierra Alhamilla to the south.

Tectonically induced dissection has removed almost all traces of fans at the faulted mountain front at the foot of the Sierra Alhamilla. Only one fan remains more or less intact there, that of the Rambla de la Sierra, and that is trenched throughout its length (Harvey, 1984a, 1987a,b). Most fans in the other two groups are currently proximally trenched but distally aggrading. Tectonically induced dissection has not reached beyond the toe areas of the most westerly of these fans. The large Sierra de los Filabres fans issue from a passive mountain front, and are back-filled into the mountain catchments. The smaller, steeper, Serrata fans issue across a faulted mountain front.

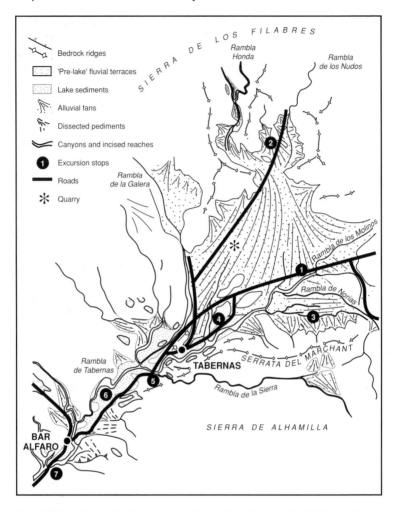

Fig. 6.3.2. Landforms of the Tabernas area: fans and 'lake' system. (Modified after Harvey *et al.* 1999a.)

Both sets of fans feed the Rambla de los Molinos and Rambla de Tabernas drainages. A series of river terraces following along these drainages record the Quaternary dissectional history of the central and lower part of the basin. During dissection, drainage was impeded and great thicknesses (*c.* 20 m) of palustrine or shallow lake sediments were deposited (Harvey & Mather, 1996; Harvey *et al.* 1999a) *c.* 150 ka (Delgado-Castilla *et al.* 1993). Two 'lakes' were formed. One, whose floor is at

380–400 m, lay within and upstream of the canyon on the lower Rambla de los Molinos, south and east of Tabernas. The other, whose floor is at 300–320 m, lay along the Rambla de Tabernas upstream from near the Bar Alfaro to the vicinity of Tabernas (Fig. 6.3.2). We believe that tectonic uplift to the south of Bar Alfaro was the primary cause of impeded drainage forming the lower 'lake' (there is evidence of Pleistocene reactivation of the Tortonian fault gouge zone described in Excursion 3.4, Stop 5), and that damming of the lower part of Tabernas canyon by tributary fans caused further impoundment of the upper 'lake' (Harvey & Mather, 1996).

This excursion deals with the alluvial fans, and the terrace and 'lake' system (Fig. 6.3.2). The first part of the excursion (Stops 1 and 2) deals with the Filabres fans, and the second part (Stop 3) with the Serrata fans, involving a several kilometre hike over gentle terrain. The third part of the excursion (Stops 4–7) deals with the terrace and 'lake' sequence.

Other excursions in the Tabernas area deal with the Neogene geology (Excursion 3.4), and the geomorphology of the badlands (Excursion 6.4). This excursion to the Tabernas fans and 'lakes' is designed as a full-day excursion, but it would be possible to cover it in a half-day by omitting Stops 3, 6 and 7 and limiting Stop 5 to a roadside view stop. This half-day could thus be combined with the first part of Excursion 6.4 to give a one-day overview of the geomorphology of the Tabernas Basin as a whole.

The excursion starts from the Bar La Molina (587 038), on the old N340 *c.* 5 km east of Tabernas (Fig. 6.3.1).

Please note that Stop 7 may have restricted access because of the construction of the new Granada–Almería autovia along the route of the N324.

Directions to Stop 1

If arriving from the east from Sorbas, follow the old N340 westwards towards Tabernas. Bar La Molina is on the left *c.* 2 km beyond the turning for Turrillas. If arriving from the west from Almería, follow the old N340 eastwards past Tabernas. Bar La Molina is on the right *c.* 5 km beyond Tabernas. Park in front of the bar. Walk *c.* 100 m east, to where you have an unimpeded view of the landscape, north and south of the road.

Stop 1: La Molina, overview of the upper part of the Tabernas Basin (587 038; Tabernas 1 : 50 000)

This location gives an overview of the coalescent alluvial fans of the upper part of the Tabernas Basin (Fig. 6.3.2). To the north of the road are the Sierra de los Filabres fans (Figs 6.3.2 and 6.3.3). The Planta Solar is built in mid-fan. These large, low-angle, coalescent, fluvially dominant fans (Harvey, 1984a,b, 1987a,b, 1990; Delgado-Castilla, 1993) are fed by large

Fig. 6.3.3. Coalescent and back-filled Filabres mountain-front fans, north margin of the Tabernas Basin. The channel of the Rambla Honda (arrowed, at bottom of photo), one of the main feeder streams, is seen here on the distal surfaces of the Honda fan (h). Mezquita fan (Stop 2) is seen in the middle distance (m). Note fan-head trench near fan apex, but untrenched distal fan surfaces.

and small catchments which drain the graphite schist terrain of the Sierra de los Filabres. These are coalescent fans (Fig. 6.3.3), unaffected by base-level change within the Tabernas Basin. To the south of the road are the much smaller, steeper, debris-flow-rich fans, fed by small catchments on the Serravallian conglomerates of the Serrata del Marchante. In the foreground is the braided channel of the Rambla de los Molinos, incised < 4 m into the toe areas of these fans. These fans are not yet affected by base-level change; however, incision of the Rambla de los Molinos is encroaching on their fan toes. This incision marks the upstream extent of the modern tectonically induced dissection of the Tabernas Basin. Depth of dissection increases markedly downstream. The sections reveal interdigitation of distal-fan fluvial sediments and shallow lake sediments. The tributary to the Rambla de los Molinos here is the Rambla de los Arcos, which drains the area behind the Serrata del Marchante and the northern slopes of the Sierra de Alhamilla. It forms a low-angle fluvial fan in the confluence zone with the Rambla de los Molinos. Hidden from view by the Serrata del Marchante are small fans at the foot of the Sierra de Alhamilla (Fig. 6.3.2), where tectonically induced incision of the Rambla de la Sierra has caused base-level induced dissection of the fan toes (Harvey, 1987a).

Transfer to Stop 2

Go west on the N340 for *c.* 4 km to the exit north of Tabernas (546 021). There, turn right and follow the minor road north, signed for Senes. Continue to follow the Senes signs, past the first road junction and past the Solar Experimental Plant entrance. Drive for *c.* 7 km along this road, to 582 088 (Stop 2), *c.* 2.5 km beyond the solar power plant to a place *c.* 200 m beyond the rock outcrop that marks the beginning of the narrowing into the valley of the Rambla de los Nudos. Park on the section of old road that forms a shoulder on the left.

En route from Tabernas the road runs up the surface of a large alluvial-fan complex, fed by two main valley systems: the Rambla Honda and the Rambla de los Nudos. This is the area that was affected by a flash flood in 1980 (Harvey, 1984c). Stop 2 is located < 1 km into the Nudos valley (Fig. 6.3.1), where a small tributary fan, the Mezquita fan (Figs 6.3.3 and 6.3.4), joins from the north.

Stop 2: Mezquita fan (Rambla de los Nudos), Filabres mountain front (578 095; Tabernas 1 : 50 000)

The Mezquita fan is a classic, proximally trenched, distally aggrading alluvial fan. It is not a simple mountain-front fan, but has back-filled into the mountain catchment, resulting in several fan segments fed from differing size source areas. Note how the individual fan surface and channel gradients reflect differing catchment sizes (Figs 6.1.2 and 6.3.4), differing processes and the sequence of fan evolution. The tributary fans to the west, fed only by small, steep, hillslope drainage areas, are steeper than the main fan. The overall profile of the main fan is concave, but note how the channel within the fan-head trench is less steep than the fan surface.

Walk up the fan over the distal surfaces to the intersection point (in the zone of man-made floodbanks, 578 095). Despite modern agricultural activity, traces of recent bar and swale sedimentation are visible on the distal fan surfaces. Walk on up the main channel. Note the tributary fan sections (575 097), and the longitudinal and transverse sections in the main fan near the fan apex (576 099).

The fan sediments, exposed in sections in the fan-head trench (2a in Fig. 6.3.4a) and in the tributary channels (2b in Fig. 6.3.4a), are dominated by fluvial rather than debris-flow deposits. This is despite the small drainage area and probably reflects the weathering characteristics of the local graphite schists (Harvey, 1984b, 1987a). The sedimentary sequences of the exposed sections suggest sedimentation in wide, shallow channels on the surfaces of an aggrading fan. Note that there are occasional

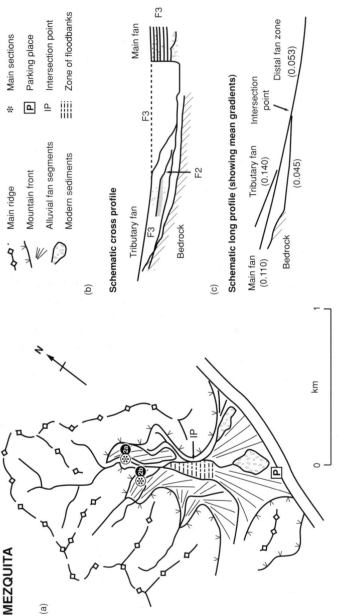

Fig. 6.3.4. Mezquita fan, Stop 2. (a) Geomorphic map; (b) schematic section; (c) schematic long profile showing mean gradients for each major segment. (From Harvey *et al.* 1999a.)

debris-flow sediments exposed locally at the base of the sequence. In places in the catchment, modern hillslope debris-flows can be seen.

Contrast the sedimentology (e.g. clast size, rounding, sorting characteristics, sedimentary structures) of the sediments derived from the side valleys with those of the main fan. Note the excellent exposure, in the apex section (see below) of shallow, braided, fluvial sediments, especially in their preservation of scour and associated turbulence structures. Contrast the deposits of the aggradational sequence with those of the modern stream bed, noting the proximity of the modern stream bed to bedrock sources.

The sections suggest a sequence with two phases of fan aggradation, separated by a soil horizon. The buried soil is visible in sections in the tributary fans, but has been eroded from the main fan. It has a similar stratigraphic position to buried soils in the Honda and Galera fans to the west. The maximum colour of the B horizon is 5YR 4/4, but there is no observable carbonate accumulation.

The soil caps deposits of an early phase of fan sedimentation (F2, Fig. 6.3.4). This was followed by a considerable period of fan surface stability, during which time the soil developed. A fan-head trench was then incised, followed by a later phase of sedimentation (F3, Fig. 6.3.4), which buried the older fan surfaces and filled the fan-head trench, back-filling into the mountain catchment, and culminating in the formation of the upper fan surfaces. Subsequent dissection created the modern fan-head trench, focusing deposition on the distal fan surfaces. There is no clear-cut dating of these events but the best estimate, made on the basis of regional correlations, would put the modern dissection as Holocene, sedimentation phase F3 as Late Pleistocene, and the gap between phases F2 and F3, represented by the buried soil, of the order of at least tens of thousands of years in the Late Pleistocene.

From the fan apex at 577 099 there is an excellent view down-fan, of the morphology of the Mezquita fan as well as of the coalescent fan surfaces of the upper Tabernas Basin as a whole. Note how fan gradient inversely reflects drainage basin area.

Transfer to Stop 3

Return to the vehicles and drive back towards Tabernas.

About 500 m beyond the solar power plant, large quarries expose sections in the mid-fan and distal-fan sediments (Fig. 6.3.2). These comprise a thick sequence of fluvial gravels, some as extensive sheet gravels, some as large channels. Buried soils cap the lower (> 6 m exposed thickness) gravels, and in turn are capped by 2 m of younger gravels.

Continue back to the old N340. If in a coach go directly to Stop 5 by turning right onto the N340, and continuing for several kilometres to the bridge over the rambla, *c.* 700 m after the junction beyond Tabernas (530 000). If proceeding to Stops 3 or 4, turn left onto the N340 and drive *c.* 3 km to the Venta El Compadre on the left. This is where the optional route for La Serrata fans (Stop 3, Fig. 6.3.1) leaves and rejoins the main excursion route. If taking this option, continue *c.* 100 m beyond the bar to KM 475.2 (579 036). Turn right here onto a single-track road that drops into the bed of the Rambla de los Molinos. Bear left along the rambla floor for *c.* 150 m.

Note the sections in the wall of the rambla, exposing distal-fan sands and gravels of the Honda/Nudos fan complex, interbedded with lake sediments comprising millimetre–centimetre laminated silts.

Next, bear right onto the track that leaves the rambla to the south. Immediately out of the rambla, turn left onto a dirt road and drive east for *c.* 2 km. There is a track on the right heading south towards La Serrata fans. Park on rough ground a little way up this track. This is the nearest point to La Serrata fans that can be reached conveniently by car.

Stop 3: Optional hike to La Serrata fans, Ceporro fan (600 014; Tabernas 1 : 50 000)

Ceporro fan, immediately in front of you, is the largest of the Serrata fans (Fig. 6.3.5). It can be identified by the ruin of a farm building on the proximal fan surface, to the west of the fan-head trench, and by the almond plantation on the west of the distal-fan surfaces.

Walk south towards the fan, keeping to the left (eastern) edge of the almond plantation. At *c.* 595 021 follow the modern distal-fan sediments to the modern channel and the intersection point at *c.* 598 020. From here it is easiest to walk up the fan surface to the east of the fan-head trench to the fan apex at *c.* 600 014 (Fig. 6.3.6).

The Serrata fans contrast markedly with Mezquita fan. They are deposited along a faulted mountain front. The sequences are more complex and include debris flows. There are several palaeosols, the more mature with red Bt horizons and Bk carbonate or K petrocalcic horizons (Harvey *et al.* 1999a). The fan surfaces in the proximal zones have well-developed calcrete crusts. The sequences show evidence of episodic aggradation followed by episodic dissection (Harvey, 1984a,b, 1987a). Intersection-point scour and distal aggradation characterize the current fan dynamics, with the fan near the mid-fan trenching threshold identified by Harvey (1987b).

On Ceporro fan, note the multiple depositional segments, including the most modern lobate deposits on the distal-fan surface. Note the terrace within the mid-fan portion of the fan-head trench. The sediments exposed in the walls of the fan-head trench show a number of palaeosols within a

Fig. 6.3.5. Ceporro fan (c), Serrata del Marchante, Tabernas Basin (Stop 3), seen from the N340 near Stop 1. Note fan-head trench (arrowed) giving way at the intersection point to the modern active distal fan surfaces (a). Note that incision of the Rambla de los Molinos (m) in the foreground, cuts into Pleistocene lake and fan sediments, but that in this location the Marchante fans are still not coupled to the stream system.

sequence that includes debris flows, especially at the fan apex and basally (Fig. 6.3.6a,b). The proximal fan surface is crusted and the intersection point scours into calcrete. Note the overall fan profile relationships (Fig. 6.3.6c). At the apex there are exposures of bedrock.

Transfer to Stop 4

Return to the vehicles and drive back to the N340 at Venta El Compadre.

Leave the Bar at Venta El Compadre towards the west and almost immediately turn left onto a narrow, metalled road that runs through a low gap into the Molinos valley. After a sharp left-hand then a right-hand turn, and *c.* 2 km from the N340, there is a place to park to the left of the road at 557 018. Walk *c.* 100 m south to a view point above the canyon of the Rambla de los Molinos (Fig. 6.3.1).

Stop 4: Tabernas lake sediments (557 018; Tabernas 1 : 50 000)

There are numerous sections in the lake sediments. The easiest to see are east of Tabernas near 557 018. From the parking place, walk *c.* 100 m to

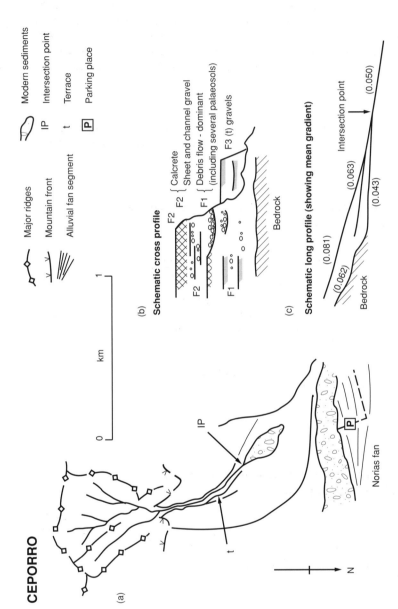

Fig. 6.3.6. Ceporro fan, Stop 4. (a) Geomorphic map; (b) schematic section; (c) schematic long profile showing mean gradients for each major segment. (From Harvey *et al.* 1999a.)

the canyon edge. Here there is a view over folded Serravallian conglomerates overlain by the Tortonian turbidites exposed in the canyon of the
Rambla de los Molinos. The Quaternary sequence (Fig. 6.3.7) starts with
two high-level, cemented gravel river terraces, followed by deep incision,
then by the aggradation of a thick sequence of lake sediments. These are
dissected by the modern canyon, with one minor later terrace stage.

Transfer to Stop 5

Return to the vehicle(s). Continue west on the minor road into Tabernas,
then through the town and onto the N340 at the western end of the town.

**Upper Tabernas 'Lake' - Schematic sequences
(Canyon section)**

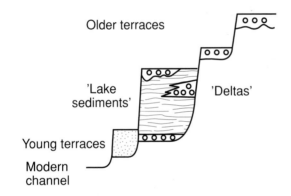

**Suggested correlation with fan sequence
(Uppermost sections)**

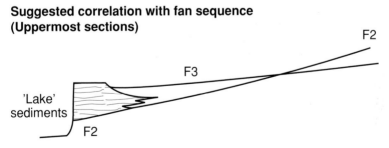

Fig. 6.3.7. Upper Tabernas 'lake' sections, Stop 4. Schematic section, illustrating the
Quaternary sequence. (From Harvey *et al.* 1999a.)

Turn left, towards Almería (take care, hazardous junction!), and after *c.* 700 m *either* park on the shoulder near the bridge over the rambla (530 000), *or* just south of the bridge, where the road takes a sharp bend to the NW, turn left onto a track into the floor of the rambla. Turn right on the floor of the rambla and park near the confluence of the two ramblas.

Stop 5: Rambla de la Sierra, confluence with the Rambla de los Molinos, distal 'upper lake' (532 998; Tabernas 1 : 50 000)

The Rambla de la Sierra is cut in a deep canyon, tributary to the canyon of the Rambla de los Molinos. The canyons are cut into folded Serravallian and Tortonian rocks (described in Excursion 3.4, Stop 1). Quaternary fluvial terraces are cut into the bedrock. There is a high terrace comprising *c.* 5 m of fluvial gravels perched *c.* 40 m above the canyon. Its constituent gravels have been deformed by a fault (at 535 999), which runs NE–SW and crosses the canyon of the Rambla de la Sierra *c.* 300 m upstream of the confluence.

Following deposition and faulting of the high terrace there was massive incision. The canyon was cut down almost to modern stream level. Then a second aggradation phase occurred, filling the canyon with fluvial gravels *c.* 25 m thick, almost up to the level of the older terrace. Along the canyon, thick, undeformed Quaternary fluvial gravels of this aggradation phase rest unconformably on the steeply dipping Serravallian and Tortonian rocks. The gravels themselves show superb channel and bar sedimentology. This massive fill has since been dissected by the modern channel to form the modern canyon (Fig. 6.3.8).

At the tributary junction the canyon of the Molinos widens out, and the tributary Rambla de la Sierra would have formed a fan into the Molinos canyon, damming or partially damming the Rambla de los Molinos. Patches of Quaternary fluvial gravels can be found on the hillslopes to the south of the rambla, between the confluence and the road bridge.

About 200 m upstream of the confluence on the Rambla de los Molinos another small tributary entering the canyon from the south forms a fan within the canyon. Tabernas 'lake' sediments are banked up on the upstream side of this fan, suggesting that both fans damming the lower part of the canyon raised the level of this upper lake to *c.* 380–400 m.

On returning to the road, take in the view NW across the Rambla de Tabernas, which is joined by the Rambla de los Molinos some way below the bridge. In the distance are mesas capped by calcrete-cemented Quaternary gravels (Nash & Smith, 1998). These, and gravels capping the hill opposite Tabernas Castle, have all been deformed to dip to the NE, suggesting an axis of uplift somewhere to the south. In the foreground, in the valley of the Rambla de Tabernas, is a near-horizontal 'terrace' surface

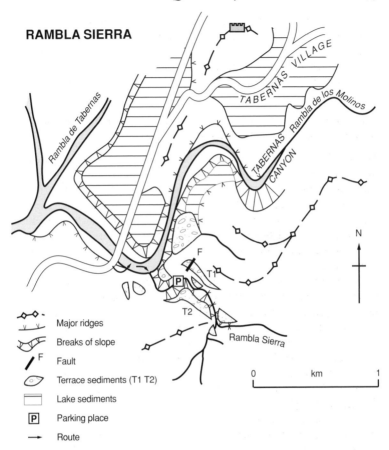

Fig. 6.3.8. Geomorphic map of the Rambla de la Sierra, Stop 5. (From Harvey *et al.* 1999a.)

at 300–320 m on fine-grained palustrine and lake sediments of the lower Tabernas 'lake'.

Transfer to Stop 6

Return to the vehicle(s). Drive back to the main road and turn left in the direction of Almería for *c.* 1.5 km, to the first track on the right (520 994), by KM 468, *c.* 400 m past the entrance on the right to the Texas Hollywood film set. If the track is chained, or you are travelling by coach, park on the hard shoulder where convenient (*c.* 300 m further on) and proceed on foot.

Follow this track for *c.* 900 m to a point above the incised meandering canyon of the Rambla de Tabernas (513 993). Park here (Stop 6).

Stop 6: Rambla de Tabernas, incised meander into Quaternary fluvial and 'lower lake' sediments (513 993; Tabernas 1 : 50 000)

The flat surface on which you are standing is the 'terrace' formed at the level of the lower Tabernas 'lake'. The view NW from the canyon edge (take care, vertical canyon walls below you, subject to collapse through piping!) shows the modern meandering canyon incised through horizontal Quaternary lake and fluvial sediments which are cut into folded Tortonian mudstones and turbidites. Small, recent, low terraces are present on the insides of the bends above the channel.

There is a tripartite division of the Quaternary sediments exposed in the canyon walls. The cemented conglomerate at the base relates to fluvial deposition following incision into the Tortonian rocks. It is cemented by a massive groundwater calcrete at the permeability interface between the Tortonian marly rocks and the overlying Quaternary sediments (Nash & Smith, 1998). Note that further downstream in the opposite wall of the canyon, this conglomerate forms a spectacular dry waterfall. Above the basal conglomerate are the palustrine and lake sediments of the lower 'lake', deposited when drainage was impeded. The sequence is locally capped by more gravels, representing the re-establishment of drainage continuity through the 'lake' environment. Incision to the present canyon floor then followed.

Above you to the south is a cemented conglomerate-capped mesa. These gravels represent an earlier fluvial terrace stage, prior to the deformation that formed the lake. Nash & Smith (1998) examined the calcretes that form the cements in the fluvial terraces of the Tabernas area, and distinguish groundwater calcretes at the bases of terrace units from pedogenic calcretes at the terrace surfaces. Furthermore, the pedogenic calcretes clearly differentiate between the older upper terrace and the younger terrace.

If you wish to examine the sediments in more detail walk SW for c. 500 m across the abandoned fields to 509 989 (the type locality for the lower lake sediments). Here the shallow surface channel drops vertically exposing c. 20 m of section through the lower 'lake'. The section reveals mass flow units at the base, typically < 1 m thick, giving way vertically to a dominance of laminated silts and sands (Fig. 6.3.9a). Reworked travertine nodules are common towards the top of the sequence and indicate carbonate precipitation around vegetation within the 'lake' unit. The background sedimentation appears to have been low energy, from suspension, and thorough bio/pedoturbation of sediments, punctuated by rare, dilute mass flows, and occasional incision of channels 1–2 m deep.

Transfer to Stop 7

Rejoin your vehicle(s). Return to the main road and drive SW towards

(a)

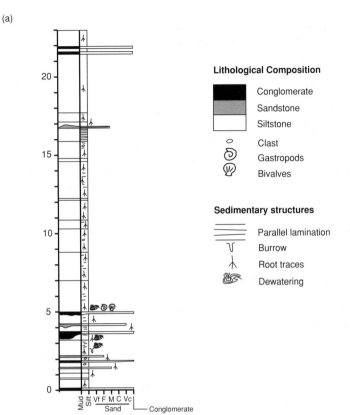

Lithological Composition

▓ (black)	Conglomerate
▒ (grey)	Sandstone
☐ (white)	Siltstone

○	Clast
◉	Gastropods
◍	Bivalves

Sedimentary structures

≡	Parallel lamination
⋎	Burrow
⋏	Root traces
⌀	Dewatering

(b)

Fig. 6.3.9. Tabernas lower lake, Stops 6 and 7. (a) Log (by A. Mather) of lower lake sediments from the type locality (509 989) near Stop 6. (b) Channels (Stop 7) inset into and infilled with 'lake' deposits. Some deformation of the sediments is evident.

Almería for *c.* 4 km to the intersection with the Guadix road, turning left towards Almería. Pass Bar Alfaro and the petrol station on your right and continue for another 1 km to a section of abandoned road on your right (486 961). Drive in here and park. Walk back to the level of the main road. *En route* notice how the dissection intensifies as you descend into the heart of the Tabernas badlands. Near Stops 5 and 6 badland erosion is confined to canyon walls and mesa margins; preservation of terrace surfaces is good, either as mesas or as the 'lake' floor forming the terrace above the modern canyons. Near Bar Alfaro incision is deeper and the dissection is more intense. Badlands occupy a far greater proportion of the landscape. A couple of kilometres up the Guadix road, the road climbs through fluvial gravels of the upper terrace, capped by a mature pedogenic calcrete. These gravels relate to a tributary from the north. At Bar Alfaro fluvial gravels from the same tributary feed into the lower 'lake' terrace.

Stop 7: West of Bar Alfaro, distal end of the Tabernas 'lake' (486 961; Tabernas 1 : 50 000)

There are magnificent views from here of the Tabernas badlands, north over the Las Salinas badlands (Excursion 6.4, Stop 2b), and SE over El Cautivo badlands (Excursion 6.4, Stop 2a). The surface here is the same 'lake' terrace surface. Note how dissected pediment surfaces in the El Cautivo badlands to the SE grade towards the centre of the valley, to levels below the high terrace. The lowest extensive pediment grades approximately to the level of the 'lake' terrace.

This zone represents the approximate downstream limit of the lake sediments. It is also just north of a zone of Quaternary tectonic deformation (corresponding to the reactivated fault gouge zone of Excursion 3.4, Stop 5). Walking NE up the road from the car, towards Bar Alfaro, *c.* 400 m from the junction with the old road, note the incised channels in the deformed lake sediments exposed on the north of the road (494 970), infilled with more lake sediments (Fig. 6.3.9b). Closer inspection of the section reveals abundant reworked carbonate rhizoliths. On the opposite side of the main road (490 965) there is abundant evidence of sediment liquefaction.

Synthesis of the 'lake' systems

East of Tabernas the distal ends of the large alluvial fans interdigitate with lake sediments, as seen *en route* to the La Serrata fans. The lake sediments can be traced for *c.* 10 km downstream from Tabernas. They appear to follow a major incisional episode in the tectonically induced dissection of the western part of the Tabernas Basin, following two earlier river terrace stages. At their western extremity (487 962) 1 km beyond Bar Alfaro (see below) they thin towards a zone of tectonic disturbance (Excursion 3.4,

Stop 5) and appear to be deformed. This indicates that tectonic uplift downstream of Bar Alfaro may have been responsible for creating the lake basin. This location coincides with a fault gouge zone, evident in the lithology of the underlying rocks (Excursion 3.4, Stop 5 and Fig. 3.4.2). However, at the confluence of the Rambla de la Sierra with the Rambla de los Molinos (Stop 5 of this excursion) and at the next small confluence upstream, fan (or fan-delta) sediments of the tributary streams appear to have dammed or partially dammed the canyon of the Rambla de los Molinos and appear, at least locally, partially responsible for modifying the upper lake basin.

The 'lake' sediments indicate a very shallow ephemeral lake environment, probably relating to two lakes rather than one large lake. The sediments range from fine, laminated silts and clays, to mottled, dark, organic muds with roots, to thin, interbedded sands and occasional gravels, sometimes channelized, indicating inflow streams. The lake sediments are up to 23 m thick, and are sometimes capped by fluvial gravels, deposited as normal drainage re-established itself. The lake sediments form extensive flat surfaces at *c.* 380–400 m in the Tabernas area and 300–320 m near Bar Alfaro. A well-developed soil has formed in places on the exposed lake floor sediments exhibiting Stage II–III carbonate accumulation. This suggests that the lake sediments equate with an older phase of fan sedimentation (F2 in Figs 6.3.4, 6.3.6, 6.3.7; *c.* 150 ka; Delgado-Castilla *et al.* 1993), rather than that which forms the most extensive fan surfaces (F3 in Figs 6.3.4, 6.3.6, 6.3.7; Harvey *et al.* 1999a). Indeed the relationships suggested here between the fan sediments, the lake sediments and the river terrace sediments are speculative. The terrace and lake sequence seen in the Tabernas Basin represents the alternation between aggradation and dissection that has resulted from interactions between Quaternary tectonic activity and climatic change. This has operated within an overall dissectional context that itself is a long-term and sustained response to tectonic uplift of the basin as a whole. The fans in the upper part of the basin appear to have been buffered from base-level change and their sedimentary and geomorphological sequences appear primarily to reflect climatic controls. The two sequences interdigitate in the upper 'lake' area.

6.4 Excursion: Badlands (one day)

ADRIAN M. HARVEY, ROY W. ALEXANDER & DIANE B. SPIVEY

Introduction

The purpose of this excursion is to examine some of the modern erosional landscapes on soft rock terrain in the badlands of the Tabernas, Sorbas and Vera Basins (Fig. 6.4.1).

(a)

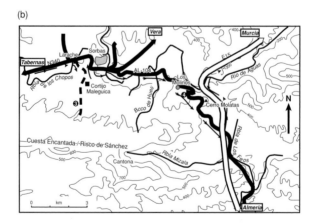

Fig. 6.4.1. Location map of the field-trip stops for Excursion 6.4. (a) Tabernas area (Stops 1 and 2). (b) Sorbas area (Stops 3 and 4).

(c)

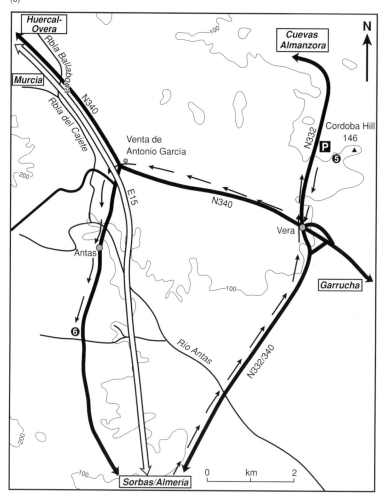

Fig. 6.4.1. (c) Vera and Antas areas (Stops 5 and 6).

In each of these areas one of the major factors leading to badland development is the creation of relief by incision through easily erodable soft rock terrain. In the Tabernas Basin this has been through sustained tectonic uplift of the thick sequence of weak Tortonian mudrocks. The river systems have cut canyons through the more resistant rocks, but the incision has created steep slopes on the marls which are prone to badland erosion.

In the Sorbas Basin, incision of the main rivers through the outcrop of the weak Abad Member marls followed the uplift of the basin. Accelerated dissection of the central part of the basin ensued after the headwards incision that resulted from the capture of the Aguas/Feos master drainage by the lower Aguas in the late Quaternary (Harvey *et al.* 1995; Mather 2000b). Where the incision has coincided with the outcrop of weak silts in the Zorreras Member and the Góchar Formation, especially in the southern part of the basin, badland development has ensued.

In the Vera Basin, badland development occurs on the steeper slopes between successive pediment surfaces, or in relation to the incision of the Antas drainage, in both cases into weak Messinian marls.

The thin vegetation cover and the semiarid climate of the region increase the effectiveness of runoff during storm rains. Although the combination of uplifted and dissected soft rock terrain would make the area naturally prone to accelerated slope erosion and gully and badland development, human activity, through unwise or ill-maintained agricultural terracing, or through the effects of overgrazing on the vegetation cover, has undoubtedly accelerated the process. The Tabernas Basin, and to a lesser extent the Aguas and Antas valleys, have undoubtedly been naturally subject to multiple generations of gully and badland development since at least the Late Pleistocene. In other areas gullying and badland development may be relatively young and triggered or accelerated by human activity.

Badland and gullying types range from valley floor gully systems of the 'arroyo' type, to hillslope gullying, induced both basally and in mid-slope, to totally dissectional landscapes (Calvo-Cases *et al.* 1991b).

Badland processes vary with: (i) position on the hillslope and relation to the drainage network; (ii) aspect; and (iii) the physical and chemical properties of the soft rock materials. Three important groups of erosional processes are evident: (i) erosion by overland flow; (ii) erosion by mass movement; and (iii) subsurface erosion by piping processes. In many badland areas interactions between these processes occur (Harvey, 1982, 1987a; Alexander & Calvo, 1990; Calvo-Cases *et al.* 1991a,b; Harvey & Calvo, 1991; Alexander *et al.* 1994; Faulkner *et al.* 2000). The style of interaction may vary spatially with lithology, topography, vegetation cover and aspect. It may vary temporally in relation to individual storms, seasons, or over a longer-term cyclicity, or progressively change with the evolution of the gully and badland morphology.

The excursion starts with the Tabernas badlands, perhaps the most impressive badland landscape in Europe, then includes two badland sites in the Sorbas Basin at La Cumbre and Los Molinos, finally dealing with two contrasting sites in the Vera Basin, at Vera and Antas.

The excursion can be treated as two half-days with the Tabernas half-day following part of the Tabernas fans excursion (see Excursion 6.3). Note that Stop 2a includes visits adjacent to the erosional experimental site run by the Desert Research Station of the Consejo Suprerior de Investigaciones Cientificas (CSIC) and access onto the site should be avoided. Allow at least 2 h for Stop 2a, and 1 h each for Stops 2b and 5 (Fig. 6.4.1b). The excursion starts at Tabernas.

Please note that Stop 2 may have restricted access because of the construction of the new Granada–Almería autovia along the route of the N324.

Directions to Stop 1

If approaching from Sorbas, follow the old N340 westwards past the Tabernas turn-offs and over the bridge over the Rambla de los Molinos. The road turns sharp right after the bridge, climbs a hill, then turns left. As the road turns left there is a wide shoulder on the right beside the turn-off to the Texas Hollywood film set. Park on this shoulder for Stop 1, a general overview of the upper part of the Tabernas badlands.

If approaching from the west, from Almería or Guadix, follow the old N340 from the intersection between the Almería and Guadix roads for *c.* 4 km, past the Mini Hollywood film set on the right, to the turning for the Texas Hollywood film set on the left. Cross the road and park on the wide shoulder on the north side of the road.

Stop 1: Upper part of the Tabernas badlands, overview (524 999; Tabernas 1 : 50 000)

Late Neogene and Quaternary uplift of the western end of the Tabernas Basin created a steep regional gradient into the Andarax drainage, towards the Rioja graben and the Almería depression, exposing the thick sequence of relatively weak Tortonian rocks to incision and erosion. The combination of tectonically induced incision into these rocks and the Pleistocene and modern dry climates has produced one of the most spectacular erosional landscapes in Europe. Where incision by the Rambla de Tabernas and its tributaries encountered more resistant rocks, such as Serravallian conglomerates or Tortonian sandstones, deep canyons have resulted. Elsewhere, rapid slope erosion of the weaker Tortonian marls and shales has produced a deeply dissected badland landscape. Away from the incising drainage network, the resistant rocks have produced caprock escarpments. The main scarp formers are the sandstone beds within the Tortonian, and Quaternary cemented calcrete-conglomerate (Nash & Smith, 1998), usually occurring at the base of terrace sediments.

The incision was not continuous throughout the Quaternary, but episodic, in response to a combination of intermittent tectonism and climatic fluctuations. These have produced a stepped landscape of steep erosional slopes alternating with near-horizontal erosional pediments and depositional surfaces on river and lake sediments. Two major and several minor terrace and/or pediment surfaces are obvious. There is a high surface developed on calcreted Quaternary gravels several kilometres to the NW of Bar Alfaro, visible along the road towards Gergal/Guadix. A surface, apparently of the same age, is formed by cemented conglomerate caps WNW of Tabernas (Nash & Smith, 1998), visible from the N340 between Tabernas and Bar Alfaro (Stop 1). There are extensive younger surfaces formed by pediments, river terraces and lake sediments at lower levels, followed by the road between Tabernas and Bar Alfaro. In the east, between Tabernas and Bar Alfaro, where the total incision is less, the landscape is dominated by these two sets of surfaces (Stop 1). To the west, where the total incision is greater, the landscape is dominated by deeply dissected badlands (Stop 2). Tabernas badlands have formed the focus of several studies of badland processes and morphology (Harvey, 1982; Alexander & Calvo, 1990; Calvo-Cases *et al.* 1991a,b; Alexander *et al.* 1994; Faulkner *et al.* 2000).

From the Stop 1 viewpoint the characteristic stepped landscape formed by the successive terrace or pediment surfaces, capped by cemented gravels, can be appreciated. The highest gravel terraces are visible to the north forming mesa caps. Another high terrace caps the hill above the road to the south. Below you to the north is the extensive terrace formed by the lower Tabernas 'lake' sediments (see Excursion 6.3), into which the modern channel of the Rambla is trenched as a canyon. Gully and badland development in this zone is restricted to steep undercut slopes and to the steep mesa-margin slopes.

Transfer to Stop 2

Rejoin the vehicle(s). Turn right onto the main road and continue west for another *c.* 4 km into the heart of Tabernas badlands and park at Bar Alfaro (494 971).

Stop 2: Tabernas badlands, Bar Alfaro (494 971; Tabernas 1 : 50 000)

Two alternative hikes are suggested from Bar Alfaro to view the geomorphology of the badlands. Alternative Stop 2a, to El Cautivo badlands, south of the N340, is the longer and more comprehensive. Alternative Stop 2b, to Las Salinas, north of the N340, is shorter but illustrates the main features of badland geomorphology and also allows inspection of extensive travertine accumulations (Mather & Stokes, 1999).

Alternative Stop 2a: El Cautivo (507 963; Tabernas 1 : 50 000)

Cross the road in front of the bar, turn left and walk east towards the bridge over the rambla (496 973). Just before the bridge, drop into the rambla bed and walk south, downstream. Note the morphology and sedimentology of the modern rambla bed, especially at the confluence at 498 968. The rambla has a wide, shallow, braided channel, with a veneer of modern gravels over a bedrock base. Note also the relationships between the two low terraces on the right bank at this point.

Leave the channel bed at 497 966, just beyond where it undercuts the marl and swings to the north side of the valley, and follow the track to the left on a modern river terrace. Note how slope angles are near vertical where the rambla is actively cutting the marl, but degrade to *c.* 45° where the channel has migrated away from the base of the slope, and badland rilling processes instead of mass movement dominate the slope evolution.

Continue along the south side of the valley floor (on the track), where the vegetated floodplain/youngest terrace surface abuts the badland margin, to the next tributary valley near an abandoned settlement at 495 964. The floodplain sediments here are young and include (undated) pottery layers. Turn up the tributary valley and after *c.* 100 m climb an abandoned and degraded track up a spur to the south towards the prominent sandstone escarpment at 500 962. There is a path that climbs onto the escarpment from below the first large sandstone outcrop, up the ledges towards the summit at 507 963 (426 m). From here there are spectacular views: south towards Alfaro mountain (another Tortonian sandstone escarpment, note the multi-stage gully and badland development on the slopes); SE towards the Sierra Alhamilla (note the belt of purple badlands picking out the faulted Triassic rocks along the mountain front); NE towards Tabernas (note the dissected pediments); and NW across deeply dissected badland terrain towards the Sierra Gádor and the Sierra Nevada in the distance. Immediately below, to the NW, are El Cautivo badlands.

El Cautivo badlands exhibit multi-stage badland development (Alexander *et al.* 1994) (Figs 6.4.2–6.4.4). The earlier stages, produced by the recession of the sandstone scarp, are represented by a series of pediment surfaces which form the interfluves between the modern gullies. The lower pediments grade to the prominent surface formed by the river terrace and lake sediments (Excursion 6.3) in the central part of the valley. Below these pediments are several younger gully and badland stages, with surfaces either feeding directly into the modern channels or grading to mini-pediments within the valley system.

Modern processes are dominated by overland flow and surface erosion, but locally mass movement occurs. Piping is rare in these badlands. There is strong aspect control, with lichens and a sporadic cover of higher plants reducing erosion rates on the north-facing slopes. Local base levels exert an important influence, with a tendency for slope stabilization to occur once a pediment has formed at the slope base. Slope stabilization occurs from the

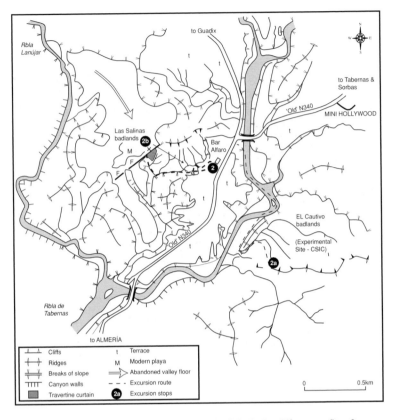

Fig. 6.4.2. Geomorphic map of the Tabernas badlands in the Bar Alfaro area, Stop 2, showing locations of the El Cautivo (alternative Stop 2a) and Las Salinas (alternative Stop 2b) badlands.

slope base up, and is accelerated as a stone cover, then a lichen cover develops, and especially when higher plants colonize (Alexander *et al.* 1994).

The vegetation within the badlands reflects the local environment, with aspect exerting a strong control and, for example, salt-tolerant plants occurring on the youngest basal pediment slopes. The older pediment surfaces show increasing maturity of vegetation with surface age, perhaps reflecting soil characteristics rather than age itself (Alexander *et al.* 1994).

El Cautivo badlands currently form an experimental site, run by the CSIC in Almería. Readers are asked to keep away from the area to the east of the route up the escarpment, avoiding any field installations that may be present. Return to the vehicle(s) at Bar Alfaro.

Fig. 6.4.3. View north from Stop 2a over Tabernas badlands from the ridge above El Cautivo badlands (foreground) towards Las Salinas badlands (middle distance, centre). Note within El Cautivo badlands, active basally undercut SW-facing badland slopes (s); much less dissected NE-facing slopes which toe out in mini-pediments (m). Larger pediment surfaces (p), dissected by the modern badlands, grade towards Late Pleistocene river terraces and lake sediments (l), forming the flat areas in the centre of the photo.

Alternative Stop 2b: Las Salinas Tabernas badlands, Bar Alfaro (492 971; Tabernas 1 : 50 000)

Walk from Bar Alfaro towards the large sandstone-capped hill behind the bar to 492 971. Note the two-stage dissection evident in the badland morphology in front of you (Fig. 6.4.2). Older stabilized gully lines, which form the bulk of the hillslope, grade towards the terrace surface on which the bar stands. These slopes are being dissected by new active hillslope gullies, which grade into the incised channels. To the SSW of the hill is a more complete badland area, dissected by an incised channel that drains to the SW. Within this badland area, note the evidence for surface processes, the influence of aspect and the evidence for base-level control of badland evolution and stabilization.

Follow the incised drainage downstream to 488 968, where the valley opens out into a wide basin, fringed by mini-pediments. Note the caprock of cemented Quaternary conglomerates, controlling slope recession and the modern evaporite deposition in depressions in the basin floor.

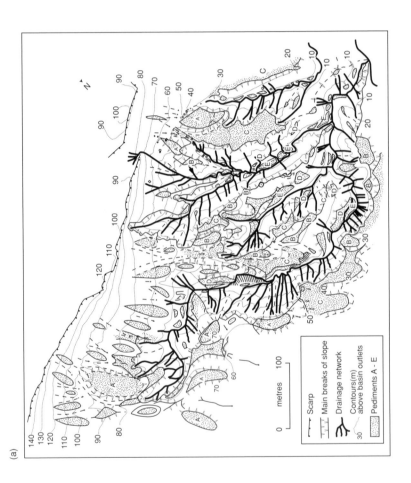

(a)

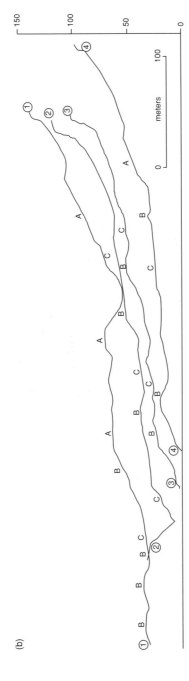

Fig. 6.4.4. El Cautivo badlands, Tabernas (Stop 2a). (a) Detailed geomorphic map. (b) Spur profiles (1–4, east–west). A–C, successive pediments. (Both modified after Alexander *et al.* 1994.)

Turn right up the main channel to 486 972, where a major fault crosses the badland surfaces. Note the impressive travertine curtains produced by seepage in the fault zone (Mather *et al.* 1997; Mather & Stokes, 1999). The fault has been active since deposition and cementation of the Quaternary conglomerates. Above the fault there is an extensive flat area that is essentially a modern evaporative playa. This is within an abandoned meander loop of the now deeply incised Rambla Lanújar, to the west.

Transfer to Stop 3

Return to the vehicle(s) at Bar Alfaro. Turn left out of the car park at Bar Alfaro (Fig. 6.4.1a), then right at the road junction, over the bridge and onto the old N340 towards Tabernas. Continue on this road past Tabernas and on towards Sorbas. On the approaches to Sorbas pass the turning to Uleila del Campo on your left. About 100 m past this junction, there is a group of white buildings on the right (Larache, 767 063), with a track between the buildings signed for La Cumbre.

Those travelling by coach should omit Stop 3 and continue beyond Sorbas directly to Stop 4. In very wet weather cars should also omit Stop 3. Otherwise, turn right here, taking the track through the houses.

Proceed along the track, bearing left at the entrance to a quarry (on the right) onto a plateau on the Sorbas Member. This was probably a previous (Stage A) river course of the Río Aguas. Bear right again (at 773 055) up the hillside, following the signs for La Cumbre. Take care; this section of the track can be in bad repair after rain. Pause at the top of the slope for the view to the north, across the old river course towards Sorbas and across the basin. Continue south along the track along the watershed. You will pass over a series of cols—former SW–NE drainage lines which have been captured by the aggressive tributaries of the south–north Barranco del Mocatán (Mather, 2000b).

Continue along the track. Note the badlands to the right developed in the basal silts and sands of the Góchar Formation. Park by the olive tree terrace to the left (Stop 3) at 796 042.

En route from Stop 2 you first climbed out of the deeply dissected Tabernas badland zone, then crossed the little dissected distal zone of the Sierra de los Filabres alluvial fans in the upper part of the Tabernas Basin (see Excursion 6.3). Quaternary alluvial-fan deposition took place after a long period of erosion which removed any basin-fill rocks above the Tortonian. You then climbed out of the Tabernas Basin and across the watershed into the western part of the Sorbas Basin. Here there has been almost no erosion since the end of basin-filling (Góchar Formation). Much of the land surface is the more or less intact end-Góchar depositional surface. As you travel east, however, notice the increasing depth of dissection (together with occasional gully development) of the Rambla de los Lobos on the right. Note also the more extensive badland development in Plio-Pleistocene red silts south of the road after the confluence of the Lobos with the Rambla de los Chopos (this is south of the abandoned road sections through deformed Góchar Formation sediments, described in Excursion 3.5b, Stop 3 and Excursion 5.3, Stop 1).

Stop 3: La Cumbre (767 042; Polopos 1 : 25 000)

These badlands (Fig. 6.4.5) are cut in red terrestrial silts of the Mocatán System of the Plio-Pleistocene Góchar Formation (Mather, 1991), and in reworked Quaternary silts on the valley floor. The Barranco de Mocatán is an incising drainage, whose incision during the Late Quaternary has been related to the capture-induced incision of the Río Aguas, working its way headwards into the centre of the Sorbas Basin (Mather, 2000b; see also Excursion 5.4, Stop 3).

In contrast with the Tabernas badlands, piping is particularly important in this area. Controls on gully and pipe form and process here are complex (Spivey, 1997), and their relative influence varies in different parts of the catchment. Much of the erosion is ultimately driven by the incision of the Barranco de Mocatán, which provides the local base level. The hillslope gullies vary in their degree of connectivity with the main drainage. There are disconnected hanging valley features, where well-vegetated, non-dissected, abandoned agricultural terraces have been left isolated from the main channel. Some valley-head and sideslope gullies toe out onto gentle vegetated slopes; some feed directly into stream channels, and some end in pipes.

The master gullies of the extensive badlands below the view point run east–west, truncating and segmenting former south–north drainage lines, which remain visible as cols in the gully interfluves (Mather 2000b). The most recent captures have been by major pipes, diverting east–west gullies into the adjacent gully to the north.

Aspect (emphasized by structural dip in parts of this area) exerts a strong control on the development of tributary gullies and pipes. Its influence on vegetation is clearly evident, especially across the main valley from the view point, with a well-developed plant cover on the north-facing slopes and extremely sparse or no vegetation on south- or SW-facing slopes (Alexander *et al.* 1999). This leads to asymmetric gully cross profiles, and process differentiation on opposing slopes. South-facing slopes tend to erode by small-scale mass movements on their steep upper parts, with the material produced choking the gully channels, causing ponding which encourages pipe formation. These pipes ultimately collapse, to give a 'notched' channel

(a)

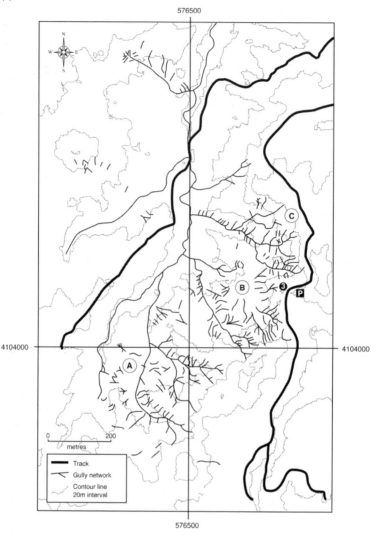

Fig. 6.4.5. La Cumbre badlands, Stop 3. (a) Geomorphic map illustrating styles of badland development at La Cumbre. (Modified after Spivey, 1997.) Area A: coupled and non-coupled gully systems, gully morphology strongly related to geology. Rilling dominant. Headcut collapse prevalent and may be associated with piping. Area B: linear headcut gullies. Block collapse along joint planes significant. Rilling and surface wash of secondary importance. Piping has no obvious role. Area C: steep linear hillslope gullies coupled to main drainage. Mature stage of gully development. Extensive piping.

(b)

Fig. 6.4.5. (b) View west over La Cumbre badlands. Note piped drainage capture (arrowed), also aspect control.

in the central section of the gully. North-facing slopes tend to have more gentle profiles and are frequently stepped, showing evidence of arcuate slump scars and occasional incipient hillslope rilling and gullying.

Local base levels exert strong controls on the form of individual gully slopes, with more gentle slope gradients occurring in areas where surface flow has been abandoned in favour of piping. The relative predominance of piping or surface flow is influenced by the physical and chemical properties of the materials and by the presence of agricultural terraces in the floors of many of the larger gullies. Recent investigation of the physical and chemical properties of the soil (Spivey, 1997; Alexander *et al.* 1999; Faulkner *et al.*, 2000) indicates that relatively high sodium concentrations cause dispersion of clays leading to piping. Piping typically occurs below a stable surface crust (of *c.* 10–20 cm) from which sodium has been leached. Pipes begin below this crust and the whole soil mass appears to retain a high hydraulic conductivity as the low percentage fraction of clay is insufficient to reduce permeability of lower levels, thus preventing autostabilization.

This catchment has undergone something of an agricultural renaissance since 1994, driven largely by the availability of EU subsidies, and the effects of this add a further layer of complexity to contemporary processes. The impact of agricultural change is most evident in the large, new, irrigated terraces to the west and north, and can also be seen in the numerous fresh tracks that zigzag through the area. Note particularly the track that cuts across a series of south–north gullies below Cerro de Juan Contreras (767 039), to the south of the view point. This is frequently rendered impassable by the development of large vertical pipes where it crosses drainage lines.

Transfer to Stop 4

Return to the vehicle(s). Drive back to Sorbas, turning right onto the main road at Larache (Fig. 6.4.1b). Beyond Sorbas, continue east on the N340 for *c.* 1 km to the intersection with the Al-104. Turn right and drive towards Los Molinos. Continue beyond Los Molinos across the bridge. After the bridge, the road bends left, then right and through a series of minor bends before climbing through a long sweeping right-hand bend, followed by a long sweeping left-hand bend. About 1.5 km from the bridge, just as the road begins to straighten out, there is a wide shoulder on the right with a track heading west between two metal posts. Park here (Stop 4).

Stop 4: South of Los Molinos (825 047; Sorbas and Polopos 1 : 25 000)

The track leads down to an abandoned farm. Walk *c.* 100 m along this track to where there is an overview of the Los Molinos badlands.

The general geomorphology of this area relating to drainage evolution, river capture, incision and slope instability, has been fully described in Excursion 6.1 (see Stops 1 and 6 on that excursion).

This stop affords a view point over the Los Molinos badlands (Fig. 6.4.6). These badlands are cut in Abad Member marl, and developed following the deep incision of the Río Aguas after the river capture at Los Molinos. Note that much of the dissected area in front of you comprises revegetated stabilized former gully systems. Modern active gullying and badland processes are restricted to south-facing slopes, especially where undercut by incising drainage. Note the differences between badlands developed on the interfluves immediately in front of you and those cutting back into the gypsum escarpment.

Human activity has also played a major part in this area. Note the gully systems developed on abandoned terraced land on the slope on the other side of the Aguas valley, opposite Los Molinos. Aspect controls are particularly evident on the gully slopes cut into the interfluve to the west. North-facing slopes carry a dense vegetation of shrubs and grasses and have more gentle gradients than the bare or sparsely vegetated south-facing slopes. The marls here are fine-grained and surface processes (wash

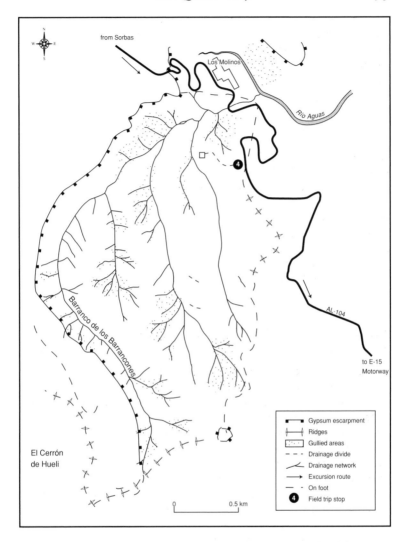

Fig. 6.4.6. Geomorphic map of Los Molinos badlands, Stop 4.

processes, rilling and shallow mass movements) predominate, especially rilling on south-facing slopes. The gully profiles are frequently stepped, where they grade into old agricultural terraces in mid-slope. Thus, in some places, a pair of drainage networks (above and below a terrace) occurs, while elsewhere the terrace may have been cut through and a single network is developing. These gully systems drain into a broad valley (behind the abandoned farm) floored by now abandoned agricultural terraces. The terraces have been deeply incised by the main channel, whereas runs down the margin of the valley floor, and undercuts the slope, initiating the basal development of steep parallel gullies. Many of these gullies have stepped profiles, caused by varying rock resistance within the Abad Member marl.

A range of single and multiple gully systems also occurs on the (south-facing) marl slope below the gypsum escarpment. The presence of large blocks of detached gypsum, as well as abandoned agricultural terraces, locally restricts the development of some gullies to the lower slopes, whereas some extend up to the base of the gypsum itself. Note the recolonization by plants in some of the lower gully channels and on side slopes where slope angles have reduced sufficiently to provide stable sites.

Transfer to Stop 5

Return to the vehicle(s). Continue south on the Al-104 to the intersection with the motorway.

As you climb up to Cerro Molatas (Fig. 6.4.1b), note the contrasting styles of the slope failure on the gypsum on the far side of the valley (topple failures) and on the Azagador Member limestones on this side (down-dip planar slides). Note also the extensive badland terrain developed on the Abad Member marls exposed down the lower Aguas valley (see Excursion 6.2 for more details). You will be passing through this area on the motorway on the way to the next stop. Note also the small-scale gullying on the Tortonian marls below the Azagador Member escarpment visible on the right as you descend from Cerro Molatas to the motorway intersection.

At the motorway intersection take the N340 north, signed for Murcia, towards Vera.

The motorway drops down into the Río Aguas canyon in the zone of the river capture, then follows the lower Aguas valley along the outcrop of the Abad Member marls. Note the extensive badland development throughout the valley past La Herrería to La Huelga, after which you begin to leave the valley to skirt the northern margins of the Sorbas then the Vera Basins. You then cross the relatively undissected terrain of the Vera Basin.

Leave the motorway at the exit for Vera and follow the old N340 into Vera. In Vera go through the town, turning right at the service station onto the N332 signed for Cuevas del Almanzora and Aguilas. Follow that road to where it crosses the new eastern bypass by a left, then a right turn. Alternatively, avoid Vera by taking the southern then the eastern bypass as far as the right turn for the N332 for Cuevas del Almanzora and Aguilas. Follow the N332 for *c.* 1 km. Either turn right into the badlands at 004 253 and part 100 m along a rough track, or continue to 005 254 and turn to the right to park in the access road to the waste treatment facility (Fig. 6.4.1c).

(a)

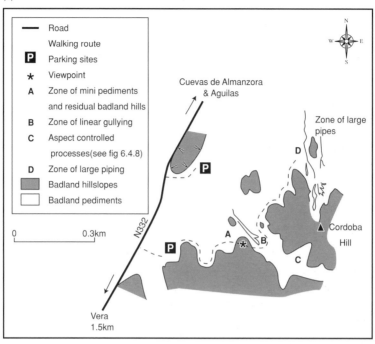

Legend:
- —— Road
- Walking route
- **P** Parking sites
- ∗ Viewpoint
- **A** Zone of mini pediments and residual badland hills
- **B** Zone of linear gullying
- **C** Aspect controlled processes (see fig 6.4.8)
- **D** Zone of large piping
- Badland hillslopes
- Badland pediments

Cuevas de Almanzora & Aguilas

Zone of large pipes

D

P

A **B**

C

Cordoba Hill

N332

0 — 0.3km

Vera 1.5km

Fig. 6.4.7. Vera badlands, Stop 5. (a) Geomorphic map. (Modified after Harvey, 1982.) (b) View within Vera badlands. Residual badland hill (s) flanked by pediment surfaces (m). Valley floor trench gully (g) cuts into Miocene marls. Note the gypsum veins within the marls (lower centre).

Stop 5: Vera badlands (008 248; Vera 1 : 50 000)

Walk east to the foot of the badlands, keeping off any cropped land, then follow the edge of the badlands north and east on the south side of the main valley. Climb the small badland hill that marks the end of the second spur (A in Fig. 6.4.7a). From here there is a view of the dissectional topography that characterizes the Quaternary of the Vera Basin.

The landforms of the Vera Basin differ from those of the Sorbas Basin. Uplift was less, and dissection was dominated by three large rivers fed from outside the basin: the Aguas in the south; the Antas in the centre; and the largest, the Almanzora, in the north. In particular the Almanzora is a transverse drainage, apparently antecedent to the uplift of the Sierra Almagro (Harvey, 1987a; Stokes, 1997). Each river crosses the basin from west to east. In the Plio-Pleistocene the Antas and the Almanzora developed large alluvial fans where they entered the Vera Basin (Stokes, 1997), but later incision produced suites of Quaternary river terraces (see Excursion 5.4). The small sizes of the other drainages feeding the basin has inhibited the development of younger, large alluvial fans, though small fans are present fringing the Sierra Lisbona to the west and the Sierra Almagro to the north. Landforms in the basin are dominated by extensive pediments (Völk, 1979) which are cut across the relatively soft Messinian rocks. Successive pediment surfaces grade towards the river terraces.

From the viewpoint in the Vera badlands at least three successive pediment surfaces can be seen. Cordoba hill (012 249), 200 m east of the view point, is capped by an Almanzora river terrace gravel, at a height to which one of the upper pediment surfaces is graded. The ridgetops into which the badlands are cutting correspond to a lower pediment, visible to the west. The modern flat valley-floor immediately to the north and in the middle distance to the west of the view point, forms the youngest pediment. This pediment is still being formed by the lateral retreat of the hillslopes, especially where slope erosion is accelerated by badland processes.

Vera badlands differ from the other badland sites visited so far, especially from the Tabernas badlands, in two main ways (Harvey, 1982, 1987a; Calvo-Cases *et al.* 1991a). First, in contrast to the Tabernas badlands, they are cut in soft gypsum-rich Messinian marls rather than the deeper, marine, gypsum-poor and more resistant Tortonian rocks of the Tabernas Basin. Secondly, in contrast to all the other sites, they are based by pediment surfaces rather than by incising channels. These two factors lead to different process interactions on the badland surfaces. Aspect, as at La Cumbre and Los Molinos, appears to be important in influencing weathering, plant cover and erosion processes. Weathering proceeds to greater depths than at the other sites, especially on north-facing slopes. The relationships between mass movement and surface erosion processes are more complex than elsewhere (except perhaps at La Cumbre), involving temporal and spatial variations in the process interactions (Calvo-Cases *et al.* 1991a), with generally high erosion rates (Alexander & Calvo, 1990; Calvo-Cases *et al.* 1991b). Piping is an important process, both on the badland surfaces (Harvey, 1982) and on abandoned agricultural terraces on the valley floors. From the viewpoint, evidence of the range of surface processes can be seen on the nearby badland surfaces. On the hillslopes aspect-controlled rilling and mass movement are evident on the south-facing slopes, with much greater stability characterizing the well-vegetated north-facing slopes. A discontinuous linear valley-floor gully (of the 'arroyo-type') dissects the valley floor in front of you. The mini-pediment surfaces in front of you (A in Fig. 6.4.7a; Fig. 6.4.7b) contrast with the steeper badland slopes of the small residual hill to the north. A little to the SW, below Cordoba hill, is a small re-entrant valley which exhibits a morphology (Fig. 6.4.8) produced by an aspect-controlled sequence of process interactions (Harvey, 1982, 1987a; Calvo-Cases *et al.* 1991a; Harvey & Calvo-Cases, 1991). Vegetated north-facing slopes are cut by widely spaced linear hillslope gullies; west-facing slopes have closer-spaced, pipe-induced gullies; south-facing slopes are the most active, dominated by surface processes involving interactions between rilling and mass movement. Erosion at this site and other parts of the Vera badlands is being monitored. Please do not walk unnecessarily on steep eroding slopes. Thank you!

From the viewpoint, walk NE across the valley floor, crossing above the headcuts of the discontinuous linear gully (B in Fig. 6.4.7a), then passing to the left (NW) of the spur on the other side of the valley. Follow the basal zone of the badland slopes around to the east until you are in a valley that drains to the north of Cordoba hill (012 252; D in Fig. 6.4.7a). Here you will see spectacular evidence of piping erosion that has followed agricultural activity and land abandonment, cutting into bedrock several metres below the surfaces of the abandoned fields. Recent agricultural restoration has destroyed some of the most spectacular features, but the scale of piping is still apparent.

Transfer to Stop 6

Return to the vehicle(s). Drive back towards Vera and turn right onto the bypass towards Murcia. After *c.* 4 km, at the Antas intersection at Venta de Antonio García (963 248), turn left towards Antas for *c.* 800 m (962 244).

For a short side-trip (by car or minibus) that affords a view over the features related to the drainage evolution of this area, turn right onto a narrow road that crosses the Rambla del Cajete (Fig. 6.4.1c). When across the rambla, bear left; follow the road up the side of the valley for *c.* 300 m to a viewpoint over the Antas valley. Park at the roadside (955 244).

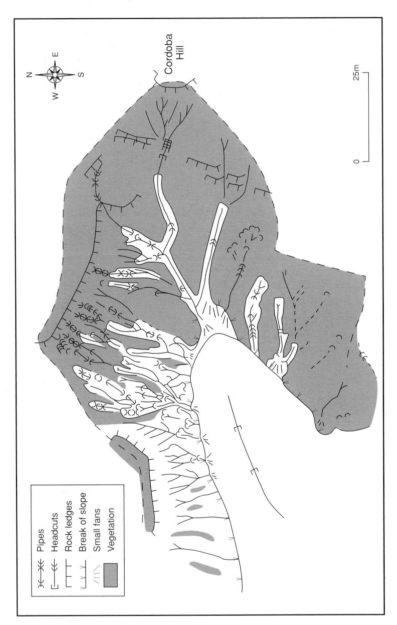

Fig. 6.4.8. Geomorphic map of small re-entrant valley in Vera badlands exhibiting an aspect-controlled sequence of process interactions. On the north-facing slope processes are dominated by linear hillslope gullies cut into stable vegetated surface. The intermediate west-facing slope is dominated by piping and the south-facing slope is dominated by surface processes with interactions between rilling and mass movement (Harvey, 1982, 1987a; Harvey & Calvo-Cases, 1991).

The Rambla de Ballabona, the main headwater of the Rambla del Cajete, formerly flowed SE through the now dry Vera valley. During the Pleistocene it was captured by the Rambla del Cajete, a short steep tributary of the Río Antas (Fig. 6.4.1c), here more deeply incised than the Ballabona. The route from Vera to this view point followed that abandoned valley as far as Venta de Antonio García. From there the broad, gentle, non-incised valley of the Ballabona could be seen to the NW. From this view point, the abandoned Vera valley is visible to the east, and the more deeply incised Antas valley to the south. There is a pronounced knickpoint at the head of the steep section in the profile of the Cajete between here and its confluence with the Antas. Also visible are the terraces of the Antas, well displayed at the confluence. Visible, to the east of Antas village, is a deeply dissected badland area.

Return to the Antas road and continue towards Antas. Turn right through the village, and continue for *c.* 2 km; stop at the roadside at 955 212 (Stop 6).

Stop 6: South of Antas (955 212; Vera 1 : 50 000)

You are now on the western side of the Vera Basin. Exposed in the hillsides above this site are limestones and marls of the Turre Formation. About 2 km to the south are the volcanics of Cabeza María (visited in Excursion 3.3b, Stop 3). The marls here are gullied by tributaries of the Río Antas.

The Antas badlands (Fig. 6.4.9) are cut into marls of the Turre Formation, in an area of abandoned agricultural terraces. In the upper and western part of the site, linear 'arroyo-type' gullies (6a and 6b in Fig. 6.4.9a) cut through agricultural terraces to depths of up to 9 m (6a in Fig. 6.4.9a), whereas the adjacent terrace surface still maintains a high cover of shrub vegetation. Nearer the road (the east of the site), badland development is much more advanced, and networks of V-shaped gullies are separated by rounded interfluves with a sparse vegetation cover (Stop 6c; Fig. 6.4.9). Shrubs and grasses dominate the vegetation here, and there is more limited development of algal crust than at Vera. Plant cover varies from *c.* 50% on the interfluves, north-facing slopes and flatter terrace remnants to < 10% on the active, dominantly south-facing gully slopes, where individual shrubs and grass clumps stand on pedestals up to 20 cm above the slope surface, indicating a rapid rate of surface lowering.

Note the multi-stage development of gullies in this lower section, with steep-sided contemporary channels actively cutting back into older, more gently sloping gully systems. Their progress is marked by well-defined knickpoints along the channels.

Antas is the most active of the badland sites described (Spivey, 1997) in terms of rates of surface lowering. Although there is some evidence of

(a)

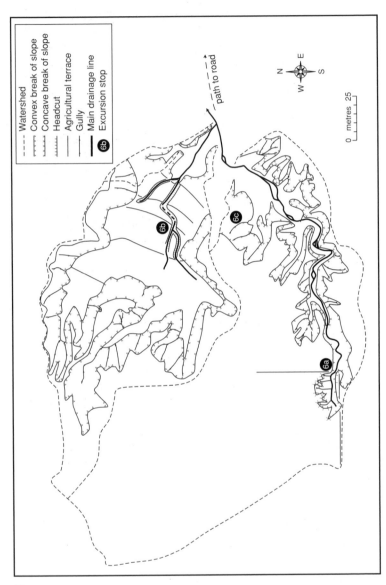

Fig. 6.4.9. Antas badlands, Stop 6. (a) Geomorphic map of main catchments. (From Mather *et al.* 1997.)

(b)

Fig. 6.4.9. (b) General view illustrating range of low erosional hills and rilling near 6c on (a) above.

piping, surface processes dominate. The high proportion of swelling clays in the geology leads to the development of both superficial and deeper cracks along which rills develop. Recent field measurements, taken over a 3-year period (Spivey, 1997), indicate an erosion rate of 9 mm a^{-1} on gully slopes and a headcut retreat rate of 150 mm a^{-1}. Mass movement is also evident as polygonal blocks of surface material are undercut by the channel. These give rise to mudball formation, often evident in the lower reaches of the channel after rainstorms (Mather *et al.* 1997).

Return to the vehicle(s). This is the end of Excursion 6.4. To return to the main road system, continue driving south along the minor road to where it rejoins the motorway.

6.5 Excursion: The landforms of the coastal zone (one day)

ADRIAN M. HARVEY, MARTIN STOKES & ANNE E. MATHER

Introduction

The purpose of this excursion is to examine the landforms and Quaternary geomorphology of the coastal zone between the Cabo de Gata and Garrucha (Fig. 6.5.1).

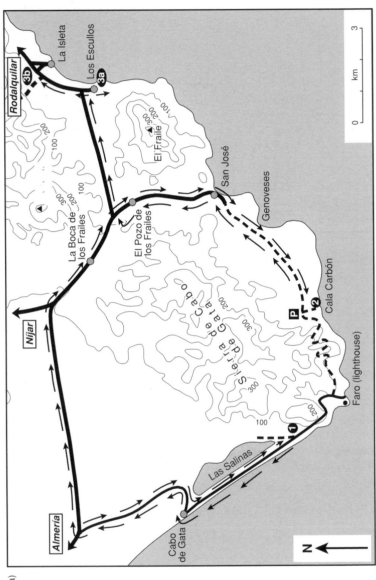

Fig. 6.5.1. Location map of the field-trip stops for Excursion 6.5. (a) Cabo de Gata (Stops 1–3).

(b)

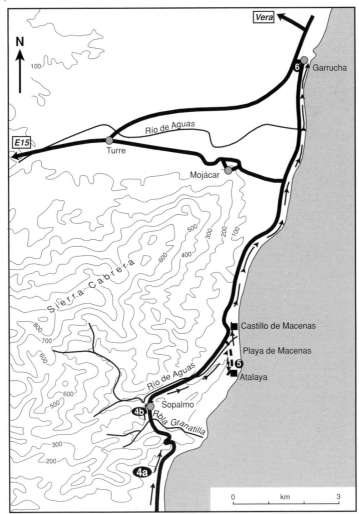

Fig. 6.5.1. (b) Carboneras to Garrucha (Stops 4–8).

The first part of this excursion deals with the Cabo de Gata coastline. The bedrock dominantly comprises the Cabo de Gata volcanics, a suite of calc-alkaline rocks dealt with in detail in Excursion 3.3a and b, Stop 1. Especially to the north, Neogene sedimentary rocks dominated by Messinian reef limestones overlie the Cabo de Gata volcanics. The sedimentary sequence fills in the volcanic topography and onlaps the volcanic slopes.

The terrain has been translated from the SW by strike-slip movement along the Carboneras faults (see Fig. 3.5.1).

The second part of the excursion follows the coast across the Carboneras Fault Zone, along the Sierra Cabrera coast and into the Vera Basin. This alignment follows that of the Palomares Fault Zone (Fig. 3.5.1).

This is a full-day excursion. If less time is available, Stops 1 or 2 could be omitted as they require most driving time to access them. The full excursion starts at Las Salinas on the road from Cabo de Gata village towards the Cabo de Gata headland.

Directions to Stop 1

If travelling from the north along the N340, from Vera, Sorbas or Carboneras, leave the motorway at the exit signed for Cabo de Gata, and continue along the old road as far as the San José/Cabo de Gata turn. Turn left onto the Cabo de Gata road.

If travelling from Almería take the main road towards the N340 motorway past the airport, to the San José/Cabo de Gata turn. Turn right onto the Cabo de Gata road.

Follow the Cabo de Gata road, turning right at the San José junction. Follow the road through Cabo de Gata village, turning left in the village. Drive along the coast road which is built on a barrier beach with the sea to the right and the Las Salinas lagoon and salt extraction pools to the left. Look for flamingos here as well as a rich variety of other wading birds.

Go through the small settlement of Las Salinas, past the cafés, and park on the right just before the road turns inland and climbs above the coast (Stop 1).

En route the road crosses the seaward end of the Almería Basin over a landscape of pedimented Neogene rocks, overlain by Quaternary rocks. In this area are a series of offlapping Pleistocene coastal sediments (Zazo *et al.* 1981). This is the one area east of Almería where the early Quaternary marine record is present, locally deformed by the Carboneras Fault Zone (see Excursion 3.5a, Stop 6).

Beyond the fault zone you cross the seaward end of the small Morales Basin, mostly on Late Pleistocene and Holocene coastal sediments (for details of the Quaternary geology see Zazo *et al.* 1981).

Stop 1: Cabo de Gata / Las Salinas (708 667; Cabo de Gata 1 : 50 000)

Stretching away to the west, the coast here is formed of sandy beaches, developed on the seaward face of a large barrier beach structure (Fig. 6.5.2). Sediments within the barrier beach structure have been radiocarbon dated to the Middle Holocene, and other older sediments by U/Th dating to the Tyrrhenian highstand (Zazo, unpublished data, quoted by Harvey *et al.*

(a)

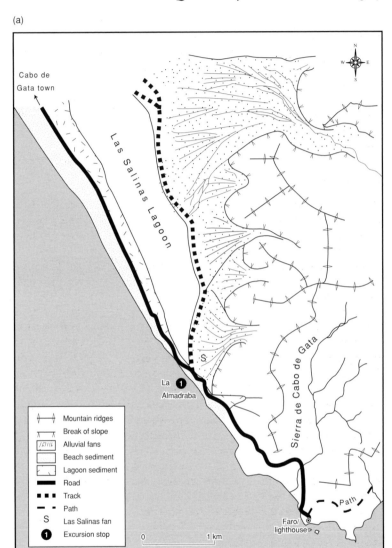

Fig. 6.5.2. (a) Geomorphic map of Cabo de Gata, Stop 1.

(b)

Fig. 6.5.2. (b) General view down La Salina fan (Stop 1). Note the untrenched distal fan surfaces and the shallow fan-head trench in the foreground (3, 5 relate to fan surfaces mapped as QF3, QF5 respectively; see text). This fan is buffered from the effects of base-level change by La Salina lagoon (s), protected from the sea now and during previous highstands by a beach–barrier complex.

1999b). Clearly this feature has been present along this coast throughout the Holocene high sea-level period (Isotope Stage 1, ≈ 7 ka), and a similar feature was present during the last interglacial highstand during the Tyrrhenian II (Isotope Stage 5, *c.* 135–110 ka). Behind the barrier beach is the Salinas lagoon (Fig. 6.5.2), now used for the production of salt. A similar feature was probably here during previous sea-level highstands. On the foreshore there are no Tyrrhenian II beach sediments *in situ* at this end of the barrier beach, but there are occasional detached blocks of cemented beach gravels that are probably of Tyrrhenian II age. Some have been incorporated into the sea defence structures.

Behind the lagoon, there are small Quaternary mountain-front alluvial fans fed by catchments in the Cabo de Gata volcanics (Fig. 6.5.2). These fans are simple fans, proximally trenched by shallow fan-head trenches, and distally prograding to the margins of the lagoon (for example of fan profiles see Fig. 6.5.3). The older proximal fan segments can be differentiated from the younger distal segments on the basis of soil development (Harvey *et al.* 1999b). Poor fan-head trench sections suggest that sedimentation on the larger fans was dominated by fluvial channel and sheet-flood

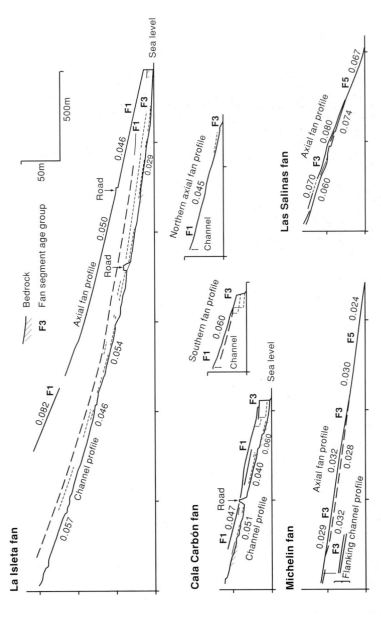

Fig. 6.5.3. Surveyed fan profiles for selected fans in the Cabo de Gata ranges. La Isleta and Cala Carbón are truncated fans on the east coast; the Cabo de Gata fans run out into the Cabo de Gata lagoon on the west coast. Figures given are mean fan or channel segment gradients. (From Harvey *et al.* 1999b.)

processes. The fans have been protected from the effects of base-level change and coastal erosion by the presence of the barrier beach–lagoon complex (Harvey *et al.* 1999b). To the south, where the road climbs above the sea, a steep debris cone is cut by the road, with good roadside sections of debris-flow deposits. This cone has been trenched throughout as the result of coastal erosion at its base.

It is worth driving to the end of the road up to a viewpoint above the lighthouse, where there are views of the erosional coast of the Cabo de Gata itself; cliffs and stacks cut in the volcanic rocks. The volcanic geology of the area is described in Excursion 3.3b, Stop 1.

Transfer to Stop 2

If time permits, it is possible to hike the *c.* 5 km to Cala Carbón (Stop 2) from this point, following the coastal track (Fig. 6.5.1a), but vehicles have to be driven the long way round. To drive to Stop 2, head back towards Almería as far as the San José turn (685 747). Turn right there towards San José. Continue beyond where the road from Níjar joins from the left and then through La Boca de los Frailes to the Los Escullos turn (792 727). If omitting Stop 2 turn left here towards Los Escullos (Stop 3). If including Stop 2 continue on the San José road, through El Pozo to the edge of San José. Turn right then right again following signs to 'playas'. At first this is a narrow road through villas above San José, then a rough, speed-restricted and, in summer, busy, gravel road. Drive slowly, continuing for *c.* 8 km to the end of the road. There is a parking area place here on top of an alluvial fan. Park here (Stop 2).

En route from Cabo de Gata the road crosses the distal areas of Quaternary alluvial fans skirting the volcanics of the Cabo de Gata ranges. At El Pozo the road crosses the Cabo de Gata ranges. From San José the road skirts the coast above large sandy bays. Note the red soils of the Quaternary valley-fills and the presence of Quaternary dunes along the coast. Also note the intense tafoni weathering on some of the volcanic lithologies. The volcanic geology of this area is described in Excursion 3.3a, Stops 5 and 6.

Stop 2: Cala Carbón (752 658; Morrón de los Geneveses 1 : 25 000)

Cala Carbón alluvial fan is a multi-surface, dissected Quaternary fan (Fig. 6.5.3) on which the oldest surfaces are capped by massive pedogenic calcrete (QF1 in Fig. 6.5.3). This is best exposed *c.* 300 m short of the first parking area, where the road crosses a side channel cut into the flanks of Cala Carbón fan.

Walk down the track from the parking area *c.* 400 m towards the beach, with the main trenched fan channel on the left (Fig. 6.5.3). Halfway down the track, in the nearest trench wall, is a large inset terrace of younger fan sediments (QF3 in Fig. 6.5.3) cut into the older fan sediments which form the main fan surface. The older sediments, exposed on the far side of the trench, and the younger sediments exposed on this side, comprise fluvial sediments, preserving channel structures, but the younger sediments are somewhat finer. Soils on the terrace surface are less red and less mature than those on the main fan surface. Downstream of here the modern channel steepens (Fig. 6.5.5) and is cut through the fan deposits into the underlying volcanic bedrock. The erosional and depositional relationships can be seen in the sections. The fan has been distally trenched in response to coastal erosion.

Descend to the beach at Cala Carbón, which is a small bay with a shingle beach backed by cliffs in bedrock and fan sediments. The next small cove to the north (Fig. 6.5.5), reached by clambering over the low bedrock ridge to the north of Cala Carbón, has a small modern sandy beach flanked by modern wavecut platforms in bedrock. At the back of the beach are boulders of cemented shelly beach conglomerates, possibly of Tyrrhenian II age (Isotope Stage 5, 135–110 ka; Harvey *et al.* 1999c). If this is so, it indicates that coastal erosion at that time had already cut the cove and foreshortened the fan. Return to the vehicle(s).

Transfer to Stop 3

Retrace the route back to San José; turn left on the main road, back through El Pozo to the Los Escullos turn (792 727, Fig. 6.5.4). Turn right towards Los Escullos, then after *c.* 4 km, past the entrance to Los Escullos camp site, turn right again (at 837 742), towards the beach at Los Escullos. Drive on for *c.* 300 m past the hostel and restaurant; park on the left of the track (Stop 3a).

Stop 3a: Los Escullos, Quaternary aeolianites (837 732; El Pozo de los Frailes 1 : 25 000)

On a small headland on the north side of Punta del Esparto (at 838 732) are flowbanded lavas with columnar jointing in places. These are overlain by richly fossiliferous calcarenites containing abundant cidaroid spines, and by soft-weathering bedded crystal lithic tuffs and agglomerates. On the north side of a small bay, this thin succession is unconformably overlain by Quaternary alluvial-fan conglomerates and a Quaternary aeolianite with preserved fossil dune morphology. Locally, the fan conglomerates are

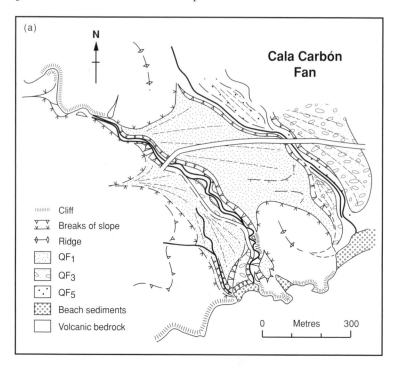

Fig. 6.5.4. Geomorphic map of Los Escullos and La Isleta fan, Stop 3. (Modified after Harvey *et al.* 1999c.)

capped by a fossil red soil, containing terrestrial gastropods. The fossil dunes extend north to beyond the hostel. Walk over the surface of the cemented dunes. Note the presence of large root casts on the surface. From the top of the dunes take in the views inland and along the coast. Inland to the west, younger fan sediments rest on the aeolianites. Behind them are older fan sediments, with red soils and calcrete crusted surfaces. Along the coast to the north, the view extends towards La Isleta, the small white settlement *c*. 2 km away across the bay.

Towards La Isleta the cliffed coast cuts a large alluvial fan (Fig. 6.5.5), whose oldest segments form the surface sloping gently seawards. The cliffs expose volcanic bedrock at the base overlain by the oldest fan deposits (QF1 in Fig. 6.5.5). Within this large fan the older sequence is predominantly composed of coarse, fluvial, cobbly sediments, but in proximal zones they are interbedded with laterally fed debris-flow deposits (Harvey *et al*. 1999b). The sequence is punctuated by several cut and fill phases clearly identifiable by marked erosion surfaces within the sections. On morphostratigraphical grounds these sediments appear to pre-date the aeolianites. The nearby lower cliff below a white villa is composed of younger fan deposits (QF3 in Fig. 6.5.5) set into the older fan. Their base is near the cliff base, and their top surface can be traced as an inset surface back into the fan. The youngest fan sediments (QF3 in Fig. 6.5.5) occur as terraces within the fan trenches.

Three main groups of fan surfaces (QF1, 3, 5 in Fig. 6.5.5) can be differentiated on the basis of soil development: with very red mature soil profiles, underlain by Stage III carbonate accumulation, characterizing the oldest surfaces; younger, less mature, less red soils with less carbonate characterizing the intermediate surfaces; and little soil development on the youngest surfaces (Harvey *et al*. 1999b).

Transfer to Stop 3b

Return to the vehicle(s). It is worth driving over the fan surfaces of La Isleta fan to Stop 3b, to see something of the sediments and the fan morphology. Drive back to the main road, and turn right (north) towards La Isleta. Within *c*. 400 m the road climbs from the youngest through the intermediate onto the oldest fan surface. Note the bright red soils forming the oldest fan surface. A little further on there is a track on the left going up-fan across the oldest surface. If you have a permit (see Chapter 1 and Appendix for details on how to obtain permission) it is possible to drive up this track to examine the fan sediments, exposed in the fan-head trench. Otherwise continue on the main road to where it crosses the main fan channel. Park on the shoulder between the channel and the turning to La Isleta.

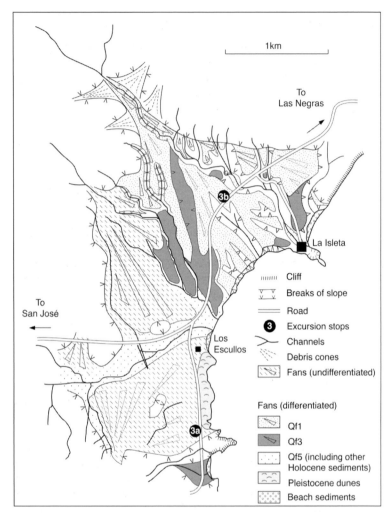

Fig. 6.5.5. (a) Geomorphic map of Cala Carbón fan, Stop 2. (Modified after Harvey *et al.* 1999.) (b) View up Cala Carbón fan (Stop 2). This fan is coupled to base levels related to Quaternary sea-level changes. At times of high sea-levels the distal parts of the fan have been dissected as a result of profile foreshortening following coastal erosion of the fan toe. Note old fan surfaces (1 relates to QF1, see text), into which intermediate fan surfaces are set (3 relates to QF3, see text). (Photo A.M. Harvey.)

Stop 3b: *La Isleta fan (843 754; El Pozo de los Frailes 1 : 25 000)*

From this location and from the side road to La Isleta (a tiny fishing settlement), the three groups of fan surfaces can be seen, with some exposure of the constituent sediments. The modern incised channel (see Fig. 6.5.5) steepens in mid-fan, which may be the result of headwards erosion from earlier profile foreshortening by marine erosion.

The contrasting morphologies of these dissected east-coast fans and the non-dissected Cabo de Gata fans illustrate the relative roles of climatic/ sediment-led proximal controls and base-level-led distal controls over fan evolution (Harvey *et al.* 1999b). On the east-coast fans, dissection has been induced both proximally, in the form of deep fan-head trenches, and distally, as expressed by profile steepenings of the fan channels (Fig. 6.5.5). This has been in response to both climatic change, controlling sediment supply, and base-level change controlling distal erosion. During times of low sea-level, coincident with global glacial phases, these fans prograded seawards. During high sea-levels the fans were subject to distal coastal erosion foreshortening their channel profiles and accentuating incisional tendencies. Interestingly, dissection occurs, not as might be expected at times of lowered base-levels, but in response to coastal erosion at times of high sea-levels (Harvey *et al.* 1999b).

Transfer to Stop 4

From La Isleta turn north along the coast road towards Las Negras. Bear right through Rodalquilar, site of a large abandoned gold mine (Cunningham *et al.* 1990), and continue to the Las Negras turn (at 869 811). Turn left towards Fernán Pérez.

En route to Fernán Pérez, the road is within the Cabo de Gata volcanics capped by Neogene sedimentary rocks. It is worth stopping at the Amatista Mirador (857 762; see Excursion 3.3a, Stop 3) on the right, beyond La Isleta, for a spectacular view of the coastline. East of Rodalquilar (at 868 794) there is access to the coast at El Playazo, where there are exposures of Tyrrhenian II (Isotope Stage 5, 135–110 ka) beach sediments and younger aeolianites (Goy & Zazo, 1986).

If travelling by car or minibus it is possible to take a shorter route *c.* 8 km along a dirt track from Fernán Pérez. In Fernán Pérez, where the main road bends to the left, near a bus shelter (835 856), bear right into a very narrow lane that soon becomes a dirt track, and widens. Follow this track, heading NE, to a large quarry. Keep the quarry on your right and join the main road leaving the quarry, continuing towards the NE. The road is now a wide gravel road, which continues for several kilometres to join the metalled side road to Agua Armaga, just past a large estancia on your left called El Molino. Near El Molino are the Tortonian temperate carbonate outcrops visited on Excursion 4.2. Turn left onto the road from Agua Amarga; drive for *c.* 3 km to the intersection with the Al-101 Carboneras road. There, turn right towards Carboneras, rejoining the main excursion route. If in a coach continue through Fernán Pérez to Campohermoso. Turn right in Campohermoso, following signs to the N340 motorway. Join the motorway, heading east (signed for Murcia) to Venta del Pobre. Exit at Venta del Pobre, following the Al-101 to Carboneras.

After Fernán Pérez the route leaves the Cabo de Gata ranges, crossing the Carboneras Fault Zone between 828 866 and 818 872. Here the zone between the main faults comprises volcanics resting on Maláguide basement. Blocky, brecciated lavas are exposed in the road cuttings. Beyond the fault zone the road descends small Quaternary alluvial fans, now obscured by plastic agriculture, to Campohermoso.

At Venta del Pobre take the exit for Carboneras (Al-101) across the Carboneras fault zone between 868 918 and 877 915. *The alternative dirt track route from Fernán Pérez rejoins this road at 903 919.* En route to Carboneras, note the rocks exposed in the roadside sections, including Pliocene pebbly conglomerates, Messinian limestones and various Cabo de Gata volcanic rocks. On the final stretch into Carboneras, along the shore, note the Pleistocene marine conglomerates capping the low cliffs on the left.

Continue through Carboneras, following the Mojácar signs through the back (landward side) of the town. Continue to follow the Mojácar signs along the coast road, crossing the Río Carboneras/Alias and climbing to the summit of the Granatilla col. It is worth making a stop to photograph the coastline at the view point at the top of the col.

For most of the way from Carboneras the route has been through volcanic rocks, but there are Neogene sedimentary rocks just south of the bridge over the Río Carboneras/Alias, and Quaternary marine cemented beach sediments along the foreshore there.

From the col, continue almost to the bridge in the valley bottom, to Stop 4 at 997 013. Here it is possible to park on a section of the old road adjacent to the new road. This might be difficult for coaches, which can park at the previous col, allowing the party to walk the *c.* 2.5 km through Stop 4a to Stop 4b at Sopalmo, however there is little space to park for long. A coach could park further along the road where space permits.

Stop 4a: Granatilla/Sopalmo, the Carboneras strike-slip Fault Zone within the Sierra Cabrera: its influence on drainage evolution (996 012; El Agua del Medio 1 : 25 000)

The Carboneras Fault Zone here involves several faults trending NE–SW, separating Cabo de Gata volcanics to the SE from black Cabrera schists to the NW (Fig. 6.5.6). Within the fault zone are slivers of a variety of rocks, including pale, fractured schists and multi-coloured Triassic phyllites. These are exposed in the fault zone at the roadside (at 997 011) and across the valley, where they have been dissected to form a small area of multi-coloured badlands (visited in Excursion 3.5a, Stop 3).

Walk back up the road to 998 010 to examine the lithology of the volcanics. The dominant lithology here is of hornblende-rich, calc-alkaline andesitic rocks, sometimes as blocky lavas, sometimes as agglomeratic rocks.

Return to the vehicle(s), and drive on to the next col at Sopalmo (996 020). Park opposite Bar Sopalmo (Stop 4b).

Stop 4b: Sopalmo (996 020; El Agua del Medio 1 : 25 000)

Sopalmo sits in a col between the drainages of the Rambla de Granatilla to the south and the Rambla de Macenas to the north. The col is floored by a thick sequence of Quaternary river gravels, suggesting that the former drainage through the col has been captured (Fig. 3.5.3). These gravels are exposed *c.* 100 m south of Sopalmo, in a road-cutting. Did the fault-aligned Macenas capture the headwaters of an original transverse Granatilla? Or did the steep, aggressive, lower Granatilla, flowing through the weak rocks of the fault zone, capture an original fault-aligned, lower-gradient Macenas? Palaeocurrent observations within the gravels indicate the latter, but the virtual absence of volcanic clasts in the gravels is not easy to explain.

Transfer to Stop 5

Drive northwards down the Macenas valley (Fig. 6.5.1b). You are now within the Cabrera schists, north of the Carboneras Fault Zone. *En route* notice the small, active, tributary junction alluvial fans, and the braided rambla channel. At the bridge over the rambla (007 028), note the two Quaternary terraces exposed downstream of the bridge.

Continue to where the road meets the coastline (023 045). Coaches can park on the shoulder beyond the track junction. Cars and minibuses take the track on the right at the Castillo de Macenas. Turn right in front of the Castillo, following the sign for the Playas de Macenas. Cross the gravel fan delta of the Rambla de Macenas, and continue south along the track, above the coast, to where one of the Carboneras faults intersects the coastline. Here (024 036), where the multi-coloured Triassic rocks are exposed, is a convenient place to park. This is the base for Stop 5.

(a)

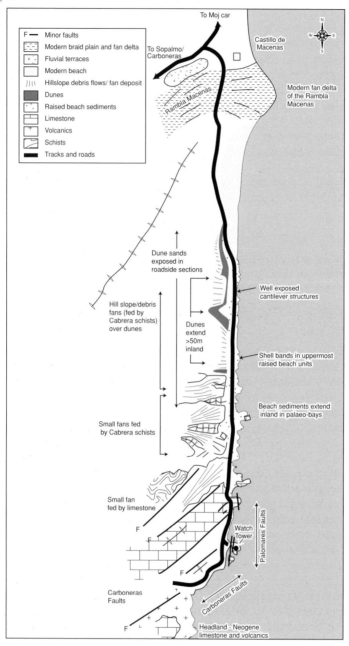

Fig. 6.5.6. (a) Stop 5: Macenas beach. (a) Sketch map of the geomorphology of Macenas

(b)

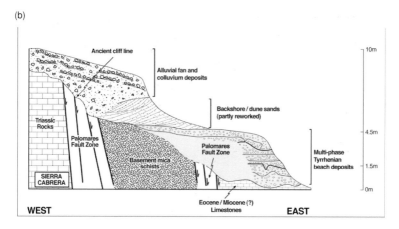

Fig. 6.5.6. (b) Interpretative sketch section of the sediment sequence exposed on the foreshore at Macenas. (Modified after Harvey, 1987a.)

Stop 5: Playa de Macenas, Carboneras and Palomares faults and Quaternary beach conglomerates (024 036; Castillo de Macenas 1 : 25 000)

The NE–SW Carboneras Fault System intersects with the north–south Palomares Fault System along this section of coast (visited as Excursion 3.5a, Stop 2).

From the watchtower on the headland (Fig. 6.5.6a) there are views along the coast. The view south, towards 024 030, shows the geology south of the Carboneras faults, with Cabo de Gata volcanics overlain by Tortonian–Messinian limestones. The view to the north (Fig. 6.5.7), towards Mojácar, is along the line of the Palomares Fault (Bousquet, 1979). In the far distance note the Sierra Almagrera, the NE extension of the Sierra Cabrera, to the east of the Palomares Fault Zone. This block has been moved more than 15 km to the north along the fault since the Miocene (Stokes, 1997).

Return to the parking place to examine the superb sequence of Quaternary shoreline deposits exposed along the foreshore between here and the Macenas valley (Harvey, 1987a; Stokes, 1997). The sequence starts with an irregular erosional topography of sea stacks, headlands and a cliffline, buried by a transgressional sequence of dominantly quartz conglomerate beach sediments. The transgressional sequence was oscillatory, involving three phases, each separated by minor regressive phases (Fig. 6.5.6b). This is clearly demonstrated by the sections, which show three successive basal conglomerates passing up into quartz beach conglomerates. Each culminates in buried cantilever slab failures (Fig. 6.5.6b) representing

Fig. 6.5.7. The coast at Macenas (Stop 5, see also Excursion 3.5a, Stop 2) looking north to the modern fan delta (d) of the Rambla de Macenas. Bedrock comprises Triassic rocks, Tertiary limestones (T) strongly faulted by the SW–NE-trending Carboneras Fault System and the coast-parallel Palomares Fault System. Cabrera schists (s) crop out beyond the Carboneras Fault Zone. The Quaternary sediments include a trangressive Tyrrhenian II sequence of cemented raised beach conglomerates (q), overlain by aeolian sands (a) and by colluvial slope and fan sediments (c) derived from the hillsopes and small catchments landward of the beach zone. Since this photo was taken (1982) erosion has removed most of the sand and shingle from the modern beach.

subaqueous cementation (J.D. Marshall, University of Liverpool, personal communication) followed by minor regression, then renewed transgression causing cantilever failure of the blocks by wave erosion. The new basal conglomerate wraps around toppled blocks. Interestingly, although these conglomerates cross one of the Carboneras faults, there is no sign of any tectonic disturbance, indicating that along that particular arm of the Carboneras Fault System there has been no movement since deposition of the beach sediments.

The third transgressional beach, exposed in approximately the position of the road, includes important shelly horizons (dominantly of bivalves, but including the rare gastropod *Strombus bubonius*) and passes up into sands, initially suggesting a back-beach environment but later clearly representing aeolian dunes. The dunes extend some way inland. Towards the top, especially in the southern part of the section, the dune sands are mixed with terrestrially derived debris-flow sediments, suggesting reworking and that aeolian sedimentation was contemporaneous with hillslope debris-flow activity. To the north of the section, where the hillslope coastal catchment areas are much smaller, there is a clearer break between the dunes and the overlying debris-flows. There, the dunes are capped by a strong pedogenic carbonate, remains of a red B horizon, and root casts in the top of the dune sands, all suggesting a time interval prior to the onset of slope processes in these locations.

The slope sediments themselves illustrate catchment controls over sedimentation styles. To the south, fed by larger catchment areas, the deposits form a small alluvial fan, and show evidence of a much more fluid environment of deposition. The sections also include palaeosols, suggesting that sedimentation was episodic. The now dissected top surface of the fan slopes at c. 10°. To the north, the sediments are typical debris flows, and form colluvial/debris-flow footslopes rather than alluvial fans, and slope seawards at c. 20°. There is much less evidence of intermittent palaeosol development here, other than the well-developed soil capping the dune sediments, suggesting that here slope processes were active later and for a shorter duration than to the south.

The whole sequence records the coastline evolution through the Late Pleistocene. The beach conglomerates are probably equivalent to dated sediments in the Mojácar area dated as Tyrrhenian II (Thurber & Stearns, 1965), and therefore relate to the transgression c. 120 ka. The dunes appear to be regressional aeolianites, related to the 'Early Würm' marine regression (Butzer & Cuerda, 1962; Butzer, 1964) c. 70 ka. The terrestrial sediments appear to be 'Würm', with the youngest relating to the 'Late Würm', perhaps c. 20–10 ka. The Holocene has seen dissection of the sequence along ravines and gullies. This is a clear illustration of the catchment rather than base-level control over terrestrial erosion and deposition even in a location so close to the sea! Rising sea-levels during the Holocene have had no influence over the sequence landwards of the modern low cliffs, cut in the Quaternary beach sediments. Local base level is and has remained the resistant top of the marine sediments.

Transfer to Stop 6

Return to the vehicle(s). Drive back to the metalled road and drive north towards Mojácar. Continue through Mojácar beach settlement, on the coast road, heading north towards Garrucha. In Garrucha, park near the cross-roads at the southern end of the town for Stop 6.

En route note several exposures of Quaternary coastal and terrestrial sediments and red soils exposed in building cuttings. Cross the Río Aguas, which here has a transverse course across the north end of the Sierra Cabrera. The valley that runs inland towards Garrucha may mark a former course of the Río Aguas. Note the hill to the north of the Río Aguas, on which Völk (1979) identified a Quaternary coastal sequence. Much seems now to have been removed by building development. Similarly, the once extensive Tyrrhenian II coastal sediments which cropped out along the shore have been bulldozed for the purpose of beach improvement.

Stop 7: Garrucha southern road section, Quaternary deformation along the Palomares Fault (045 151; Garrucha 1 : 25 000)

Exposed in the road-cutting, running at right angles to the sea (Fig. 6.5.8), is a complex set of deformed (?)Pliocene and Quaternary sediments, adjacent to the Palomares Fault Zone. Our interpretation of the sequence (Stokes, 1997) is of a first phase of Quaternary beach sedimentation, transgressive over an eroded Neogene (probably Pliocene?) base, which culminates in a lagoonal phase. This was followed by complex deformation involving faulting and slumping. This in turn was followed by marine regression, stabilization and soil formation. The final sedimentation phase was a further marine transgression, presumably the widely represented Tyrrhenian II sequence, reaching its landward limit near the seaward end of this section, followed by regression and a second soil. There appears to have been some minor tilting since then.

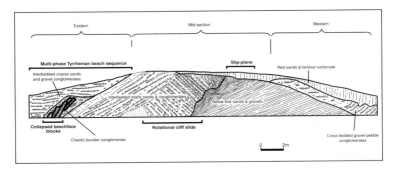

Fig. 6.5.8. Stop 6: Garrucha southern road section. (Modified after Stokes, 1997.)

Appendix

Permission to walk/sample in the National Parks

Issued via:
El Director Conservador
Parque de Cabo de Gata
Conserjería de Medio Ambiente
Junta de Andalucía
Rodalquilar
Almería
Spain
Tel.: +34 950 38 97 42
Fax: +34 950 38 97 54

Aerial photographs and maps

Centro Nacional de Información Geográfico
Ministerio de Obras Publicas
Transportes y Medio Ambiente c/General Ibáñez de Ibero, 3, E-28003
Madrid, Spain
Tel.: +34 91 536 06 36
Fax: +34 91 553 29 13
http://www.oan.es/servicios/e/CNIG.html

Military topographic maps on CD ROM

Servicio Geográfico del Ejército
Calle Darío Gazapo n°. 8
28024 Madrid
Tel.: +34 91 711 50 43

Information on geological maps and related resources

Instituto Geológico y Minero de España (IGME)
Ríos Rosas
23-28003 Madrid
España
Tel.: +34 91 349 5700
Fax: +34 91 442 6216
Email: webmaster@igmes.es
http://www.igme.es/internet/itge.htm

Accommodation

Cortijo Urra Field Study Centre
Sorbas 04270
Almería
Spain
Tel.: (020)7 359 7087
Fax: (020)7 503 9912
Email: anna@urra.globalnet.co.uk

Useful texts

ALLEN, P.A. (1997) *Earth Surface Processes.* Blackwell Science, Oxford.
EMERY, D. & MYERS, K. (1996) *Sequence Stratigraphy.* Blackwell Science, Oxford.
FRY, N. (1991) *The Field Description of Metamorphic Rocks.* Wiley, Chichester.
JONES, A.P., TUCKER, M.E. & HART, J.K., eds. *The Description and Analysis of Quaternary Stratigraphic Sections.* Technical Guide 7. Quaternary Research Association.
THOMAS, D.S.G. (1997) *Arid Zone Geomorphology* (2nd edn). Wiley, Chichester.
THORPE, R. & BROWN, G. (1991) *The Field Description of Igneous Rocks.* Wiley, Chichester.
TUCKER, M. (1991) *Sedimentary Rocks in the Field.* Wiley, Chichester.

References

Acosta, A. (1998) *Estudio de los fenómenos de fusión cortical y generación de granitoides asociados a las peridotitas de Ronda.* PhD thesis, University of Granada, Spain.

Addicot, W.O., Snavely, P.D., Bukry, D. Jr & Poore, R.Z. (1977) Neogene stratigraphy and paleontology of southern Almería Province, Spain: an overview. *US Geol. Surv. Open-File Rep.* **77**, 1–49.

Aguirre, J. (1995) *Tafonomía y evolución sedimentaria del Plioceno marino en el litoral sur de España entre Cádiz y Almería.* PhD thesis, University of Granada, Spain.

Aldaya, F., Alvarez, F., Galíndo-Zaldívar, J., González-Lodeiro, J., Jabaloy, A. & Navarro-Vila, F. (1991) The Maláguide–Alpujárride contact (Betic Cordilleras, Spain): a brittle extensional detachment. *C. R. Acad. Sci. Paris* **313**, 1447–1453.

Aldaya, F., García-Dueñas, V. & Navarro-Vila, F. (1979) Los Mantos Alpujárrides del tercio central de las Cordilleras Béticas: ensayo de correlación tectónica de los Alpujárrides. *Acta geol. Hispánica* **14**, 154–166.

Alexander, R.W. & Calvo, A. (1990) The influence of lichens on slope processes in some Spanish badlands. In: *Vegetation and Erosion* (Ed. Thornes, J.B.), pp. 385–398. Wiley, Chichester.

Alexander, R.W., Harvey, A.M., Calvo, A., James, P.A. & Cerda, A. (1994) Natural stabilisation mechanisms on badland slopes: Tabernas, Almería, Spain. In: *Environmental Change in Drylands: Biogeographical and Geomorphological Perspectives* (Eds Millington, A.C. & Pye, K.), pp. 85–111. Wiley, Chichester.

Alexander, R.W., Spivey, D.B., Faulkner, H. & Willshaw, K. (1999) Badland morphology and geoecology. Mocatán Systems: processes and patterns. In:

BSRG/BGRG SE Spain Field Meeting: Field Guide. (Eds Mather, A.E. & Stokes, M.), pp. 134–151. University of Plymouth.

Álvarez, F. (1987) *La tectónica de la Zona Bética en la región de Aguilas.* PhD thesis, University of Salamanca, Spain.

Amor, J.M. & Florschutz, F. (1964) Results of the preliminary palynological investigation of samples from a 50-m boring in Southern Spain. *Bol. R. Soc. Esp. Hist. Nat. (Geol.)* **62**, 251–255.

Angelier, J., Cadet, J.P., Delibrias, G. *et al.* (1976) Les déformations du Quaternaire marin, indicateurs néotectoniques: quelques exemples Méditerranéans. *Rev. Geogr. phys. Geol. Dynam.* **18**, 427–448.

Araña, V. & Vegas, R. (1974) Plate tectonics and volcanism in the Gibraltar arc. *Tectonophysics* **24**, 197–212.

Arribas, A. (1993). *Mapa Geológico del Distrito Minero de Rodalquilar, Almería.* Instituto Tecnológico Geominero de España, Madrid.

Arribas, A., Cunningham, C.G., Rytuba, J.J. *et al.* (1995) Geology, geochronology, fluid inclusions, and isotope geochemistry of the Rodalquilar gold alunite deposit, Spain. *Econ. Geol.* **90**, 795–822.

Azañón, J.M., Crespo-Blanc, A. & García-Dueñas, V. (1997) Continental collision, crustal thinning and nappe forming during the pre-Miocene evolution of the Alpujárride Complex (Alborán Domain, Betics). *J. struct. Geol.* **19**, 1055–1071.

Azañón, J.M., Crespo-Blanc, A., García-Dueñas, V. & Sánchez-Gómez, M. (1996) Folding of metamorphic isograds in the Adra extensional unit (Alpujárride Complex, Central Betics). *C. R. Acad. Sci. Paris* **323**, 949–956.

Azañón, J.M. & Goffe, B. (1997) Ferro- and magnesiocarpholite assemblages as record of high-*P*, low-*T* metamorphism in the Central Alpujárrides, Betic Cordillera (SE Spain). *Eur. J. Mineral.* **9**, 1035–1051.

Azema, J. (1961) Etude géologique des abords de Málaga (Espagne). *Estud. Geol.* **17**, 131–160.

Bakker, H.E., De Jong, K., Helmes, H. & Biermann, C. (1989) The geodynamic evolution of the internal zone of the Betic Cordilleras (south-east Spain): a model based on structural analysis and geothermobarometry. *J. metam. Geol.* **7**, 359–381.

Balanyá, J.C., Azañón, J.M. & García-Dueñas, V. (1997) Alternating contractional and extensional events in the Alpujarride nappes of the Alborán Domain (Betics, Gibraltar Arc). *Tectonics* **16**, 226–238.

Balanyá, J.C., Azañón, J.M., Sánchez-Gómez, M. & García-Dueñas, V. (1993) Pervasive ductile extension, isothermal decompression and thinning of the Jubrique unit during the Paleogene times (Alpujárride Complex, western Betics). *C. R. Acad. Sci. Paris* **316**, 1595–1601.

Barragan, G. (1986–87) Una nueva interpretación de la sedimentación neógena en el sector suroccidental de la Cuenca de Vera. *Acta geol. Hispánica* **21–22**, 449–457.

Barragán, G. (1994) Algunos datos sobre la actividad hidrotermal Pliocena al oeste de Cuevas del Almanzora. Encuadre geológico y cronológico de las manifestaciones magmáticas e hidrotermales de la depresión de Vera (Provincia de Almería). Simposia de Recursos Naturales y Medio Ambiente en el Sureste Peninsular. Investigacion y Aprovechamiento. Cuevas del Almanzora (Almería). Almería Instituto de Estudios Almerianses: Cuevas del Amanzora Ayuntamiento.

Barragan, G. (1997) *Evolución geodinámica de la Depresión de Vera, Provincia de Almería, Cordilleras Béticas.* PhD thesis, University of Granada, Spain.

Bell, J.W., Amelung, F. & King, G.C.P. (1997) Preliminary late Quaternary slip history of the Carboneras Fault, Southeastern Spain. *J. Geodyn.* **24**, 51–66.

Bellon, H., Bordet, P. & Montenat, C. (1983) Le magmatisme néogène des Cordillères Bétiques (Espagne): chronologie et principaux caractères géochimiques. *Bull. Soc. geol. France* **25**, 205–218.

Bennett, M., Doyle, P. & Mather, A.M. (1996) Dropstones: their origin and significance. *Palaeogeogr. Palaeoclimatol. Palaeoecol.* **121**, 331–339.

Bennett, M., Doyle, P., Mather, A.E. & Woodfin, J.L. (1994) Testing the climatic significance of dropstones: an example from southeast Spain. *Geol. Mag.* **131**, 845–848.

Berggren, W.A., Kent, D.V., Swisher, C.C. III, & Aubry, M.P. (1995) A revised Cenozoic geochronology and chronostratigraphy. In: *Geochronology, Time Scales and Global Stratigraphic Correlation* (Eds Berggren, W.A., Kent, D.V., Aubry, M.P. & Hardenbol, J.), Spec. Publ. Soc. econ. Paleont. Miner., Tulsa, **54**, 129–212.

Betzler, C., Brachert, T., Braga, J.C. & Martín, J.M. (1997) Nearshore, temperate, carbonate depositional systems (lower Tortonian, Agua Amarga Basin, southern Spain): implications for carbonate sequence stratigraphy. *Sediment. Geol.* **113**, 27–53.

Betzler, C., Brachert, T., Braga, J.C. & Martín, J.M. (1998) Temperate carbonates in the Agua Amarga Basin (Miocene, Almería, SE Spain). In: *Fieldguide Book*, pp. 277–292. 15th International Sedimentological Congress, Alicante, Spain.

Blanco, M.J. & Spakman, W. (1993) The P-wave velocity structure of the mantle below the Iberian Peninsula: evidence for subduction lithosphere below southern Spain. *Tectonophysics* **221**, 13–34.

Blumenthal, M. (1927) Versuch einer tektonischen Gliederung der betischen Cordilleren von Central- und Südwest-Andalusien. *Eclogae geol. Helv.* **20**, 487–532.

Blumenthal, M. (1935) Reliefüberschiebungen in den westlichen Betischen Cordilleren. *Geología Del Mediterraneo Occident* **4**, 3–28.

Bodinier, J.L., Morten, L., Puga, E. & Díaz De Federico, A. (1987) Geochemistry of metabasites from the Nevado–Filábride Complex, Betic Cordilleras, Spain: relics of a dismembered ophiolitic sequence. *Lithos* **20**, 235–245.

Boorsma, L.J. (1992) Syn-tectonic sedimentation in a Neogene strike-slip basin containing a stacked Gilbert-type delta (SE Spain). *Sediment. Geol.* **81**, 105–124.

Bordet, P. (1985) *Le volcanisme miocène des Sierras de Gata et de Carboneras (Espagne du Sud-Est).* Doc. et Trav. IGAL, **8**.

Bousquet, J.C. (1979) Quaternary strike-slip faults in southeastern Spain. *Tectonophysics* **52**, 277–286.

Brachert, T.C., Betzler, C., Braga, J.C. & Martín, J.M. (1996) Record of climatic change in neritic carbonates: turnover in biogenic associations and depositional modes (late Miocene, southern Spain). *Geol. Rundsch.* **85**, 327–337.

Brachert, T.C., Betzler, C., Braga, J.C. & Martín, J.M. (1998) Micro-taphofacies of a warm-temperate carbonate ramp (uppermost Tortonian–lower Messinian, Southern Spain). *Palaios* **13**, 459–475.

Braga, J.C. & Martín, J.M. (1987) Distribución de las algas dasycladáceas en el Trías Alpujárride. *Cuad. Geol. Ibérica* **11**, 475–489.

Braga, J.C. & Martín, J.M. (1992) Messinian carbonates of the Sorbas basin: sequence stratigraphy, cyclicity and facies. In: *Late Miocene Carbonate Sequences of Southern Spain: A Guidebook for the Las Negras & Sorbas Area*, in conjunction with the SEPM/IAS Research Conference on Carbonate Stratigraphic Sequences: Sequence Boundaries & Associated Facies, August 30–September 3, La Seu, Spain, pp. 78–108.

Braga, J.C. & Martín, J.M. (1996a) Geometries of reef advance in response to relative sea-level changes in a Messinian (uppermost Miocene) fringing reef (Cariatiz reef, Sorbas Basin, SE Spain). *Sediment. Geol.* **107**, 61–81.

Braga, J.C. & Martín, J.M. (1996b) Messinian 'reefs' of the Sorbas basin, Almería, SE Spain. In: *2nd Cortijo Urra Field Meeting, SE Spain: Field Guide* (Ed. Mather, A.E. & Stokes, M.), pp. 4–13. University of Plymouth, England.

Braga, J.C. & Martín, J.M. (1997) Evolución paleogeográfica de la cuenca de Sorbas y relieves adyacentes (Almería, SE España) en el Mioceno superior. In: *Avances en el Conocimiento del Terciario Ibérico* (Eds Calvo, J.P. & Morales, J.), pp. 61–64. Universidad Complutense de Madrid–Museo Nacional Ciencias Naturales, Spain.

Braga, J.C. & Martín, J.M. (2000) Subaqueous siliciclastic stromatolites: a case history from beach deposits in the Sorbas Basin, SE Spain. In: *Microbial Sediments* (Eds Riding, R. & Awramik, S.W.), pp. 226–232. Springer, Berlin.

Braga, J.C., Martín, J.M. & Alcalá, B. (1990) Coral reefs in coarse terrigenous sedimentary environments (upper Tortonian, Granada basin, S. Spain). *Sediment. Geol.* **66**, 135–150.

Braga, J.C., Martín, J.M., Betzler, C. & Brachert, T.C. (1996b) Miocene temperate carbonates in the Agua Amarga Basin (Almería, SE Spain). *Rev. Soc. geol. España* **9**, 285–296.

Braga, J.C., Martín, J.M., Betzler, C.H., Brachert, T., Civis, J. & Sierro, F.J. (1994) Carbonatos templados Tortonienses de la cuenca de Agua Amarga (Almería, SE España). *II Congreso del Grupo Español del Terciario*, Jaca (Huesca), Septiembre 1994. Comunicaciones, pp. 69–72.

Braga, J.C., Martín, J.M. & Riding, R. (1995) Controls on microbial dome fabric development along a carbonate–siliciclastic shelf–basin transect, Miocene, SE Spain. *Palaios* **10**, 347–361.

Braga, J.C., Martín, J.M. & Riding, R. (1996a) Internal structure of segment reefs: *Halimeda* algal mounds in the Mediterranean Miocene. *Geology* **24**, 35–38.

Buforn, E., Udías, A. & Mezcua, J. (1988) Seismicity and focal mechanisms in south Spain. *Bull. Seism. Soc. Am.* **78**, 2008–2024.

Butzer, K.W. (1964) Climatic–geomorphologic interpretation of Pleistocene sediments in the Eurafrican sub tropics. In: *African Ecology and Human Evolution* (Eds Howell, F.C. & Bourliere, F.), pp. 1–25. Methuen, London.

Butzer, K.W. & Cuerda, J. (1962) Coastal stratigraphy of Southern Mallorca and its implications for the Pleistocene chronology of the Mediterranean Sea. *J. Geol.* **70**, 398–416.

Calvo-Cases, A., Harvey, A.M. & Paya-Serrano, J. (1991a) Process interactions and badland development in SE Spain. In: *Soil Erosion Studies in Spain* (Eds Sala, M., Rubio, J.L. & Garcia-Ruiz, J.M.), pp. 75–90. Geoforma Ediciones, Logroño, Spain.

Calvo-Cases, A., Harvey, A.M., Paya-Serrano, J. & Alexander, R.W. (1991b) Response of badland surfaces in southeast Spain to simulated rainfall. *Cuarternario y Geomorfología* **5**, 3–14.

Cesare, B., Salvioli Mariani, E. & Venturelli, G. (1997) Crustal anatexis and melt extraction during deformation in the restitic xenoliths at El Joyazo (SE Spain). *Mineral. Mag.* **61**, 15–27.

Cita, M.B. (1991) Development of a scientific controversy. In: *Controversies in Modern Geology: Evolution of Geological Theories in Sedimentology, Earth History and Tectonics* (Eds Müller, D.W. McKenzie, J.A. & Weissert, H.), pp. 13–23. Academic Press, London.

Civis, J., Martinell, J. & de Porta, J. (1977) Precisiones sobre la edad del miembro Zorreras (Sorbas, Almería) In: *Messinian Seminar No. 3*, Abstracts, Málaga, Spain.

Clark, J.D. & Pickering, K.T. (1996) *Submarine Channels, Processes and Architecture*. Vallis Press, London.

Claypool, G.E., Holser, W.T., Kaplan, I.R., Sakai, H. & Zak, I. (1980) The age curves of sulfur and oxygen isotopes in marine sulfate and their mutual interpretation. *Chem. Geol.* **28**, 199–260.

Comas. M.C., García-Dueñas, V. & Jurado, M.J. (1992) Neogene tectonic evolution of the Alborán basin from MCS data. *Geo-Marine Lett.* **12**, 157–164.

Comas, M.C., Zahn, R., Klaus, A. *et al.* (1996). *Proceedings of the ODP, Initial Reports*, **161**. Ocean Drilling Program, College Station, TX.

Crespo-Blanc, A., Orozco, M. & García-Dueñas (1994) Extension versus compression during the Miocene tectonic evolution of the Betic Chain: late folding of normal fault systems. *Tectonics* **13**, 78–88.

Cronin, B.T. (1994) *Channel-fill architecture in deep water sequences: variability, quantification and applications*. PhD thesis, University of Wales, UK.

Cronin, B.T. (1995) Structurally-controlled deep sea channel courses: examples from the Miocene of southeast Spain and the Alborán Sea, southwest Mediterranean. In: *Characterisation of Deep Marine Clastic Systems* (Eds Hartley, A.J. & Prosser, D.J.), Spec. Publ. geol. Soc. London, No. 94, pp. 115–135. Geol. Soc. London, Bath.

Cronin, B.T., Ivanov, M.K., Limonov, A.F. *et al.* & Shipboard Scientific Party TRR-5 (1997) New discoveries of mud volcanoes on the Eastern Mediterranean Ridge. *J. geol. Soc. London* **154**, 173–182.

Cuevas, J. (1990) Microtectónica y metamorfismo de los Mantos Alpujárrides del tercio central de las Cordilleras Béticas (entre Motril y Adra), *Publicaciones especiales del Boletín Geológico y Minero de España.*

Cuevas, J. & Tubía, J.M. (1990) Quartz fabric evolution within the Adra Nappe (Betic Cordilleras, Spain). *J. struct. Geol.* **12**, 823–833.

Cunningham, C.G., Arribas, A., Jr, Rytuba, J.J. & Arribas, A. (1990) Mineralized and unmineralized calderas in Spain: Part I, evolution of the Los Frailes Caldera. *Mineral. Deposita* **25**, S21–S28.

Dabrio, C.J. (1990) *Fan-Delta Facies Associations in the Late Neogene and Quaternary Basins of Southeastern Spain*. In: *Coarse-grained Deltas* (Eds Colella, A & Prior, D.B.), Spec Publs int. Ass. Sediment., No. 10, pp. 91–111. Blackwell Scientific Publications, Oxford.

Dabrio, C.J., Esteban, M. & Martín, J.M. (1981) The coral reef of Níjar, Messinian (uppermost Miocene), Almería province, SE Spain. *J. sediment. Petrol.* **51**, 521–539.

Dabrio, C.J., Fortun, A.R., Polo, M.D. & Roep Th. B. (1998) Seaward and landward jumping barrier islands and high-frequency sea-level changes in the Sorbas Basin. In: *Fieldguide Book*, 15th Internaional Sedimentological Congress, Alicante, Spain, 255–274.

Dabrio, C.J. & Martín, J.M. (1978) Los arrecifes Messinienses de Almería (SE de España). *Cuad. Geol. University Granada* **8–9**, 85–100.

Dabrio, C.J., Martín, J.M. & Megías, A.G. (1985) The tectosedimentary evolution of Mio-Pliocene reefs in the Province of Almería. In: *6th European Regional Meeting of Sedimentologists, Excursion Guidebook* (Eds Milá, M.D. & Rosell, J.), pp. 269–305. Lleida, Spain.

Dabrio, C.J. & Polo, M.D. (1995) Oscilaciones eustáticas de alta frecuencia en el Neógeno superior de Sorbas (Almería, sureste de España). *Geogaceta* **18**, 75–78.

Davies, G.R., Nixon, P.H., Pearson, D.G. & Obata, M. (1993) Tectonic implication of graphitized diamonds from the Ronda peridotite massif, southern Spain. *Geology* **21**, 471–474.

De Dekker, P. & Chivas, A.R. (1988) Palaeoenvironment of the Messinian Mediterranean 'Lago Mare' from strontium and magnesium in the ostracode shells. *Palaios* **3**, 352–358.

De Jong, K. (1991) *Tectono-metamorphic studies and radiometric dating in the Betic Cordilleras (SE Spain), with implications for the dynamics of extension and compression in the western Mediterranean area.* PhD thesis, University of Amsterdam, the Netherlands.

De Jong, K. & Bakker, H.E. (1991) The Mulhacén and Alpujárride Complex in the Sierra de los Filabres, SE Spain: lithostratigraphy. *Geol. Mijnbouw* **70**, 93–103.

De Larouzière, F.D., Bolze, J., Bordet, P., Hernandez, J., Montenat, C. & Ott D'Estevou, P. (1988) The Betic segment of the lithospheric trans-Alborán shear zone during the Late Miocene. *Tectonophysics* **152**, 41–52.

Delgado, F., Estevez, A., Martin, J.M. & Martín-Algarra, A. (1981) Observaciones sobre la estratigrafía de la formación carbonatada de los mantos Alpujárrides (Cordilleras Béticas). *Estud. Geol.* **37**, 45–57.

Delgado-Huertas, A. (1993) *Estudio isotópico de los procesos diagenéticos e hidrotermales relacionados con la génesis de bentonitas (Cabo de Gata, Almería).* PhD thesis, University of Granada, Spain.

Delgardo-Castilla, L. (1993) Estudio sedimentológico de los cuerpos sedimentarios Pleistocenos en la Rambla Honda, al N de Tabernas, provincia de Almería (SE de España). *Cuaternario Y Geomorfología* **7**, 91–100.

Delgardo-Castilla, L., Pascual-Molina, A. & Ruiz-Bustos, A. (1993) Geology and micromammals of the Serra-1 site (Tabernas Basin, Betic Cordilera). *Estud. Geol.* **49**, 361–366.

Di Battistini, G., Toscani, L., Iaccarino, S. & Villa, I.M. (1987) K/Ar ages and the geological setting of calc-alkaline volcanic rocks from Sierra de Gata, SE Spain. *Neues Jahrb. Mineral. Monatsh.* **H8**, 369–383.

Doyle, P., Mather, A.E., Bennett, M.R. & Bussell, M.A. (1997) Miocene barnacle assemblages from southern Spain and their palaeoenvironmental significance. *Lethaia* **29**, 267–274.

Dronkert, H. (1977) The evaporites of the Sorbas basin. *Rev. Instit. Inv. Geol. Dip. Provincial Univ. Barcelona* **32**, 55–76.

Dupuy, C., Dostal, J. & Boivin, P.A. (1986) Geochemistry of ultramafic xenoliths and their host alkali basalt from Tallante, southern Spain. *Mineral. Mag.* **50**, 231–239.

Durand-Delga, M. (1968) Coup d'oeil sur les unités Malaguides des Cordillères Bétiques (Espagne). *C. R. Acad. Sci. Paris* **266**, 190–193.

Egeler, C.G. (1963) On the tectonics of the eastern Betic Cordilleras (SE Spain). *Geol. Rundsch.* **53**, 260–269.

Egeler, C.G. & Simon, O.J. (1969) Orogenic evolution of the Betic Zone (Betic Cordilleras, Spain), with emphasis on the nappe structures. *Geol. Mijnbouw* **48**, 296–305.

Ekdale, A.A., Bromley, R.G. & Pemberton, S.G. (1984) *Ichnology: Trace Fossils in Sedimentology and Stratigraphy.* Soc. Econ. Paleont. Mineral., Tulsa, Short Course 15.

Esteban, M. (1979/80) Significance of the upper Miocene coral reefs of the western Mediterranean. *Palaeogeogr. Palaeoclimatol. Palaeoecol.* **29**, 169–188.

Esteban, M., Braga, J.C., Martín, J.M. & Santisteban, C. (1996) Western Mediterranean reef complexes. In: *Models for Carbonate Stratigraphy from Miocene Reef Complexes of Mediterranean Regions* (Eds Franseen, E.K., Esteban, M., Ward, W.C. & Rouchy, J.M.), Concepts in Sedimentology and Paleontology Series, Soc. econ. Paleont. Miner., Tulsa, **5**, 55–72.

Esteban, M. & Giner, J. (1980) Messinian coral reefs and erosion surfaces in Cabo de Gata (Almería, SE Spain). *Acta geol. Hispánica* **4**, 97–104.

Fallot, P., Faure-Muret, A., Fontboté, J.M. & Solé-Sabaris, L. (1960) Estudios sobre las series de Sierra Nevada y de la llamada Mischungzone. *Bol. Inst. Geol. Min. España* **71**, 345–557.

Faulkner, H., Spivey, D. & Alexander, R. (2000) The role of some site geochemical processes in the development and stabilisation of three badland sites in Almería, Southern Spain. *Geomorphology* **35**, 87–99.

Fernández-Soler, J.M. (1992) *El volcanismo calco-alcalino de Cabo de Gata (Almería): estudio volcanológico y petrológico.* PhD thesis, University of Granada, Spain. (Published by the Sociedad Almeriense de Historia Natural and Consejería de Medio Ambiente, Junta de Andalucía, Almería, 1996.)

Foley, S.F. & Venturelli, G. (1989) High K_2O rocks with high MgO, high SiO_2

affinities. In: *Boninites and Related Rocks* (Ed. Crawford, A.J.), pp. 72–88. Unwin Hyman, London.

Foley, S.F., Venturelli, G., Green, D.H. & Toscani, L. (1987) The ultrapotassic rocks: characteristics, classification and constraints for petrogenetic models. *Earth Sci. Rev.* **24**, 81–124.

Fontboté, J.M. (1984) La Cordillera Bética. In: *Libro Jubilar J.M. Ríos, Geología de España II* (Ed. Instituto Geológico y Minero de España), pp. 251–343.

Fortuin, A.R., Kelling, J.M.D. & Roep, T.B. (1995) The enigmatic Messinian– Pliocene section of Cuevas del Almanzora (Vera basin, SE Spain) revisited: erosional features and strontium isotope age. *Sediment. Geol.* **97**, 177–201.

Franseen, E.K. (1989) Depositional sequences and correlation of Middle to Upper Miocene carbonate complexes, Las Negras area, southeastern Spain. PhD thesis, University of Wisconsin–Madison, USA.

Franseen, E.K. & Goldstein, R.H. (1996) Palaeoslope, sea-level, and climate controls on Upper Miocene platform evolution, Las Negras area, southeastern Spain. In: *Models for Carbonate Stratigraphy from Miocene Reef Complexes of Mediterranean Regions* (Eds Franseen, E.K., Esteban, M., Ward, W.C. & Rouchy, J.M.), Concepts in Sedimentology and Paleontology Series, Soc. econ. Paleont. Miner., Tulsa **5**, 159–176.

Franseen, E.K. & Mankiewicz, C. (1991) Depositional sequences and correlation of middle (?) to late Miocene carbonate complexes, Las Negras and Níjar areas, southeastern Spain. *Sedimentology* **38**, 871–898.

Franz, G., Gómez-Pugnaire, M.T. & López Sánchez-Vizcaíno, V. (1994) Retrograde formation of NaCl-scapolite in high pressure metaevaporites from the Cordilleras Béticas (Spain). *Contrib. Mineral. Petrol.* **116**, 448–461.

Fúster, J.M. (1956) Las erupciones delleníticas del Terciario superior de la fosa de Vera (provincia de Almería). *Bol. R. Soc. Esp. Hist. Nat.* **54**, 53–88.

Fúster, J.M. & De Pedro, F. (1953) Estudio petrológico de las rocas volcánicas lamproíticas de Cabezo María (Almería). *Estud. Geol.* **9**, 477–508.

Fúster, J.M., Gastesi, P., Sagredo, J. & Fermoso, M.L. (1967) Las rocas lam-

proíticas del SE de España. *Estud. Geol.* **23**, 35–69.

Galíndo-Zaldívar, J. (1993) *Geometría y cinemática de las deformaciones neógenas en Sierra Nevada (Cordilleras Béticas)*. PhD thesis, University of Granada.

Galíndo-Zaldívar, J., González-Lodeiro, F. & Jabaloy, A. (1989) Progressive extensional shear structures in a detachment contact in the western Sierra Nevada (Betic Cordilleras, Spain). *Geod. Acta* **3**, 73–85.

García Dueñas, V., Martínez-Martínez, J.M. & Soto, J.I. (1988) Los Nevado-Filábrides, una pila de pliegues-mantos separados por zonas de cizalla. In: *II Congreso Español de Geología* (Ed. Sociedad Geológica de España), pp. 17–26.

García-Casco, A. & Torres-Roldan, R.L. (1996) Disequilibrium induced by fast decompression in St–Bt–Grt–Ky–Sill–And metapelites from the Betic Belt (Southern Spain). *J. Petrol.* **37**, 1207–1239.

García-Dueñas, V., Balanyá, J.C. & Martínez-Martínez, J.M. (1992) Miocene extensional detachments in the outcropping basement of the northern Alborán basin (Betics) and their implications. *Geo-Marine Lett.* **12**, 88–95.

García-Dueñas, V. & Comas, M.C. (1971) Estructuras de colapso en la vertiente occidental de Sierra Nevada (Sector de Nigüelas, Granada). *Bol. Inst. Geol. Min. España* **82**, 507–511.

García-Hernández, A.C., López-Garrido, A.C., Rivas, P., Sanz De Galdeano, C. & Vera, J.A. (1980) Mesozoic paleogeographic evolution of the External Zones of the Betic Cordillera. *Geol. Mijnbouw* **59**, 155–168.

Gautier, F., Clauzon, G., Suc, J.-P., Cravatte, J. & Violanti, D. (1994) Age et durée de la crise de salinité messinienne. *C. R. Acad. Sci. Paris* **318**, 1103–1109.

Geel, T. (1973) The geology of the Betic of Malaga: the Subbetic and the zone between these two units in the Vélez Rubio Area (Southern Spain), *GUA Pap. Geology*, 5, pp. 181.

Gilbert, G.K. (1885) The topographic features of lake shores. *US Geological Survey, 5th Annual Report*, pp. 69–123.

Gile, L.H., Peterson, F.F. & Grossman, R.B. (1966) Morphological and genetic sequences of carbonate accumulation in desert soils. *Soil Sci.* **101**, 347–360.

Gómez-Pugnaire, M.T. (1981) La evolución del metamorfismo alpino en el Complejo Nevado–Filábride de la Sierra de Baza (Cordilleras Béticas, España). *Tecniterrae*, 1–130.

Gómez-Pugnaire, M.T., Braga, J.C., Martín, J.M., Sassi, J.P. & Del Moro, A. (2000) The age of the Nevado–Filábride cover (Betic Cordilleras, S Spain): regional implications. *Schweiz. Mineral. Petrograph. Mitt.* **80**, 45–52.

Gómez-Pugnaire, M.T., Chacón, J., Mitrofanov, F. & Timofeev, V. (1982) First report on Pre-Cambrian rocks in the graphite-bearing series of the Nevado–Filábride complex (Betic Cordilleras, Spain). *Neues Jahrb. Geol. Palaeontol. Monatsh.* **3**, 176–180.

Gómez-Pugnaire, M.T. & Fernández-Soler, J.M. (1987) High-pressure metamorphism in metabasites from the Betic Cordilleras (SE Spain) and its evolution during the Alpine orogen. *Contrib. Mineral. Petrol.* **95**, 231–234.

Gómez-Pugnaire, M.T. & Franz, G. (1988) Metamorphic evolution of the Palaeozoic series of the Betic Cordilleras (Nevado–Filábride Complex, SE Spain) and its relationship with the Alpine orogeny. *Geol. Rundsch.* **77**, 619–640.

Gómez-Pugnaire, M.T., Franz, G. & López-Sánchez-Vizcaíno (1994) Retrograde formation of NaCl-scapolite in high pressure metaevaporites from the Cordilleras Béticas (Spain). *Contrib. Min. Petrol.* **116**, 448–461.

Gómez-Pugnaire, M.T. & Muñoz, M. (1991) Al-rich xenoliths in the Nevado–Filábride metabasites: evidence for a continental setting of this basic magmatism in the Betic Cordilleras (SE Spain). *Eur. J. Mineral.* **3**, 193–198.

Gómez-Pugnaire, M.T. & Sassi, F.P. (1983) Pre-Alpine metamorphic features and Alpine overprints in some parts of the Nevado–Filábride basement (Betic Cordilleras, Spain). *Mem. Sci. Geol.* **36**, 49–72.

González-Lodeiro, F., Aldaya, F., Galíndo-Zaldivar, J. & Jabaloy, A. (1996) Superposition of extensional detachments during the Neogene in the internal zones of the Betic cordilleras. *Geol. Rundsch.* **85**, 350–362.

Goy, J.L. & Zazo, C. (1986) Synthesis of the Quaternary in the Almería littoral,

neotectonic activity and its morphologic features. *Tectonophysics* **130**, 259–270.

Goy, J.L., Zazo, C., Hillaire-Marcel, C. & Causse, C. (1986) Stratigraphie et chronologie (U/Th) du Tyrrhenian du Sud-Est de l'Espagne. *Z. Geomorphol. (Suppl.)* **62**, 71–82.

Hall, S.H. (1983) *Post Alpine tectonic evolution of SE Spain and the structure of fault gouge.* PhD thesis, University of London, UK.

Haq, B.U., Hardenbol, J. & Vail, P.R. (1987) Chronology of fluctuating sea levels since the Triassic. *Science* **235**, 1156–1167.

Haq, B.U., Hardenbol, J. & Vail, P.R. (1988) Mesozoic and Cenozoic chronostratigraphy and cycles of sea-level change. In: *Sea-Level Changes: an Integrated Approach* (Eds Wilgus, C.K., Hastings, B.S., Kendall, C.G.St.C., Posamentier, H.W., Ross, C.A. & Van Wagoner, J.C.), Spec. Publ. Soc. econ. Paleont. Miner., Tulsa, **42**, 71–108.

Hart, A. (1999) An introduction to the landslides of the Sorbas basin. In: *BSRG/BGRG SE Spain Field Meeting: Field Guide* (Eds Mather, A.E. & Stokes, M.), pp. 124–133. University of Plymouth, UK.

Harvey, A.M. (1982) The role of piping in the development of badlands and gully systems in south-east Spain. In: *Badland Geomorphology and Piping* (Eds Bryan, R. & Yair, A.), pp. 317–335. Geobooks, Norwich.

Harvey, A.M. (1984a) Aggradation and dissection sequences on Spanish alluvial fans: influence on morphological development. *Catena* **11**, 289–304.

Harvey, A.M. (1984b) Debris flow and fluvial deposits in Spanish Quaternary alluvial fans: implications for fan morphology. In: *Sedimentology of Gravels and Conglomerates* (Eds Koster, E.H. & Steel, R.), Mem. Can. Soc. petrol. Geol., Calgary, No. 10, McAra, Calgary, Alberta, pp. 123–132.

Harvey, A.M. (1984c) Geomorphological response to an extreme flood: a case from southeast Spain. *Earth Surf. Process. Landf.* **9**, 267–279.

Harvey, A.M. (1987a) Patterns of Quaternary aggradational and dissectional landform development in the Almería region, southeast Spain: a dry-region tectonically active landscape. *Die Erde* **118**, 193–215.

Harvey, A.M. (1987b) Alluvial fan dissection: relationships between morphology and sedimentation. In: *Desert Sediments, Ancient and Modern* (Eds Frostick, L. & Reid, I.), Spec. Publ. Geol. Soc. London, No. 35, pp. 87–103. Blackwell Scientific Publications, Oxford.

Harvey, A.M. (1990) Factors influencing Quaternary alluvial fan development in southeast Spain. In: *Alluvial Fans: a Field Approach* (Eds Rachocki, A. & Church, M.), pp. 247–269. Wiley, Chichester.

Harvey, A.M. (1992a) Controls on sedimentary style on alluvial fans. In: *Dynamics of Gravel-Bed Rivers* (Eds Billi, P., Hey, R.D., Thorne, C.R. & Tacconi, P.), pp. 519–535. Wiley, Chichester.

Harvey, A.M. (1992b) The influence of sedimentary style on the morphology and development of alluvial fans. *Israel J. Earth Sci.* **41**, 123–134.

Harvey, A.M. (1997) The role of alluvial fans in arid zone fluvial systems. In: *Arid Zone Geomorphology: Process, Form and Change in Drylands* (Ed. Thomas, D.S.G.), 2nd edn, pp. 231–259. Wiley, Chichester.

Harvey, A.M. & Calvo-Cases, A. (1991) Process interactions and rill development on badland and gully slopes. *Z. Geomorphol. (Suppl.)* **83**, 175–194.

Harvey, A.M. & Mather, A.E. (1996) Tabernas Quaternary alluvial fans and 'lake' system. In: *2nd Cortijo Urra Field Meeting, Southeast Spain: Field Guide* (Eds Mather, A.E. & Stokes, M.), pp. 39–42. University of Plymouth, UK.

Harvey, A.M. & Wells, S.G. (1987) Response of Quaternary fluvial systems to differential epeirogenic uplift: Aguas and Feos river systems, southeast Spain. *Geology* **15**, 689–693.

Harvey, A.M., Foster, G., Hannam, J. & Mather, A.E. (1999a) Mineral magnetic characteristics of the soils and sediments of the Tabernas alluvial fan and 'lake' system. In: *BSRG/BGRG SE Spain Field Meeting: Field Guide* (Eds Mather, A.E. & Stokes, M.), pp. 43–61. University of Plymouth, UK.

Harvey, A.M., Mather, A.E., Stokes, M., Silva, P.G., Goy, J. & Zazo, C. (1999c) Influence of Quaternary sea-level change on the alluvial fans of the Cabo de Gata coastal zone. In: *BSRG/BGRG SE Spain Field Meeting: Field Guide.* (Eds Mather, A.E. & Stokes, M.), pp. 84–96. University of Plymouth, UK.

Harvey, A.M., Miller, S.Y. & Wells, S.G. (1995) Quaternary soil and river terrace sequences in the Aguas/Feos river systems: Sorbas basin, southeast Spain. In: *Mediterranean Quaternary River Environments* (Eds Lewin, J., MacKlin, M.G. & Woodward, J.C.), pp. 263–281. Balkema, Rotterdam.

Harvey, A.M., Silva, P.G., Mather, A.E., Goy, J.L., Stokes, M. & Zazo, C. (1999b) The impact of Quaternary sea-level and climatic change on coastal alluvial fans in the Cabo de Gata ranges, southeast Spain. *Geomorphology* **28**, 1–22.

Haughton, P.D.W. (1994) Deposits of deflected and ponded turbidity currents, Sorbas Basin, southeast Spain. *J. sediment Res.* **64**, 233–246.

Haughton, P.D.W. (2000) Evolving turbidite systems on a deforming basin floor, Tabernas, SE Spain. *Sedimentology*, **47**, 497–518.

Hebeda, E.M., Boelrijk, N.A.I.M., Priem, H.N.A. & Vendurmen, R.H. (1980) Excess radiogenic Ar and undisturbed Rb–Sr systems in basic intrusives subjected to Alpine metamorphism in SE Spain. *Earth planet. Sci. Lett.* **47**, 81–90.

Herbig, H.G. (1983) The Carboniferous of the Betic Cordillera. In: *X Congreso Internacional Estratigrafia y Geologia del Carbonífero* (Ed. Martínez-Díaz, C.), pp. 343–356.

Herbig, H.G. (1984) Reconstruction of a lost sedimentary realm: the limestone boulders in the Carboniferous of the Maláguides (Betic Cordillera, Southern Spain). *Facies* **11**, 1–108.

Hermes, J.J. (1978) The stratigraphy of the Subbetic and southern Prebetic of the Velez–Rubio–Caravaca area and its bearing on the transcurrent faulting in the Betic Cordilleras in southern Spain, *Proc. K. Ned. Akad. Wet.* **81**, 1–54.

Hernandez, J. (1983) *Le volcanisme miocène du Rif oriental (Maroc): géologie, pétrologie et minéralogie d'une province shoshonitique.* PhD thesis, Université Paris.

Hernandez, J., De Larouzière, F.D., Bolze, J. & Bordet, P. (1987) Le magmatisme néogène bético-rifain et le couloir de décrochement trans-Alborán. *Bull. Soc. geol. France* **(8)**, 257–267.

Hernandez-Pacheco, A. & Ibarrola, E. (1970) Nuevos datos sobre la petrología y geoquímica de las rocas volcánicas de la isla de Alborán (Mediterráneo occidental, Almería). *Estud. Geol.* **26**, 93, 103.

Hodell, D.A., Mueller, P.A., McKenzie, J.A. & Mead, G.A. (1989) Strontium isotope stratigraphy and geochemistry of the late Neogene ocean. *Earth planet. Sci. Lett.* **92**, 165–168.

Hoernle, K., Van Den Bogaard, P., Duggen, S., Mocek, B. & Garbe-Schönberg, D. (1999) Evidence for Miocene subduction beneath the Alborán Sea (Western Mediterranean) from ^{40}Ar/^{39}Ar age dating and the geochemistry of volcanic rocks from Holes 977A and 978A. *Proceedings of the ODP, Scientific Results* **161**, pp. 357–374. Ocean Drilling Program, College Station, TX.

Hsü, K.J., Montadert, L., Bernoulli, D. *et al.* (1978) *Initial Reports of the Deep Sea Drilling Project 42A.* (US Govt. Printing Office, Washington, DC.

Hsü, K.J., Montadert, L., Bernoulli, D. *et al.* (1977) History of the Messinian salinity crisis. *Nature* **267**, 399–403.

Hurst, V.J. (1977) Visual estimates of iron in saprolites. *Geol. Soc. Am. Bull.* **88**, 174–176.

Iaccarino, S., Morlotti, E., Papani, G., Pelosio, G. & Raffi, S. (1975) Litho-stratigrafia e biostratigraphia di alcune serie Neogeniche della provincia di Almeria (Andalusia orientale, Spagna). *Ateneo Parmenje, Acta Nat.* **11**, 237–313.

Jabaloy, A. (1993) *La estructura de la región occidental de la Sierra de los Filabres.* PhD thesis, University of Granada.

Jabaloy, A., Galíndo-Zaldivar, J. & González-Lodeiro, F. (1993) The Alpujárride–Nevado–Filábride extensional shear zone, Betic Cordillera, SE Spain. *J. struct. Geol.* **15**, 555–569.

Jimenez, A.P. & Braga, J.C. (1993) Occurrence and taphonomy of bivalves from the Níjar reef (Messinian, Late Miocene, SE Spain). *Palaeogeogr. Palaeoclimatol. Palaeoecol.* **102**, 239–251.

Johnson, C., Harbury, N. & Hurford, A.J. (1997) The role of extension in the Miocene denudation of the Nevado–Filabride Complex, Betic Cordillera (SE Spain). *Tectonics* **16**, 189–204.

Junta de Andalucia (1985) *Mapa Geológico-Minero de Andalucia.* ISBN 84-398-5259-2.

Kelly, M., Black, S. & Rowan, J.S. (2000) A calcrete-based U/Th chronology for landform evolution in the Sorbas Basin, southeast Spain. *Quat. Sci. Rev.* **19**, 995–1010.

Kendall, G.S.C. & Lerche, I. (1988) The rise and fall of eustasy. In: *Sea-Level Changes: an Integrated Approach* (Eds Wilgus, C.K., Hasting, B.S., Kendall, C.G.St., Posamentier, H.W., Ross, C.A. & Van Wagoner, J.C.). Spec. Publ. Soc. econ. Paleont. Miner., Tulsa, **42**, 3–18.

Kleverlaan, K. (1987) Gordo megabed: a possible seismite in a Tortonian submarine fan, Tabernas basin, Province Almería, southeast Spain. *Sediment. Geol.* **51**, 165–180.

Kleverlaan, K. (1989a) Three distinctive feeder-lobe systems within one time slice of the Tortonian Tabernas fan, SE Spain. *Sedimentology* **36**, 25–45.

Kleverlaan, K. (1989b) *Tabernas fan complex: a study of a Tortonian fan complex in a Neogene basin, Tabernas, Province of Almería, SE Spain.* PhD thesis, University of Amsterdam, the Netherlands.

Kleverlaan, K. (1989c) Neogene history of the Tabernas basin, (SE Spain) and its Tortonian submarine fan development. *Geol. Mijnbouw* **68**, 421–432.

Kornprobst, J. (1976) Signification structurale des péridotites dans l'orogène bético-rifain: arguments tirés de l'étude des détritus observés dans les sédiments paléozöiques. *Bull. Soc. geol. France* **18**, 607–618.

Kozur, H., Mulder-Blanken, C.W.H. & Simon, O.J. (1985) On the Triassic of the Betic Cordilleras (southern Spain), with special emphasis on holothurian sclerites. *Stratig. Paleontol.* **88**, 83–110.

Lafuste, M.L.J. & Pavillon, M.J. (1976) Mise en évidence d'Eifélien daté au sein des terrains métamorphiques des zones internes des Cordilléres bétiques: intérêt de ce nouveau repère stratigraphique. *C. R. Acad. Sci. Paris* **283**, 1015–1018.

Leone, G., Reyes, E., Cortecci, G., Pochini, A. & Linares, J. (1983) Genesis of bentonites from Cabo de Gata, Almería, Spain: a stable isotope study. *Clay Min.* **18**, 227–238.

Lodder, W. (1966) *Gold–alunite deposits and zonal wall-rock alteration near Rodalquilar, SE Spain.* PhD thesis, University of Amsterdam, the Netherlands.

Loomis, T.P. (1972) Contact metamorphism of pelitic rocks by the Ronda ultramafic intrusion, southern Spain. *Geol. Soc. Am. Bull.* **83**, 2449–2474.

Loomis, T.P. (1975) Tertiary mantle diapirism, orogeny, and plate tectonics east of the Strait of Gibraltar. *Am. J. Sci.* **275**, 1–30.

López-Arroyo, A., Martín-Martín, A.J., Mezcua-Rodríguez, J., Muñoz, D. & Udías, A. (1980) *El terremoto de Andalucía de 25 de Diciembre de 1884.* Presidencia del Gobierno, Instituto Geográfico Nacional, Madrid.

López-Ruiz, J. & Rodríguez-Badiola, E. (1980) La región volcánica Neógena del sureste de España. *Estud. Geol.* **36**, 5–63.

López-Ruiz, J. & Wasserman, M.D. (1991) Relación entre la hidratación/desvitrificación y el δ18O en las rocas volcánicas neógenas del SE de España. *Estud. Geol.* **47**, 3–11.

López Sánchez-Vizcaíno, V., Gómez-Pugnaire, M.T. & Fernández-Soler, J.M. (1991) Petrological features of some Alpujárride, mafic igneous bodies from the Sierra de Almagro (Betic Cordilleras, Spain). *Rev. Soc. Geol. España* **4**, 321–335.

López Sánchez-Vizcaíno, V. (1994) *Evolución petrológica y geoquímica de las rocas carbonáticas en el área de Macaél–Cóbdar, Complejo Nevado–Filábride, SE España.* PhD thesis, University of Granada, Spain.

MacHette, M.N. (1985) Calcic soils of the southwestern United States. In: *Soils and Quaternary Geology of the Southwestern United States* (Ed. Weide, D.L.), Geol. Soc. Am. Spec. Paper, Boulder, **203**, 12–21.

Mankiewicz, C. (1987) *Sedimentology and calcareous algal paleoecology of middle and upper Miocene reef complexes near Fortuna (Murcia Province) and Níjar (Almería Province), southeastern Spain.* PhD thesis, University of Wisconsin–Madison, USA.

Mankiewicz, C. (1988) Occurrence and paleoecologic significance of *Halimeda* in late Miocene reefs, southeastern Spain. *Coral Reefs* **6**, 271–279.

Mankiewicz, C. (1996) The Middle to Upper Miocene carbonate complex of Níjar, Almería Province, southeastern Spain. In: *Models for Carbonate Stratigraphy from Miocene Reef Complexes of Mediterranean Regions* (Eds Franseen, E.K., Esteban, M., Ward, W.C. & Rouchy, J.M.), Concepts in Sedimentology and Paleontology Series, Soc. econ. Paleont. Miner., Tulsa **5**, 141–157.

Marjanac, T. (1987) Ponded megabeds and some characteristics of the Eocene Adriatic basin (Middle Dalmatia, Yugoslavia). *Mem. Soc. geol. Ital.* **40**, 241–249.

Martín, J.M. (1980) *Las dolomías de las Cordilleras Béticas.* PhD thesis, University of Granada, Spain.

Martín, J.M. & Braga, J.C. (1987) Alpujárride carbonate deposits (southern Spain): marine sedimentation in a Triassic Atlantic. *Palaeogeogr. Palaeoclimatol. Palaeoecol.* **59**, 243–260.

Martín, J.M. & Braga, J.C. (1989) Algae in the Níjar Messinian coral reef (Almería, SE Spain). In: *Algae in Reefs* (Eds Braga, J.C. & Martín, J.M.), *Field Trip Guidebook*, pp. 45–57. University of Granada, Spain.

Martín, J.M. & Braga, J.C. (1994) Messinian events in the Sorbas Basin in southeastern Spain and their implications in the recent history of the Mediterranean. *Sediment. Geol.* **90**, 257–268.

Martín, J.M. & Braga, J.C. (1996) Tectonic signals in the Messinian stratigraphy of the Sorbas basin (Almería, SE Spain). In: *Tertiary Basins of Spain: the Stratigraphic Record of Crustal Kinematics* (Eds Friend, P.F. & Dabrio, C.J.), pp. 387–391. Cambridge University Press, Cambridge.

Martín, J.M., Braga, J.C., Betzler, C. & Brachert, T. (1994) Modelo sedimentario y asociaciones de organismos en playas conglomeráticas Plio-Cuaternarias de caracter templado del SE de España. *II Congreso del Grupo Español del Terciario*, Jaca (Huesca), Septiembre 1994. *Comunicaciones*, pp. 153–156.

Martín, J.M., Braga, J.C., Betzler, C. & Brachert, T. (1996) Sedimentary model and high-frequency cyclicity in a Mediterranean, shallow-shelf, temperate carbonate environment (uppermost Miocene, Agua Amarga Basin, Southern Spain). *Sedimentology* **43**, 263–277.

Martín, J.M., Braga, J.C. & Riding, R. (1993) Siliciclastic stromatolites and thrombolites, late Miocene, S.E. Spain. *J. sediment. Petrol.* **63**, 131–139.

Martín, J.M., Braga, J.C. & Riding, R. (1997) Late Miocene *Halimeda* alga–microbial segment reefs in the marginal Mediterranean Sorbas Basin, Spain. *Sedimentology* **44**, 441–456.

Martín, J.M., Braga, J.C. & Riding, R. (1998) Messinian reefs and stromatolites of the Sorbas Basin (Almería, SE Spain). In: *Fieldguide Book*, pp. 111–125. 15th International Sedimentological Congress, Alicante, Spain.

Martín, J.M., Braga, J.C. & Rivas, P. (1989) Coral successions in Upper Tortonian reefs in SE Spain. *Lethaia* **22**, 271–286.

Martín, J.M., Braga, J.C. & Sánchez-Almazo, I.M. (1999) The Messinian record of the outcropping marginal Alborán Basin deposits: significance and implications. *Proceedings of the ODP, Scientific Results* **161**, pp. 543–551. Ocean Drilling Program, College Station, TX.

Martín-Algarra, A. & Vera, J.A. (1982) Penibético, las unidades del Campo de Gibraltar, las Zonas Internas y las unidades implicadas en el contacto entre Zonas Internas y Zonas Externas. In: *El Cretácico de España* (Ed. University Complutense of Madrid), pp. 603–632.

Martínez-Martínez, J.M. (1986) Evolución tectonometamórfica del complejo Nevado–Filábride en el sector de unión entre Sierra Nevada y Sierra de los Filabres (Cordilleras Béticas). *Cuad. Geol. Iberica* **13**, 1–19.

Mather, A.E. (1991) *Late Caenozoic drainage evolution of the Sorbas basin, southeast Spain*. PhD thesis, University of Liverpool, UK.

Mather, A.E. (1993a) Basin inversion: some consequences for drainage evolution and alluvial architecture. *Sedimentology* **40**, 1069–1089.

Mather, A.E. (1993b) Evolution of a Pliocene fan delta: links between the Sorbas and Carboneras basins, SE Spain. In: *Tectonic Controls and Signatures in Sedimentary Successions* (Eds Frostick, L.E. & Steel, R.J.). Spec. Publs int. Ass. Sediment., No. 20, pp. 277–290. Blackwell Scientific Publications, Oxford.

Mather, A.E. (1999) Alluvial fans: a case study from the Sorbas Basin, southeast Spain. In: *The Description and Analysis of Quaternary Stratigraphic Sections* (Eds Jones, A.P., Tucker, M.E. & Hart, J.K.), Technical Guide 7, pp. 77–110. Quaternary Research Association.

Mather, A.E. (2000a) Impact of headwater river capture on alluvial system development: an example from SE Spain. *J. geol. Soc. London* **157**, 957–966.

Mather, A.E. (2000b) Adjustment of a drainage network to capture induced base-level change: an example from the Sorbas Basin, SE Spain. *Geomorphology* **34**, 271–289.

Mather, A.E. & Harvey, A.M. (1995) Controls on drainage evolution in the Sorbas basin, southeast Spain. In: *Mediterranean Quaternary River Environments* (Eds Lewin, J., MacKlin, M.G. & Woodward, J.C.), pp. 65–76. Balkema, Rotterdam.

Mather, A.E., Harvey, A.M. & Brenchley, P.J. (1991) Halokinetic deformation of Quaternary river terraces in the Sorbas Basin, South-East Spain. *Z. Geomorphol. (Suppl.)* **82**, 87–97.

Mather, A.E., Harvey, A.M. & Stokes. M. (2000) Quantifying long term catchment changes of alluvial fan systems. *GSA Bull.* **112**, 1825–1833.

Mather, A.E. & Stokes, M. (1999) Pleistocene travertines and lakes of Tabernas: evidence for a wetter climate? In: (Eds Mather, A.E. & Stokes, M.), *BSRG/BGRG SE Spain Field Meeting: Field Guide*, pp. 63–71. University of Plymouth, UK.

Mather, A.E., Stokes, M. & Pirrie, D. (1997) Marl braid bars and their significance: an example from SE Spain. 6th Int. Conf. Fluvial Sedimentol., University of Cape Town, South Africa, Abstracts Volume.

Mather, A.E. & Westhead, R.K. (1993) Plio/Quaternary strain of the Sorbas Basin, SE Spain: evidence from sediment deformation structures. *Quaternary Proc.* **3**, 57–65.

McPhie, J., Doyle, M. & Allen, R. (1993). *Volcanic Textures: a Guide to the Interpretation of Textures in Volcanic Rocks*. University of Tasmania, Centre for Ore Deposit and Exploration Studies, Hobart.

Meyers, W.J., Lu, F.H. & Zachariah, J.K. (1997) Dolomitization by mixed evaporative brines and freshwater, Upper Miocene carbonates, Níjar, Spain *J. sediment. Res. A* **67**, 898–912.

Michard, A., Chalouan, A., Montigny, R. & Ouazzani-Touhami, A. (1983) Les nappes Cristallophylliennes du Rif (Sebtides, Maroc). *C. R. Acad. Sci. Paris* **296**, 1337–1340.

Mitchell, R.H. & Bergman, S.C. (1991). *Petrology of Lamproites*. Plenum, New York.

Molin, D. (1980) *Le volcanisme miocène du Sud-Est de l'Espagne (Provinces de*

Murcia et d'Almeria). Thesis 3e cycle, Université Paris VI, France.

Mon, R. (1969) Rapports entre la nappe de Málaga et les unités Alpujarrides à l'Ouest de Málaga (Espagne). *C. R. Acad. Sci. Paris* **268**, 1008–1011.

Monié, P., Galíndo-Zaldivar, J., González-Lodeiro, F., Goffe, B. & Jabaloy, A. (1991) $^{40}Ar/^{39}Ar$ geochronology of alpine tectonism in the Betic Cordilleras (Southern Spain). *J. geol. Soc. London* **148**, 289–297.

Montenat, Ch. (1990) *Les bassins Néogènes du domaine bétique oriental (Espagne)*. Documents et Travaux IGAL, Paris, pp. 12–13.

Montenat, C., Ott D'Estevou, P., Larouziére, F.D. & Bedu, P. (1987b) Origin geodynamique des bassins Neogenes du domaine Betique Oriental (Espagne). *Total Compagnie Francaise Des Petroles, Notes et Memoires* **21**, 11–49.

Montenat, C., Ott D'Estevou, P. & Masse, P. (1987a) Tectonic–sedimentary characters of the Betic Neogene basins evolving in a crustal transcurrent shear zone (SE Spain). *Bull. Cent. Rech. Explor. Prod. Elf Aquitaine*, **11**, 1–22.

Morten, L., Bargossi, G.M., Martínez-Martínez, J.M., Puga, E. & Díaz de Federico, A. (1987) Metagabbro and associated eclogites in the Lubrín area, Sierra Nevada complex, Spain. *J. metamorph. Geol.* **5**, 155–174.

Müller, D.W., Hodell, D.A. & Ciesielki, P.F. (1991) Late Miocene to earliest Pliocene (9.8–4.5) paleoceanography of the Subantarctic Southeast Atlantic: stable isotopic, sedimentologic, and microfossil evidence, Leg 114. In: *Proceedings of the ODP, Sci Results*, (Eds Ciesielki, P.F., Kristoffersen, Y., Clement, B. *et al*.), **114**, pp. 459–474. Ocean Drilling Program, College Station, TX.

Müller, D.W. & Mueller, P.A. (1991) Origin and age of the Mediterranean Messinian evaporites: implications from strontium isotopes. *Earth planet. Sci. Lett.* **107**, 1–12.

Munksgaard, N.C. (1984) High $\delta^{18}O$ and possible pre-eruptional Rb–Sr isochrons in cordierite-bearing Neogene volcanics from SE Spain. *Contrib. Mineral. Petrol.* **87**, 351–358.

Muñoz, M. (1986) Estudio comparativo de los cuerpos intrusivos básicos asociados a los materiales de edad triásica de los

dominios Subbético y Nevado–Filábride del sector centro-oriental de las Cordilleras Béticas. *Geogaceta* **1**, 35–37.

Mutti, E. & Normark, W.R. (1987) Comparing examples of modern and ancient turbidite systems: problems and concepts. In: *Marine Clastic Sedimentology* (Eds Legget, J.K. & Zuffa, G.G.), pp. 1–38. Graham and Trotman, London.

Nachite, D. (1993) *Los ostrácodos y la evolución paleoambiental del Neógeno reciente del NO de Marruecos y del SE de España*. PhD thesis, University of Granada, Spain.

Nash, D.J. & Smith, R.F. (1998) Multiple calcrete profiles in the Tabernas basin, southeast Spain: their origins and geomorphic implications. *Earth Surf. Process. Landf.* **23**, 1009–1029.

Nelson, D.R., McCulloch, M.T. & Sun, S.S. (1986) The origin of ultrapotassic rocks as inferred from Sr, Nd and Pb isotopes. *Geochim. Cosmochim. Acta* **50**, 231–245.

Nieto, J.M. (1995) *Petrología y geoquímica de los gneises del Complejo del Mulhacén, Cordilleras Béticas*. PhD thesis, University of Granada, Spain.

Nieto, J.M., Puga, E., Díaz de Federico, A. *et al.* (1997) Petrological, geochemical and geochronological constraints on the geodynamic evolution from the Hercynian to the Alpine orogeny in the Mulhacén Complex (Betic Cordillera, Spain). *Quad. Geodinam. Alpina Quat.* **4**, 85–86.

Niggli, P. (1923). *Gesteins- und Mineralprovinzen*. Borntraeger, Berlin.

Nijhuis, H.J. (1964) *Plurifacial alpine metamorphism in the south-eastern Sierra de los Filabres, south of Lubrín*. PhD thesis, University of Amsterdam, the Netherlands.

Nobel, F.A., Andriessen, P.A.M., Hebeda, E.H., Priem, H.N.A. & Rondeel, H.E. (1981) Isotopic dating of the post-alpine Neogene volcanism in the Betic Cordilleras, Southern Spain. *Geol. Mijnbow* **60**, 209–214.

Obata, M. (1980) The Ronda peridotites garnet–spinel and plagioclase-lherzolite facies and the $P–T$ trajectories of a high-temperature mantle intrusion. *J. Petrol.* **21**, 533–572.

Ott D'Estevou, P. (1980) *Evolution dynamique du bassin Néogene de Sorbas (Cordillères Bétiques Orientales, Espagne)*. Doc. et Trav. IGAL, Paris, 1.

Ott D'Estevou, P. & Montenat, C. (1990) *Le bassin de Sorbas–Tabernas*. Documents et Travaux IGAL, Paris, **12–13**, 101–128.

Ovejero, G. & Zazo, C. (1971) Niveles marinos Pleistocenos en Almería (SE de Espana). *Quaternaria* **15**, 145–159.

Pascual-Molina, A.M. (1997) *La Cuenca Neógena de Tabernas (Cordilleras Béticas)*. PhD thesis, University of Granada, Spain.

Pearson, D.G., Davies, G.R., Nixon, P.H. & Milledge, H.J. (1989) Graphitized diamond from a peridotite massif in Morocco and implications for anomalous diamond occurrences. *Nature* **338**, 60–62.

Pichler, H. (1965) Acid hyaloclastites. *Bull. Volcanol.* **28**, 293–310.

Pickering, K.T. & Hiscott, R.N. (1985) Contained (reflected) turbidity currents from the Middle Ordovician Cloridorme Formation, Quebec, Canada: an alternative to the antidune hypothesis. *Sedimentology* **32**, 373–394.

Platt, J.P., Behrmann, H.H., Martínez-Martínez, J.M. & Vissers, R.L.M. (1984) A zone of mylonite and related ductile deformation beneath the Alpujárride Nappe Complex, Betic Cordilleras, S Spain. *Geol. Rundsch.* **73**, 773–785.

Platt, J.P., Soto, J.I. & Comas, M.C. (1996) Decompression and high-temperature–low-pressure metamorphism in the exhumed floor of an extensional basin, Alborán Sea, western Mediterranean. *Geology* **24**, 447–450.

Platt, J.P., Van Den Eekhout, B., Janzen, E., Konert, G., Simon, O.J. & Weifermars, R. (1983) The structure and tectonic evolution of the Aguilón fold–nappe. Sierra Alhamilla, Betic Cordilleras, SE Spain. *J. struct. Geol.* **5**, 519–538.

Platt, J.P. & Vissers, R.L.M. (1980) Extensional structures in anisotropic rocks. *J. struct. Geol.* **2**, 397–410.

Platt, J.P. & Vissers, R.L.M. (1989) Extensional collapse of thickened continental lithosphere: a working hypothesis for the Alborán Sea and Gibraltar arc. *Geology* **17**, 540–543.

Playá, E., Rosell, L. & Ortí, F. (1997) Geoquímica isotópica (δ^{34}S, ^{87}Sr/^{86}Sr) y contenidos en estroncio de las evaporitas messinienses de la cuenca de Sorbas (Almería). In: *Avances en el Conocimiento del Terciario Ibérico* (Eds Calvo,

J.P. & Morales, J.), pp. 161–164. Universidad Complutense de Madrid–Museo Nacional Ciencias Naturales, Madrid.

Postma, G. (1979) Preliminary note on a significant sequence in conglomeratic flows of a mass-transport dominated fan-delta (Lower Pliocene, Almería Basin, SE Spain). *Ned. Akad. Wet. Proc. Ser.* **B82**, 465–471.

Postma, G. (1984) Mass-flow conglomerates in a submarine canyon: Abrioja fan-delta, Pliocene Southeast Spain. In: *Sedimentology of Gravels and Conglomerates* (Eds Koster, K.H. & Steel, R.), Mem. Can. Soc. petrol. Geol., Calgary, **10**, 237–258.

Postma, G. (1995) Sea-level related architectural trends in coarse grained delta complexes. *Sediment. Geol.* **98**, 3–12.

Postma, G. & Roep, T.B. (1985) Resedimented conglomerates in the bottomsets of Gilbert-type gravel deltas. *J. sediment. Petrol.* **55**, 874–885.

Priem, H.N.A., Boelrijk, N.A.J.M., Hebeda, E.H., Oen, I.S., Verdurmen, E.A.T.H. & Verschure, R.H. (1979) Isotopic dating of the emplacement of the ultramafic masses in the Serranía de Ronda, southern Spain. *Contrib. Mineral. Petrol.* **70**, 103–109.

Priem, H.N., Boelruk, N.A., Hebeda, E.H. & Verschuren, R.H. (1966) Isotopic age determinations on tourmaline granite-gneiss (South-Eastern Sierra de los Filabres). *Geol. Mijnbouw* **45**, 184–187.

Prior, D.B., Bornhold, B.D. & Johns, M.W. (1986) Active sand transport along a fjord-bottom channel, Bute Inlet, British Columbia. *Geology* **14**, 581–584.

Puga, E. (1976) *Investigaciones petrológicas en Sierra Nevada occidental (Cordilleras Béticas)*. PhD thesis, University of Granada, Spain.

Puga, E. (1980) Hypothèses sur la genèse des magmas calco-alcalins intra-orogéniques alpins dans les Cordillères bétiques. *Bull. Soc. geol. France* **22**, 243–250.

Puga, E., Díaz De Federico, A. & Fontboté, J.M. (1974) Sobre la individualización y sistematización de las unidades profundas de la Zona Bética. *Estud. Geol.* **30**, 543–548.

Puga, E., Nieto, J.M., Diaz De Federico, A., Bodiner, J.L. & Morten, L. (1999) Petrology and metomorphic evolution of ultramafic rocks and dolerite dykes of the Betic Ophiolite Association (Mulhacén Complex, SE Spain): evidence of eo-Alpine subduction following an ocean-floor metasomatic process. *Lithos* **49**, 23–56.

Rhodenburg, H. & Sabelberg, U. (1980) Northwest Sahara margin: terrestrial stratigraphy of the Upper Quaternary and some palaeoclimatic implications. In: *Palaeocology of Africa and the Surrounding Islands* (Eds Van Zinderen Bakker, E.M. & Coetsee, J.A.), No. 12, pp. 267–276.

Riding, R., Braga, J.C. & Martín, J.M. (1991b) Oolite stromatolites and thrombolites, Miocene, Spain: analogues of Recent giant Bahamian examples. *Sediment. Geol.* **71**, 121–127.

Riding, R., Braga, J.C. & Martín, J.M. (1999) Late Miocene Mediterranean desiccation: topography and significance of the 'salinity crisis' erosion surface onland in southeast Spain. *Sediment. Geol.* **123**, 1–7.

Riding, R., Braga, J.C., Martín, J.M. & Sánchez-Almazo, I.M. (1998) Mediterranean Messinian Salinity Crisis: constraints from a coeval marginal basin, Sorbas, SE Spain. *Mar. Geol.* **146**, 1–20.

Riding, R., Martín, J.M. & Braga, J.C. (1991a) Coral–stromatolite reef framework, Upper Miocene, Almería, Spain. *Sedimentology* **38**, 799–818.

Roep, T.B. & Beets, D.J. (1977) An excursion to coastal and fluvial sediments of Messinian–Pliocene age (Sorbas and Zorreras Members) in the Sorbas Basin, SE Spain. In: *Messinian Seminar, 3, Field Trip Guide, Field Trip, 2, Málaga*, pp. 22–36.

Roep, ThB., Beets, D.J., Dronkert, H. & Pagnier, H. (1979) A prograding coastal sequence of wave-built structures of Messinian age, Sorbas, Almería, Spain. *Sediment. Geol.* **22**, 135–163.

Roep, T.B., Dabrio, C.J., Fortuin, A.R. & Polo, M.D. (1998) Late highstand patterns of shifting and stepping coastal barriers and washover fans (Later Messinian, Sorbas Basin, SE Spain). *Sediment. Geol.* **116**, 27–56.

Rouchy, J.M. & Saint-Martin, J.P. (1992) Late Miocene events in the Mediterranean as recorded by carbonate–evaporite relations. *Geology* **20**, 629–632.

Ruegg, G.J.H. (1964) *Geologische Onderzoekingen in Het Bekken Van Sorbas, S Spanje.* Amsterdam Geological Institute, University of Amsterdam, the Netherlands.

Rytuba, J.J., Arribas, A., Jr, Cunningham, C.G. *et al.* (1990) Mineralized and unmineralized calderas in Spain: Part II, evolution of the Rodalquilar caldera complex and associated gold–alunite deposits. *Mineral. Deposita* **25**, S29–S35.

Sabelberg, U. (1977) The stratigraphic record of late Quaternary accumulation series in southwest Morocco and its consequences concerning the pluvial hypothesis. *Catena* **4**, 204–214.

Sánchez-Almazo, I., Braga, J.C. & Martín, J.M. (1997) Palaeotemperature and sealevel control on carbonate deposition (Late Miocene, Sorbas Basin, SE Spain). *International Conference on Neogene Mediterranean Paleoceanography*, Erice, Sicily, Italy, September 1997, (Abstract), p. 70.

Sánchez-Rodríguez, L. (1998) *Pre-Alpine and Alpine evolution of the Ronda Ultramafic Complex and its countryrocks (Betic chain, Southern Spain): U–Pb SHRIMP zircon and fission-track dating.* PhD thesis, ETH, Zurich.

Sánchez-Rodríguez, L. & Gebauer, D. (2000) Mesozoic formation of pyroxenites and gabbros in the Ronda area (southern Spain), followed by Early Miocene subduction metamorphism and emplacement into the middle crust: U–Pb sensitive high resolution ion microprobe dating of zircon. *Tectonophysics* **316**, 19–44.

Sanger-Von Oepen, P., Friedrich, G. & Vogt, G.H. (1989) Fluid evolution, wallrock alteration, and ore mineralisation associated with the Rodalquilar epithermal deposits in Southeast Spain. *Mineral. Deposita* **24**, 235–243.

Santisteban, C. (1981) *Petrología y sedimentología de los materiales del Mioceno superior de la cuenca de Fortuna (Murcia) a la luz de la teoría de la crisis de salinidad.* PhD thesis, University of Barcelona, Spain.

Santisteban, C. & Taberner, C. (1988) Sedimentary models of siliciclastic deposits and coral reefs interrelation. In: *Carbonate–Clastic Transitions* (Eds Doyle, L.J. & Roberts, H.H.), Developments in Sedimentology, 42, pp. 35–76. Elsevier, Amsterdam.

Sanz De Galdeano, C. (1987) Strikeslip faults in the southern border of the Vera Basin (Almería, Betic Cordilleras). *Estud. Geol.* **43**, 435–443.

Sanz De Galdeano, C. (1990) Geologic evolution of the Betic Cordilleras in the Western Mediterranean, Miocene to the present. *Tectonophysics* **172**, 107–119.

Sanz De Galdeano, C. & Vera, J.A. (1992) Stratigraphic record and palaeogeographical context of the Neogene basins in the Betic Cordillera, Spain. *Basin Res.* **4**, 21–36.

Sendra, J., Stokes, M. & De Renzi, M. (1996) Consecuencias tafonómicas de la evolución de un fan-delta: el Konservat Fossil-Lagerstätte de Cuevas del Almanzora (Plioceno, Almería, España). In: *Comunicaciones de la II Reunión de Tafonomía y Fosilización* (Eds Hevia, G., Sancho Blasco, M.F.Y. & Pérez Urresti, I.), pp. 357–362. Institución Fernando el Católico,

Serrano, F. (1979) *Los foraminíferos planctónicos del Mioceno superior de la cuenca de Ronda y su comparación con los de otras áreas de las Cordilleras Béticas.* PhD thesis, University of Málaga, Spain.

Serrano, F. (1990) El Mioceno Medio en el área de Níjar (Almería, España). *Rev. Soc. geol. España* **3**, 65–77.

Shackleton, N.J. (1984) Oxygen isotope evidence for Cenozoic climatic cooling. In: *Fossils and Climate* (Ed. Brenchley, P.J.), pp. 27–34. Wiley, Chichester.

Sierro, F.J., Flores, J.A., Civis, J., González-Delgado, J.A. & Frances, G. (1993) Late Miocene globorotaliid event-stratigraphy and biogeography in the NE-Atlantic and Mediterranean. *Mar. Micropaleontol.* **21**, 143–168.

Silva, P.G., Goy, J.L., Somoza, L., Zazo, C. & Bardají, T. (1993) Landscape response to strike-slip faulting linked to collisional settings: Quaternary tectonics and basin formation in the Eastern Betics, southeastern Spain. *Tectonophysics* **224**, 289–303.

Soediono, H. (1971) *Geological investigations in the Chirivel area, province of Almería, Southeastern Spain.* PhD thesis, University of Amsterdam, the Netherlands.

Soto, J.I. (1993) *Estructura y evolución metamórfica del Complejo Nevado–Filábride en la terminación oriental de la Sierra de los Filabres, Cordilleras Béticas.* PhD thesis, University of Granada, Spain.

Spivey, D.B. (1997) *Scale, process and badland development in Almería Province,* *southeast Spain.* PhD thesis, University of Liverpool, UK.

Stokes, M. (1997) *Plio-Pleistocene drainage evolution of the Vera Basin, SE Spain.* PhD thesis, University of Plymouth, UK.

Stokes, M. & Griffiths, J.S. (1999) Small-scale mass movement failures in Early to Mid Pleistocene fluvial deposits in the Vera Basin, South-east Spain. In: *Proceedings of the 9th International Conference and Field Trip on Landslides* (Eds Griffiths, J.S., Stokes, M. & Thomas, R.), pp. 135–147. Balkema, Rotterdam.

Stokes, M. & Mather, A.E. (2000) Response of Plio-Pleistocene alluvial systems to tectonically induced base-level changes, Vera Basin, SE Spain. *J. geol. Soc. London* **157**, 303–316.

Stokes, M. & Sendra, J.R. (1996) Stratigraphical, sedimentological and palaeontological consequences of Pliocene fan-delta evolution, SE Spain. In: *2nd Cortijo Urra Field Meeting, Southeast Spain: Field Guide* (Eds Mather, A.E. & Stokes, M.), pp. 49–57. University of Plymouth, UK.

Tapia Garrido, J.A. (Ed.) (1980) *Los Baños de Sierra Alhamilla.* Cajal, Almería.

Thompson, R. & Oldfield, F. (1986). *Environmental Magnetism.* Allen & Unwin, London.

Thornes, J.B. (1974) The rain in Spain. *Geog. Mag.* **47**, 337–343.

Thurber, D.L. & Stearns, C.E. (1965) Th^{230}–U^{234} dates of late Pleistocene fossils from the Mediterranean and Moroccan littorals. *Quaternaria* **7**, 29–42.

Toelstra, S.R., Van De Poel, H.M., Huisman, C.H.A., Geerlings, L.P.A. & Dronkert, H. (1980) Paleoecological changes in the latest Miocene of the Sorbas Basin, SE Spain. *Géol. Méditerranée* **7**, 115–126.

Torres-Roldán, R.L. (1979) *La evolución tectonometamórfica del macizo de los Reales (extremo occidental de la Zona Bética): un ensayo sobre el origen de gradientes anómalos y de alta temperatura en el dominio cortical Alpujárride–Maláguide de la Zona Bética (Cordilleras Béticas, Andalucia).* PhD thesis, University of Granada, Spain.

Torres-Roldán, R.L. (1981) Plurifacial metamorphic evolution of the Sierra Bermeja peridotite aureole (southern Spain). *Estud. Geol.* **37**, 115–133.

Torres-Roldán, R.L., Poli, G. & Peccerillo, A. (1986) An early Miocene arc-tholeiitic magmatic event from the Alborán Sea: evidence from precollisional subduction and back-arc crustal extension in the westernmost Mediterranean. *Geol. Rundsch.* **75**, 219–234.

Toscani, L., Contini, S. & Ferrarino, M. (1995) Lamproitic rocks from Cabezo Negro de Zeneta: brown micas as a record of magma mixing. *Mineral. Petrol.* **55**, 281–292.

Toscani, L., Venturelli, G., Barbieri, M., Capedri, S., Fernández-Soler, J.M. & Oddone, M. (1990) Geochemistry and petrogenesis of two-pyroxene andesites from Sierra de Gata (SE Spain). *Mineral. Petrol.* **41**, 199–213.

Trommsdorff, V., López Sánchez-Vizcaíno, V., Gómez-Pugnaire, M.T. & Müntener, O. (1998) High pressure breakdown of antigorite to spinifex-textures olivine and orthopyroxene, SE Spain. *Contrib. Mineral. Petrol.* **132**, 139–148.

Tubía, J.M. (1988) Estructura de los Alpujárrides occidentales: cinemática y condiciones de emplazamiento de las peridotitas de Ronda. *Bol. Inst. Geol. Min. España* **99**, 165–212.

Tubía, J.M., Cuevas, J. & Gil-Ibarguchi, J.I. (1997) Sequential development of the metamorphic aureole beneath Ronda peridotites and its bearing on the evolution of the Betic Cordillera. *Tectonophysics* **279**, 227–252.

Tubía, J.M., Cuevas, J., Navarro-Vila, F., Álvarez, F. & Aldaya, F. (1992) Tectonic evolution of the Alpujárride Complex (Betic Cordilleras, southern Spain). *J. struct. Geol.* **14**, 193–203.

Tubía, J.M. & Gil-Ibarguchi, J.I. (1991) Eclogites of the Ojén nappe: a record of subduction of the Alpujárride Complex (Betic Cordillera, southern Spain). *J. geol. Soc. London* **148**, 801–804.

Turner, S.P., Platt, J.P., George, R.M.M., Keley, S.P., Pearson, D. & Nowell, C.M. (1999) Magmatism associated with orogenic collapse of the Betic–Alborán Domain, SE Spain. *J. Petrol.* **40**, 1011–1036.

Udías, A., Lopez Arroyo, A. & Mezcua, J. (1976) Seismotectonics of the Azores–Alborán Region. *Tectonophysics* **31**, 259–289.

Van Bemmelen, R.W. (1927) *Bijdrage de geologie der Betische Ketens in de pro-*

vincie Granada. PhD thesis, University of Delft, the Netherlands.

Van Bemmelen, R.W. (1969) Origin of the Western Mediterranean Sea. *Geol. Mijnbouw* **26**, 13–52.

Van De Poel, H.M. (1991) Messinian stratigraphy of the Níjar basin (SE Spain) and the origin of its gypsum-ghost limestones. *Geol. Mijnbouw* **70**, 215–234.

Van De Poel, H.M., Roep, T.B. & Pepping, N. (1984) A remarkable limestone breccia and other features of the Mio-Pliocene transition in the Agua Amarga Basin. *Géol Méditerranéenne* **11**, 265–276.

Van Der Wal, D. & Vissers, R.L.M. (1993) Uplift and emplacement of upper mantle rocks in the western Mediterranean. *Geology* **21**, 1119–1122.

Van Morkhovan, F.P.C.M. (1962). *Post-Palaeozoic Ostracoda.* Elsevier, London.

Venturelli, G., Capedri, S., Barbieri, M., Toscani, L., Salvioli Mariani, E. & Zerbi, M. (1991a) The Jumilla lamproite revisited: a petrological oddity. *Eur. J. Mineral.* **3**, 123–145.

Venturelli, G., Capedri, S., Di Battistini, M., Crawford, A., Kogarko, L.N. & Celestini, S. (1984) The ultrapotassic rocks from southeastern Spain. *Lithos* **17**, 37–54.

Venturelli, G., Salvioli Mariani, E., Foley, S.F., Capedri, S. & Crawford, A.J. (1988) Petrogenesis and conditions of crystallization of Spanish lamproitic rocks. *Can. Mineral.* **26**, 67–79.

Venturelli, G., Toscani, L. & Salvioli Mariani, E. (1991b) Mixing between lamproitic and dacitic components in Miocene volcanic rocks of SE Spain. *Mineral. Mag.* **55**, 282–285.

Vissers, R.L.M. (1981) A structural study of the central Sierra de los Filabres (Betic Zone, SE Spain) with emphasis on deformational processes and their relation to the Alpine metamorphism. *GUA Pap. Geol.* **15**, 1–154.

Vissers, R.L.M., Platt, J.P. & Van Der Val, D. (1995) Late orogenic extension of the Betic Cordillera and the Alborán Domain: a lithospheric view. *Tectonics* **14**, 786–803.

Voet, H.W. (1967) *Geological investigations in the northern Sierra de los Filabres around Macaél and Cóbdar, south-eastern Spain.* PhD thesis, University of Amsterdam, the Netherlands.

Völk, H.R. (1966) Aggradational directions and biofacies in the youngest postorogenic deposits of Southeastern Spain: a contribution to the determination of the age of the East Mediterranean coast of Spain. *Palaeogeogr. Palaeoecol. Palaeoclimatol.* **2**, 313–331.

Völk, H.R. (1967) *Zur geologie und stratigraphie des Neogenbecken von Vera, Südost-Spanien.* PhD thesis, University of Amsterdam, the Netherlands.

Volk, H. (1979) *Quatare Reliefentwicklung in Sudost-Spanien.* Heidelberger Geographische Arbieten, **58**, 146.

Völk, H.R. & Rondeel, H.E. (1964) Zur Gliederung des Jungtertiars im Becken von Vera, Sudostspanien. *Geol. Mijnbouw* **43**, 310–315.

Wagner, C. & Velde, D. (1986) The mineralogy of K-richterite-bearing lamproites. *Am. Mineral.* **71**, 17–37.

Watts, A.B., Platt, J.P. & Buhl, P. (1993) Tectonic evolution of the Alborán Sea basin. *Basin Res.* **5**, 153–177.

Weijermars, R. (1985) Uplift and subsidence history of the Alborán Basin and a profile of the Alborán diapir (W Mediterranean). *Geol. Mijnbouw* **64**, 349–356.

Weijermars, R. (1987) The Palomares brittle–ductile shear zone of southern Spain. *J. struct. Geol.* **9**, 139–157.

Weijermars, R. (1988) Neogene tectonics in the Western Mediterranean may have caused the Messinian Salinity Crisis and an associated glacial event. *Tectonophysics* **148**, 211–219.

Weijermars, R. (1991) Geology and tectonics of the Betic Zone, SE Spain. *Earth Sci. Rev.* **31**, 153–236.

Weijermars, R., Roep, T.H.B., Van Den Eeckhout, B., Postma, G. & Kleverlaan, K. (1985) Uplift history of a Betic fold nappe inferred from the Neogene–Quaternary sedimentation and tectonics (in the Sierra Alhamilla and Almeria, Sorbas and Tabernas Basins of the Betic Cordillera, SE Spain). *Geol. Mijnbouw* **64**, 379–411.

Williams, D.F. (1988) Evidence for and against sea-level changes from the stable isotopic record of the Cenozoic. In: *Sea-Level Changes: an Integrated Approach* (Eds Wilgus, C.K., Hasting, B.S., Kendall, C.G. St., Posamentier, H.W., Ross, C.A. & Van Wagoner, J.C.). Spec. Publ. Soc. econ. Paleont. Miner., Tulsa, **42**, 31–36.

Wood, J. (1996) An introduction to the lower Messinian temperate water facies of the Sorbas basin (Abad Member and Azagador Member). In: *2nd Cortijo Urra Field Meeting, SE Spain: Field Guide* (Eds Mather, A.E. & Stokes, M.), pp. 14–23. University of Plymouth, UK.

Zazo, C., Goy, J.L., Hoyes, M. *et al.* (1981) Ensayo de sintesis sobre el Tirreniense Peninsular Español. *Estud. Geol.* **37**, 257–262.

Zeck, H.P. (1970) An erupted migmatite from Cerro del Hoyazo, SE Spain. *Contrib. Mineral. Petrol.* **26**, 225–246.

Zeck, H.P. (1992) Restite–melt and mafic–felsic magma mixing and mingling in an S-type dacite, Cerro del Hoyazo, southeastern Spain. *Trans. R. Soc. Edinburgh: Earth Sci.* **83**, 139–144.

Zeck, H.P. (1996) Betic–Rif orogeny: subduction of Mesozoic Tethys lithosphere under eastward drifting Iberia, slab detachment shortly before 22 Ma, and subsequent uplift and extensional tectonics. *Tectonophysics* **254**, 1–16.

Zeck, J. & Whitehouse, M.J. (1999) Hercynian, Pan-African, Proterozoic and Archean ion-microprobe zircon ages for a Betic–Rif core complex, Alpine belt, Mediterranean: consequences for its P–T–t path. *Contrib. Mineral. Petrol.* **134**, 134–149.

Zeng, J., Lowe, D.R., Prior, D.B., Wiseman, J.R., W.J. & Bornholm, B.D. (1991) Flow properties of turbidity currents in Bute Inlet, British Columbia. *Sedimentology* **38**, 975–996.

Index

342